Praise for Joseph LeDoux's *Synaptic Self*

"LeDoux's work, covering an avalanche of neuroscientific research, is surely the most accessible contemporary work for those interested in the brain's effect on personality."

—Gilbert Taylor, *Booklist*

"LeDoux offers a fascinating view into that 'most unaccountable of machinery,' the brain."

—*Kirkus Reviews*

"*Synaptic Self* represents a brilliant manifesto at the cutting edge of psychology's evolution into a brain science. Joseph LeDoux is one of the field's pre-eminent, most important thinkers."

—Daniel Goleman, author of *Emotional Intelligence*

"In this pathbreaking synthesis, Joseph LeDoux draws on dazzling insights from the cutting edge of neuroscience to generate a new conception of an enduring mystery: the nature of the self. Enlightening and engrossing, LeDoux's bold formulation will change the way you think about who you are."

—Daniel L. Schacter, Chairman of Psychology at Harvard University, author of *The Seven Sins of Memory*

"*Synaptic Self* is a wonderful tour of the brain circuitry behind some of the critical aspects of the mind. LeDoux is an expert tour guide and it is well worth listening. His perspective takes you deep into the cellular basis of what it is to be a thinking being."

—Antonio R. Damasio, neuroscientist, author of *The Feeling of What Happens*

"A clear, up-to-date, and impressively fair-minded account of what neuroscience has established about human nature."

—Howard Gardner, author of *Frames of Mind* and *Intelligence Reframed*

ABOUT THE AUTHOR

Joseph LeDoux, Henry and Lucy Moses Professor of Science at New York University's Center for Neural Science, is the author of *The Emotional Brain: The Mysterious Underpinnings of Emotional Life* and the coauthor, with Michael Gazzaniga, of *The Integrated Mind.*

SYNAPTIC SELF

HOW OUR BRAINS BECOME WHO WE ARE

JOSEPH LeDOUX

PENGUIN BOOKS

PENGUIN BOOKS
Published by the Penguin Group
Penguin Putnam Inc., 375 Hudson Street, New York, New York 10014, U.S.A.
Penguin Books Ltd, 80 Strand, London WC2R 0RL, England
Penguin Books Australia Ltd, 250 Camberwell Road, Camberwell, Victoria 3124, Australia
Penguin Books Canada Ltd, 10 Alcorn Avenue, Toronto, Ontario, Canada M4V 3B2
Penguin Books India (P) Ltd, 11 Community Centre,
Panchsheel Park, New Delhi - 110 017, India
Penguin Books (N.Z.) Ltd, Cnr Rosedale and Airborne Roads, Albany, Auckland, New Zealand
Penguin Books (South Africa) (Pty) Ltd, 24 Sturdee Avenue,
Rosebank, Johannesburg 2196, South Africa

Penguin Books Ltd, Registered Offices:
Harmondsworth, Middlesex, England

First published in the United States of America by Viking Penguin,
a member of Penguin Putnam Inc. 2002
Published in Penguin Books 2003

10 9 8 7 6 5 4 3 2 1

THE LIBRARY OF CONGRESS HAS CATALOGED THE HARDCOVER EDITION AS FOLLOWS:
LeDoux, Joseph E.
Synaptic self : how our brains become who we are / Joseph LeDoux.
p. cm.
ISBN 0-670-03028-7(hc)
ISBN 0 14 20.0178 3 (pbk)
1. Personality. 2. Self. 3. Neuropsychology. I. Title.
QP402 .L43 2002
612.8'2—dc21 2001045356

Printed in the United States of America
Set in Adobe Garamond with ITC Symbol
Designed by Carla Bolte

FOR NANCY

CONTENTS

ACKNOWLEDGMENTS

The bottom-line point of this book is "You are your synapses." Synapses are the spaces between brain cells, but are much more. They are the channels of communication between brain cells, and the means by which most of what the brain does is accomplished.

To propose a synaptic explanation of the self, I had to discuss in some detail how the brain works. I've tried to do this without trivializing the facts—it's not a pop psychology, how-to, or self-help book. Although writing about the brain in a way that will be clear to lay readers and at the same time not insulting to other scientists is tough, I'm pleased with the result.

But I didn't achieve this goal alone. My wife, Nancy Princenthal, is an art critic, and a fantastic writer. She constantly urged me to use words economically, and to avoid repetition. She read and reread, each time (often to my annoyance) with a sharpened pencil in hand. And when I had finally met her critical approval, I turned the manuscript over to Rick Kot, my editor at Viking. It was great to have a pro like Rick on my side.

I also want to thank present and past members of my lab. Without their creativity and hard work, the science that inspired me to write *Synaptic Self* would not exist. And the work could not have been done without the generous support of the National Institute of Mental Health, New York University, the W. H. Keck Foundation, and the Henry and Lucy Moses Fund.

I'm also grateful to colleagues who read certain chapters and commented on them. Thanks to Yadin Dudai, Jacek Debiec, Karim Nader, Greg Sullivan, and Rafi Lamprecht. In addition, several colleagues consulted with me on specific topics, including Tony Movshon, Carla Shatz, Dan Sanes, Jerome Kagan, Tim Wilson, Stephen Happel, Nancey Murphy, Barry Everitt, Norm White, Sandra File, Justine Kent, David Silbersweig, Jack Gorman, Amy Arnsten, Liz Phelps, and Jonathan Cohen.

Will Chang went above and beyond on the Works Cited list. And Mian Hou and Claudia Farb made indispensable contributions to the illustrations.

Brett Kelly, Rick Kot's assistant at Viking, was also very helpful throughout. Special thanks to Barbara Campo and Carla Bolte at Viking for their thoroughness and their patience with my corrections and to Don Homolka, an eagle-eyed copy editor.

I also want to express my gratitude to my agents, Katinka Matson and John Brockman. I greatly appreciate their advice and support.

Finally, I thank my children, Jacob and Milo. They inspire my synapses to change every day.

Writing a book is a humbling experience. You come to realize how many things you thought you understood but really didn't, at least not well enough to explain them clearly. I learned a tremendous amount while writing *Synaptic Self,* and hope you do too while reading it.

Note: Internet-equipped readers can visit the LeDoux Lab Home Page (*www.cns.nyu.edu/home/ledoux*) for more information about my reasearch. Once there, it will be possible to navigate to the Synaptic Self page, where information relevant to the book is presented, including links to reviews and a full bibliographic listing.

CHAPTER ONE

THE BIG ONE

DAD, WHAT IS THE MIND? IS IT JUST A SYSTEM OF IMPULSES OR SOMETHING TANGIBLE?

—Bart Simpson

"I don't know, so maybe I'm not," thc T-shirt said. Had it been another time or place, I probably would've chuckled and moved on. But I was on Bourbon Street when this postmodern homage to Descartes's proclamation, "I think, therefore I am," appeared on a young man weaving toward me. I had just had dinner in the French Quarter after a long day of data digestion and schmoozing at the Society of Neuroscience Conference, an annual get-together of twenty thousand or so brain researchers from around the world. The sound of Dixieland tunes, the aroma of stale beer, and the sight of scantily clad women dancing on runways inside dark bars had me contemplating the years and life changes that had come and gone since I myself had weaved down Bourbon Street during my college days in Louisiana. I headed back to my hotel, reflecting on my past and present life and wondering about what, all of a sudden, seemed to me to be the big question brain researchers should be asking: "What makes us who we are?"

Neuroscience hasn't yet delved deeply into this puzzling issue.[1] It has, for good reason, focused on how specific processes, like perception, memory, or emotion, work in the brain, but much less on how our brains make us who we are. I'd venture a guess that if a random sample of neuroscientists were asked, "What do we know about the brain mechanisms of the self and personality?" the predominant answer would be "Not much."

But maybe we know more than we think. Maybe some or even many pieces of the puzzle have already been discovered, and just have to be assembled into a coherent whole. Actually, I believe this might be the case. A lot of information is available about how the brain works, and while it may not yet be sufficient to fully explain persons, it should certainly encourage us to begin thinking about the problem.

My notion of personality is pretty simple: it's that your "self," the essence of who you are, reflects patterns of interconnectivity between neurons in your brain. Connections between neurons, known as synapses, are the main channels of information flow and storage in the brain (fig. 1.1). Most of what the brain does is accomplished by synaptic transmission between neurons, and by calling upon the information encoded by past transmission across synapses.

Given the importance of synaptic transmission in brain function, it should practically be a truism to say that the self is synaptic. What else could it be?[2] Not everyone, however, will be happy with this conclusion. Many will surely

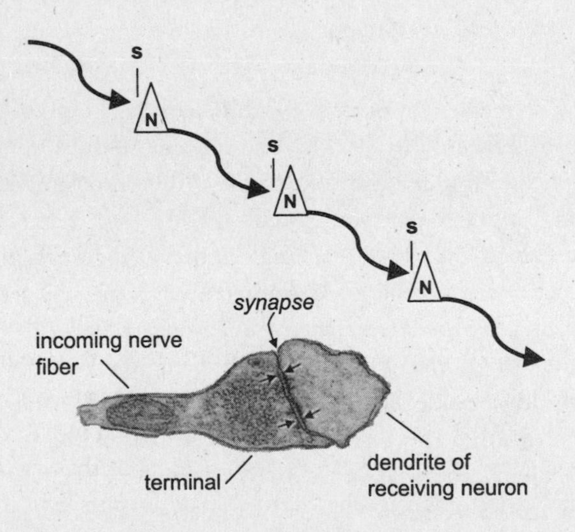

FIGURE 1.1 WHAT IS A SYNAPSE?

Synapses are small gaps between neurons (*top:* s, synapse; N, neuron). When a neuron is active, an electrical impulse travels down its nerve fiber and causes the release of a chemical neurotransmitter from its terminal. The transmitter drifts across the synaptic space and binds to a dendrite on the receiving neuron, thus closing the gap. Essentially everything the brain does is accomplished by the process of synaptic transmission. A picture (taken on an electron microscope) of an actual synaptic connection between two neurons is shown (*bottom*).

counter that the self is psychological, social, moral, aesthetic, or spiritual, rather than neural, in nature. My synaptic theory of the self is not proposed as an alternative to these views. It is, rather, an attempt to portray the way the psychological, social, moral, aesthetic, or spiritual self is realized.

I'll state unashamedly from the start that we can't, at this point, go all the way in formulating a complete synaptic theory of personality. But even a partial understanding of the synaptic basis of who we are is, for me, an acceptable goal. For seeking knowledge about the brain is not only a valid scientific pursuit; it can also improve the quality of life, as when it uncovers new ways of treating neurological or psychiatric disorders.

SYNAPTIC NATURE

Let's start with a fact: People don't come preassembled, but are glued together by life. And each time one of us is constructed, a different result occurs. One reason for this is that we all start out with different sets of genes; another is that we have different experiences. What's interesting about this formulation is not that nature and nurture both contribute to who we are, but that they actually speak the same language. They both ultimately achieve their mental and behavioral effects by shaping the synaptic organization of the brain (fig. 1.2). The particular patterns of synaptic connections in an individual's brain, and the information encoded by these connections, are the keys to who that person is.

The genetic blueprint begins to unfold in the newly fertilized egg. Genes actually do two things in the broadest biological sense: they make us all the same (we're all humans), and they also distinguish us from one another (each of us has a unique genetic makeup that contributes to our individuality). When two people get together and make a baby, they always end up with a creature that looks and acts like a human and not like a monkey, dog, or fish. The common genetic heritage of our species dictates that the basic systems and molecules in my brain are the same as those in yours, and the basic mental and behavioral repertoire available to me is also available to you. We all walk upright, speak through our mouths, laugh, cry, and learn from experiences. But as children of a particular set of parents, who are themselves products of a particular genealogical history, we each also have genes that give our brains unique qualities and direct the specific manner in which the general mental and behavioral characteristics of our species are expressed. Genetic factors are in fact known to influence a variety of individual or personality

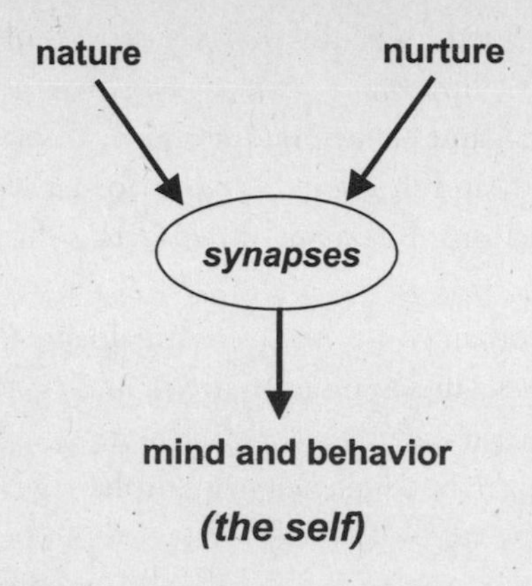

FIGURE 1.2 NATURE, NURTURE, SYNAPSES
Nature and nurture are not different things but instead are different ways of doing the same thing—wiring synapses in the brain. Synapses encode who we are.

characteristics, including how outgoing, fearful, or aggressive one is, as well as the likelihood that one will develop depression, anxiety, or schizophrenia.

Just how do genes affect individual behavior? In the simplest terms, they do so by making proteins that shape the way neurons get wired together. There are many steps, to say the least, between genes and their expression in actual behavior, but genetic sculpting of the synaptic organization of neural systems is the key to the process. Animal breeders have long known that it takes only a few generations of mating among carefully selected individuals to begin influencing behavioral traits of their offspring, such as how tame or fierce a given line of dogs is. When breeders try to customize behavior in this fashion, they in fact are often working with the synaptic organization of the brain. A few extra connections here, a little more or a little less neurotransmitter there, and animals begin to act differently. Once we realize that the basic wiring plan of the brain is under genetic influence, it's easy to see how not only animals but also people can have very similar brains and yet be so different, right from the start of their lives. Genetic forces, operating on the synaptic arrangement of the brain, constrain, at least to some extent, the way we act, think, and feel. The hopes of some, and the fears of others, surrounding the issue of human cloning obviously hinge on the importance of genes in regulating not only how we look, but also who we are.

For good reason, then, quite a lot of attention is being paid to genes these days. Widely read authors, including Richard Dawkins, E. O. Wilson, and

Steven Pinker, have argued forcefully that key aspects of mind and behavior are inherited.[3] Still, it's important to recognize that genes only shape the broad outline of mental and behavioral functions, accounting for at most 50 percent of a given trait, and in many instances for far less.[4] Inheritance may bias us in certain directions, but many other factors dictate how one's genes are expressed.

For example, if a woman consumes excessive alcohol during pregnancy, or a child has a diet deficient in certain nutrients, a brain genetically destined for brilliance can instead turn out to be cognitively impaired. Likewise, a family history of extraversion can be squelched in an orphanage run with an iron fist, just as a natural tendency to be shy and withdrawn can be compensated for to some degree by the supportive encouragement of parents.[5] Even if it becomes possible to clone a child who has died at a tender age, it's probable that the look-alike, having his own set of experiences, is going to act, think, and feel differently.

We hear a lot these days about how identical twins, reared apart by separate adoptive parents, can have similar habits and traits.[6] We hear less about the many ways they differ. The main outcome of Judith Rice Harris's controversial 1998 book, *The Nurture Assumption,* in which she proposed that parents hardly matter, was probably the emergence of clearer ideas about just how important, and under what conditions, parents *do* matter.[7] The personality disorders of children raised in brutal Romanian orphanages are a shocking testimony to the fact that experiences can have profound effects on behavior.[8] Genes are important, but not all-important.

NURTURING NATURE

The puzzle of how nature and nurture shape who we are is simplified by the realization that synapses are the key to the operations of both. Whether your paycheck is deposited to your bank account automatically or you hand it over to the teller in person, it goes to the same place. Nature and nurture function similarly: they are simply two different ways of making deposits in the brain's synaptic ledgers.

For example, in life (and not just in cartoons) rodents really are afraid of cats.[9] This is true of rats and mice living in the wild, but also of the rats studied in laboratories. Although lab rats come from breeding colonies that have been isolated from cats for many, many generations, any one of them, upon seeing a cat for the first time, will "freeze" dead in its tracks. This is Mother

Nature at work, since the rat (and his immediate ancestors) never had the opportunity to learn from personal experience that cats are dangerous.[10] This behavior is not just specific to rats or even animals, for people also freeze when faced with danger. Recall the amateur videotape of the bombing at the 1996 Atlanta Summer Olympics shown over and over on CNN. Within a second of the explosion, everyone in the vicinity was crouching motionless.[11]

Why did freezing become a genetically hardwired behavior? Obviously, there's no way to know for sure, since we have no fossil record tracing the evolution of behavioral or mental capacities.[12] However, a reasonable hypothesis is that freezing is a beneficial response when faced with a predator. Predators, the primal danger for most animals, respond to and are excited by movement. Keeping still in the face of danger is often the best thing for the prey to do. Because millions of years ago animals who did so were more likely to survive, today it's what most animals do, at least as an initial line of defense. Freezing is not a choice but an automatic response, a preprogrammed way of dealing with danger. It sometimes backfires, however, as when a deer is frozen in the headlights of an oncoming car. Like most evolutionarily based strategies, it's good for many animals much of the time, but not for all animals all of the time.

What's interesting is that freezing also occurs if a rat (or other animal) merely hears a sound that preceded an aversive stimulus (a mild electrical shock of its feet) on some prior occasion.[13] There's no predator around in this case, so how is the connection formed? The sound is a warning signal. Any rat that survives an encounter with a cat or other predator should store in its brain as much about the situation as possible so that the next time the sounds or sights or smells that preceded the arrival of the cat occur, those stimuli can be attended to in order to increase its chances of staying alive. In the case of the electrical shock, pain receptors are activated and this substitutes for a close encounter of the harmful kind with a predator, making it possible for the stimuli that preceded the shock to be stored as if they were those that preceded a cat.

There are two basic ways in which these kinds of processes might have been wired up in the brain. First, a two-system operation might exist, with one system for responding to species-typical (innate or genetically programmed) dangers and another for learning about novel ones experienced by individuals in their lives. Or there might just be one system that takes care of both situations. In fact, experiments show that the latter is the way the brain actually works (fig. 1.3).

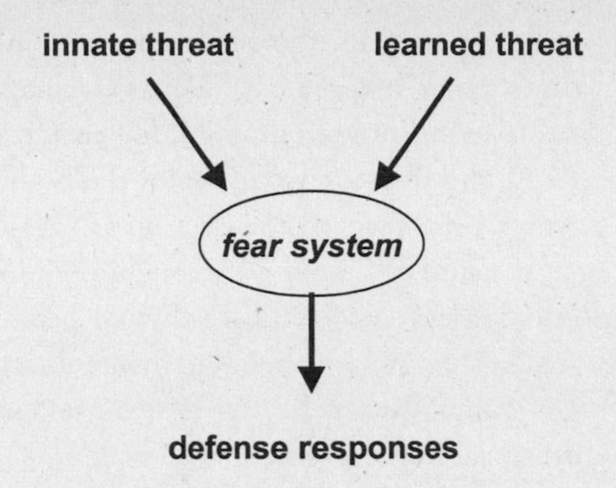

FIGURE 1.3 THE FEAR SYSTEM
The brain's fear or defense system determines whether danger (innate or learned) is present and, if so, produces protective responses.

Damage to a region of the brain called the amygdala eliminates the tendency for rats to freeze to both the cat and to the sound.[14] The amygdala is part of the brain system that controls freezing behavior and other defensive responses in threatening situations. Its synapses are wired by nature to respond to the cat, and by experience to respond in the same way to dangers that are learned about (fig. 1.4). It's a wonderfully efficient way to do things: rather than create a separate system to accommodate learning about new dangers, just enable the system that is already evolutionarily wired to detect danger to be modifiable by experience. The brain can, as a result, deal with novel dangers by taking advantage of evolutionarily fine-tuned ways of responding. All it has to do is create a synaptic substitution whereby the new stimulus can enter the circuits that the prewired ones used.

For the past twenty years, I've been trying to figure out how brains learn about dangers. In my previous book, *The Emotional Brain,* I described much of what is known about the organization of the brain system involved, and laid out the implications for understanding emotions.[15] The basic wiring plan is simple: it involves the synaptic delivery of information about the outside world to the amygdala, and the control of responses that act back on the world by synaptic outputs of the amygdala. If the amygdala detects something dangerous via its inputs, then its outputs are engaged. The result is freezing, changes in blood pressure and heart rate, release of hormones, and lots of other responses that either are preprogrammed ways of dealing with

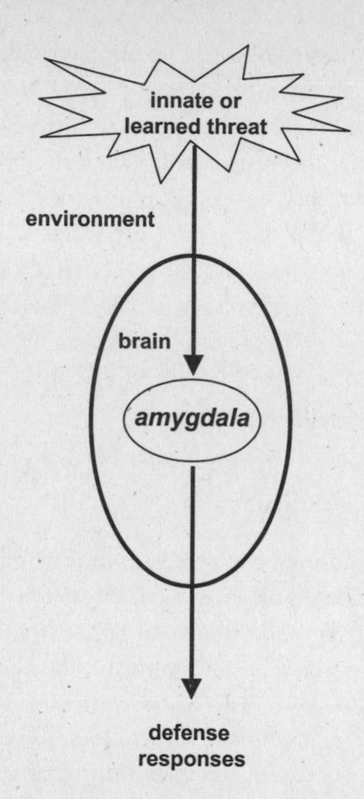

FIGURE 1.4 THE AMYGDALA: CENTERPIECE OF THE DEFENSE SYSTEM

The amygdala determines whether danger (innate or learned) is present and, if so, initiates bodily responses that were designed by evolution to deal with dangers.

danger or are aspects of body physiology that support defensive behaviors.[16] Throughout my past research, I've made use of the fact that the fear system can learn and store information about stimuli that warn of impending bodily harm or other dangers. Recently, though, my lab has turned to studies of exactly how fear learning takes place at the synaptic level. This new work, described in this book, sheds light on how synaptic modifications in the fear system, particularly in the amygdala, allow us to benefit from past encounters with danger.

Most systems of the brain are plastic, that is, modifiable by experience, which means that the synapses involved are changed by experience. But, as the fear example shows, learning is not the function that those systems originally were designed to perform. They were built instead to accomplish certain tasks (like detecting danger, finding food and mates, hearing sounds, or moving a limb toward some desired object). Learning (synaptic plasticity) is just a feature that helps them do their job better.

Plasticity in all the brain's systems is an innately determined characteristic. This may sound like a nature-nurture contradiction, but it is not. An innate capacity for synapses to record and store information is what allows systems to encode experiences. If the synapses of a particular brain system cannot change, this system will not have the ability to be modified by experience and to maintain the modified state. As a result, the organism will not be able to learn and remember through the functioning of that system. All learning, in other words, depends on the operation of genetically programmed capacities to learn. Learning involves the nurturing of nature.

LEARNING WHO WE ARE

The study of learning and memory processes in the brain has advanced rapidly in the last few decades. This book is in large part based upon this new understanding of encoding and storage and their origin in synaptic function. Learning, and its synaptic result, memory, play major roles in gluing a coherent personality together as one goes through life. Without learning and memory processes, personality would be merely an empty, impoverished expression of our genetic constitution. Learning allows us to transcend our genes, or, as the novelist Salman Rushdie said, "Life teaches us who we are."[17] Our genes may bias the way we act, but the systems responsible for much of what we do and how we do it are shaped by learning. Although a rat is by nature afraid of cats, it will live longer if it learns where in its particular world cats are most likely to be encountered and what kinds of sounds and smells are present when cats are nearby. A rat that has stored this information in its fear system is more worldly, and much better off, than one that has not. Something similar occurs in many if not most other brain systems: the information they encode and store today will contribute importantly to how they function tomorrow.

Our knowledge of who we are, of the way we think about ourselves, of what others think of us, and of how we typically act in certain situations is in large part learned through experience, and this information is accessible to us through memory. Without learning and memory, we wouldn't know if the person we are today jibes with the one we were yesterday or the one we expect to be tomorrow. Without learning and memory, a person would have the bare-bones personality provided by genes, but wouldn't know much about it.

But, as we will see throughout this book, learning and memory also contribute to personality in ways that exceed explicit self-knowledge. The brain, in other words, learns and stores many things in networks that function outside of conscious awareness. These learned tendencies affect all aspects of mind and behavior, and are probably at least as important for day-to-day functioning as what we know about ourselves consciously.

THE BIG ONE, OR IS IT?

My walk down Bourbon Street got me thinking about the origins of the self as the key overarching problem that neuroscience should aim to solve. Many other brain scientists, however, might be inclined to say that consciousness is the big one.

Neuroscientists traditionally have avoided confronting consciousness. The topic was one that retired neuroscientists, facing their own mortality, would talk about, but young brain researchers knew better. Even joking about it could give you a bad reputation. But times have changed and discussions of consciousness by neuroscientists are on the rise.[18] They have even been the basis of an indictment of science. In *The End of Science,* John Horgan, a social critic turned science journalist turned science critic, announced that neuroscience was, like other sciences, dead. But unlike some other disciplines, he argued, neuroscience was ending not because it had solved its big question—how does consciousness work?—but because it never would.[19]

I think it's good that scientists are now interested in consciousness, but I also believe it's being overemphasized. Suppose next week we find out that after decades of false starts and failed promises, an indefatigable neuroscientist finally has solved the consciousness problem. Would that really tell us what makes people tick? Would we now understand why schizophrenia emerges in one person but not his twin brother; why, when two people are faced with bodily harm, one is paralyzed by his fear while the other fights back; why an excessively shy child is likely to become an anxious adult; why some people are vegetarians and others enjoy red meat; why a meat-eater sometimes orders a vegetarian dish; why my kids can't stand the music I listen to; or why I like theirs? The answer to each of these is clearly No!

The question *Synaptic Self* asks is not "How does consciousness come out of the brain?" but rather "How does our brain make us who we are?"

What a person is, and what he or she thinks, feels, and does, is by no stretch of the imagination influenced only by consciousness. Many of our thoughts, feelings, and actions take place automatically, with consciousness only coming to know them as they happen, if at all. Figuring out the mechanism of consciousness would surely be a major scientific coup, but it wouldn't explain how the brain works, or how our brains make us the individuals we are.

An understanding of the mystery of personality crucially depends on figuring out the unconscious functions of the brain. *Unconscious* is actually a notoriously ambiguous word. Some people may think it refers to Freudian repressed memories, others that it is what happens in coma, or after being hit on the head, or drinking too much. None of these definitions is what I have in mind. Instead, what I mean by the term is the many things that the brain does that are not available to consciousness. (We can be thankful this is the way we function, for if we had to consciously plan each muscle contraction, our brain would be so busy we would probably never end up actually taking a step or uttering a sentence.)

Consciousness, at least the kind of consciousness we mean when we talk about our own mental states, very likely developed in the brain recently in evolutionary history, layered on top of all the other processes that already existed.[20] Unconscious operation of the brain is thus the rule rather than the exception throughout the evolutionary history of the animal kingdom.[21] It's a linguistic quirk, or a revealing cultural assumption, that the older (unconscious) processes are defined as negations of the newer one (consciousness). Language isn't perfect.

What, then, are all these unconscious processes? Actually, they include almost everything the brain does, from standard body maintenance like regulating heart rate, breathing rhythm, stomach contractions, and posture, to controlling many aspects of seeing, smelling, behaving, feeling, speaking, thinking, evaluating, judging, believing, and imagining.[22] We can be and often are aware of what we are doing when these things happen, but much of the time consciousness is informed after the fact. When someone speaks to you, for example, you decode sentence meaning on the basis of the sound of the words (phonology), the meaning of the words (semantics), the grammatical relations between the words (syntax), and your knowledge about the world (pragmatics). You usually are not aware of performing these operations, but simply do them. While you end up consciously knowing what the person said, you don't have conscious access to the processes that allowed you to comprehend the sentence. Similarly, when you yourself utter sentences, you go through the same processes, often without a conscious thought, but this time as the generator rather than the receiver. Our abilities to perceive the world, attend to objects and events, remember, imagine, and think all operate pretty much in this fashion. Collectively, these processes have been called the psychological or cognitive unconscious,[23] and they account for much of mental life.

Does that mean we'll know what a person is when we figure out the operations of these various conscious and unconscious functions? While unlocking the synaptic mechanisms underlying each of these processes is itself going to be quite a challenge, we need to go beyond the mere explanation of how each works in isolation. We need to understand how the many processes interact, and how the particular interactions that take place inside an individual's brain give rise to and maintain who he or she is. My aim in this book is to show how it is possible, at least in principle, to begin to understand the self in terms of such synaptic interactions.

Previous attempts to relate the self to the brain have mostly done so via the conscious self.[24] Recently, though, it appears the tide is beginning to turn. Antonio Damasio's book, *The Feeling of What Happens,* discusses the *protoself,* a kind of core self that exists outside of consciousness, and in *The Mind's Past* Michael Gazzaniga emphasizes the importance of unconscious processes in the production of consciousness.[25] Thus, consciousness, after long being neglected in brain science, is finally getting the attention it deserves, but it is also being put in its place—as part of, but not comprising the entirety of, the mental terrain. Although these other books deal with the self and the brain, they do not explore the biological mechanisms by which the brain makes the self. This is what *Synaptic Self* does.

As we begin to understand ourselves in synaptic terms, we don't have to sacrifice other ways of understanding existence. The idea that the self is created and maintained by arrangements of synaptic connections, in other words, doesn't diminish who we are. It instead provides a simple and plausible explanation for how the enormously complex psycho-spiritual-sociocultural package of protoplasm we call our self is possible.

CHAPTER TWO

SEEKING THE SELF

KNOW THYSELF.

—Oracle at Delphi

SELF-KNOWLEDGE IS A DANGEROUS THING.

—Lou Reed

Before we go looking for the essence of a person in the brain, it would help to have some conception of what we are seeking. There's certainly no shortage of opinion about what terms like *personality* or *the self* might mean. William James, for example, proposed: "*In its widest possible sense . . . a man's Self is the sum total of all that he CAN call his,* not only his body and his psychic powers, but his clothes and his house, his wife and children, his ancestors and friends, his reputation and works, his lands and horses, and yacht and bank-account. . . . If they wax and prosper, he feels triumphant; if they dwindle and die away, he feels cast down,—not necessarily in the same degree for each thing, but in much the same way for all."[1]

In fact, a whole area of psychology is devoted to the study of personality and the self.[2] And theologians, philosophers, novelists, and poets have also had much to say on the subject. While seemingly deep truths have sometimes emerged from these musings, it is not clear how, if at all, these insights might relate to the workings of the brain. After all, for many people, the brain and the self are quite different. I hope to show here that this is not the case. To do so, I first need to describe a way of thinking about the self that is compatible with current understanding of brain function.

"SOUL SERENADE"[3]

A few months after starting this book, I attended a conference on the relation between the brain and the soul, sponsored by (fittingly enough) the Vatican.[4] The specific topic was "Neuroscience and Divine Action," and the theologians who organized this meeting were trying to reconceptualize Church teachings in a way that would make sense in light of current scientific understanding of how the world works. In particular, they were attempting to determine how it is possible for God to influence people's lives without violating the laws of physics. I can't present the full range of views expressed, but one that stood out was the notion that God *interacts* but doesn't *intervene.*

The basic idea goes something like this: In the beginning, God set the universe up in a certain way (that is, he created the laws of physics) and has subsequently left it alone, at least for the most part.[5] On a typical day, therefore, God doesn't control the position of stars and planets, move mountains, part seas, change weather conditions, or make people do things they wouldn't otherwise do (in other words, he doesn't usually intervene), but he does communicate with people (he interacts).

Our concern here is not with the theological arguments for and against a noninterventionist view of God but rather with the possibility (or impossibility) of a scientific view of interaction. Given that people live a physical existence in the physical world, and that God is not part of the physical world, the question is: How can God interact with people? If you believe in the existence of a nonmaterial soul, then all you need to assume is that when God was creating the universe, he worked out some way of interacting with the soul. Since both God and the soul are nonmaterial, that interaction would also be nonmaterial, and the laws of physics would therefore be unviolated when interactions occur.

Much to my surprise, however, many of the theologians attending this meeting didn't believe in a classic nonmaterial soul (this would probably be an even bigger surprise to the faithful they represent). Instead, they seemed to accept the principle that the mind is inexorably tied to the brain, and they consequently believed in a soul that is pretty much one and the same as the neurally mediated mind, a part of the physical world that must by its nature obey the laws of physics.

Theologians who link the soul to the physical world actually have history on their side. While many Christians today continue to believe in a soul that is separate from the body and that survives death, this idea didn't really become prominent in Christianity until the Middle Ages. Early Christian teachings emphasized the resurrection of the body itself on Judgment Day rather than merely the survival of a nonmaterial soul. According to Mark 9:47, Jesus said, "It is better for you to get into the Kingdom of God with but one eye than to be thrown into Gehenna with both eyes." This lesson was apparently not meant symbolically but instead reflected the early Judaic notion that we take into the afterlife the bodily state with which we leave this life.[6] (This explains why Jews and Christians, and Muslims, have cemeteries on the western slopes of Mount Olive, facing the Eastern Gate of the Old City in Jerusalem: the closer their bodies are buried to the Eastern Gate, which is where the Final Judgment is expected, the sooner they will be raised.) Ancient Egyptians also seem to have believed that the body, and not just the spirit, carries on in some way after death. In the Tomb of Menena, for example, a foe of the royal family chiseled off their faces to ensure that they went into the next life physically challenged.[7]

If the soul is indeed physical in nature, part of the dilemma about how to sustain a belief in both physics and God would be solved (the part about how the soul meshes with the body). However, the thoroughly modern theologians would still be in a bit of a quantum pickle. If the soul is equivalent to the mind, and the mind depends on the functioning of the brain, how can God interact with people without physically affecting their neurons and, thus, intervening? And where is the soul hanging out while the body decays in the interim between death and Judgment Day? Not surprisingly, the Vatican conference ended inconclusively. No matter how all the pieces of the puzzle were moved around, they didn't fit together to make a coherent picture. As the philosopher David Hume said long ago, logic and reasoning (and presumably science) cannot explain the immortality of the soul.[8] Either you believe or you don't.

My reason for discussing this conference and the issues it raised is not so much to argue the point that it would be difficult, and maybe impossible, to find scientific solutions to theological riddles, but rather to demonstrate that a spiritual view of the self isn't (or doesn't have to be) *completely* incompatible with a biological one. Whatever else we are and aren't, much of what we are is accounted for by what goes on in our brains. Some theologians, as we've seen, have come to accept this. But even people who believe in an immaterial soul that survives death have acknowledged the fact that the normal functioning of the soul depends on the brain. Shakespeare embraced this notion when he called the brain the soul's frail dwelling.[9] A few minutes with my mother, a devout Catholic with Alzheimer's disease, makes it painfully clear just how fragile the soul's dwelling is.

THE LIMITS OF LOGIC

Theologians aren't the only ones who have concerned themselves with interactions between body and soul.[10] The question is also one of the major puzzles that has occupied philosophers through the ages. In the seventeenth century, the French mathematician René Descartes devised a way of thinking about body and soul that has shaped philosophical debate on this topic ever since.[11] Like the contemporary theologians described above, Descartes sought a means of reconciling science and faith. His solution was to propose that "the mental" and "the physical" were separate substances that met and interacted at a special place in the brain.

Historians seem to agree that the earliest Greek philosophers did not have clearly distinguished notions of body and soul.[12] Later, though, some philosophers came to view body and soul as separate. Plato, for example, believed that the intellectual essence of an individual—his psyche or soul—survives death.[13] In fact, Plato looked forward to death so that he could be free of his body and all its needs and passions and finally be capable of pure thought.[14] For Aristotle, in contrast, body and soul were so integrally related that they could not be separated, though they could be distinguished conceptually.[15] By the Middle Ages, philosophers adopted a combination of these notions, viewing body and soul as two unified 'substances' (like Aristotle) but regarding the soul as eternal (like Plato). Aquinas, for example, believed that the intellectual, nonmaterial qualities of mind gave the soul immortality, and that the body was resurrected and reunited with the soul on Judgment Day.[16]

This was the intellectual backdrop against which Descartes played his influential discussion of body and mind. Like Plato, he viewed the mental and physical as separate substances: "My soul, by which I am what I am . . . is entirely and absolutely distinct from my body, and can exist without it." In a fusion of faith and psychological theory, Descartes equated the soul with consciousness, and said only humans have conscious control of their behavior. Therefore, only human souls can gain or lose access to heaven by their actions. The behaviors of other animals were, in Descartes's scheme, reflexive or automatic, and carried out without thought. So, for Descartes, if it wasn't conscious, it wasn't mental. Descartes didn't exactly deny the existence of unconscious processes, but simply relegated them to the physical world, proposing that they function in humans the way they do in mindless (soulless) animals.

But if 'the physical' and 'the mental' are completely different entities, how can the conscious soul (the mental) be responsible for the physical body? Descartes's solution was that the conscious soul substance can interact with the material body by means of a small region of the brain called the pineal gland. While most parts of the brain exist in duplicate, the pineal gland is singular and centrally located, which suggested to Descartes that it must be the seat of mind-body interaction—a place where commands from the soul can influence the body, and where information from the body (either about the outside world or the body itself) can enter the soul as perceptions, emotions, and knowledge.

In Descartes's scheme, the nonphysical soul substance actually served the dual function of communicating with the physical body as well as with God. This interplay between physical and nonphysical substances is precisely the kind of solution to the problem of how God interacts with people that the theologians at the Vatican conference were trying to move beyond. Our interest here, however, is not in the theological question (How does God interact with the soul?), but in the philosophical one (How does the mind interact with the body?). Descartes's framing of the philosophical question, and his particular answer to it (a mind-body interaction in the brain), set up the conundrum known as the mind-body problem, which philosophers have struggled with ever since.[17] I want to make two points about this subject that are relevant to the present discussion.

First, in equating the mind with consciousness, Descartes framed the mind-body problem in terms of the relation between consciousness and the brain. As such, the mind-body problem, in its traditional conception, concerns an *aspect* of the mind rather than the whole mind. Most of what the brain does is not, in fact, part of the traditional mind-body debate. Some contemporary philosophers take a broader view than Descartes and accept that certain nonconscious aspects of brain function also contribute to mental life.[18] However, they regard such nonconscious aspects as "easy problems" that are not really the concern of philosophy. As will become apparent later in this chapter, I believe that these implicit or unconscious aspects of the self also play an important role, in fact an essential one, in shaping who we are and explaining why we do what we do.

Second, it's important to distinguish the philosophical mind-body problem from the neuroscientific problem about how the brain creates the mind.

Philosophers, by definition, seek philosophical solutions to problems (including the mind-body problem) and discuss the possible relations, the logic, that might exist between fundamental substances in nature (matter and mind). Neuroscientists, by contrast, typically start with the assumption that the materialist view of the mind-body problem is correct (that the mind is a product of the brain), and then try to understand how the brain makes the mind possible.[19] In fact, many philosophers today accept some version of materialism, but even if the tides should shift in the coming years and dualism (the belief that mind and body are separate substances) should take over philosophy, neuroscientists will not be out of jobs. Brain researchers are, after all, studying the brain, not philosophy. This does not mean that the paths of the neuroscientist and philosopher never cross. They often do, and when they do, each group has sometimes been enlightened (and sometimes enraged) by the other.[20] But, ultimately, because philosophers and brain scientists are pursuing different concerns, progress in one field does not necessarily signal an advance or defeat in the other.

In spite of my own contention that consciousness is not the be-all and end-all of mind and behavior, I nevertheless have considerable sympathy for the belief that neuroscience will come to explain consciousness. Descartes was correct in thinking about unconscious mental processes in physical terms; he erred, however, in conceiving of consciousness as nonphysical. That the brain mechanisms underlying conscious experience haven't been figured out yet doesn't mean that they will remain obscure forever. In fact, recent research has begun to make some headway in understanding the brain mechanisms of consciousness, and we'll take a look at this work later in the book.

OUR BODIES, OUR SELVES

Although the mind-body problem is the favorite topic of philosophers who work in the area called the philosophy of the mind, some of these philosophers have other things on their minds. One that is particularly relevant to us, and that is closely intertwined with the mind-body problem, is the issue of what constitutes a person. Is a person a body, a mind, a mind *in* a body? Does a person have to be human? Are all humans persons? Could a creature from another planet be a person? Can a human lose personhood as a result of brain damage, insanity, or moral transgressions? When during life does personhood start and stop? Is an embryo or an infant a person? What about someone who lingers for months in a coma from which he will, by medical prediction, never recover? The latter questions have wide-ranging social and legal implications, but only if the former ones can be answered in some reasonable way. If we can't establish precisely what a person is, it matters little whether we are one or not. John Locke had something like this in mind when he said, hundreds of years ago, that *person* is a "forensic term, appropriating actions and their merit; and so belongs only to intelligent agents, capable of a law, and happiness and misery. . . . This personality extends itself beyond present existence to what is past, only by consciousness."[21]

Peter Strawson is perhaps the best-known modern philosopher in this area. His much-cited paper "Persons"[22] starts with Ludwig Wittgenstein's assertion that bodies are not in possession of the states of consciousness that come out of them: "The I occurs in philosophy through the fact that the 'world is my world.' The philosophical I is not the man, not the human body, or the human soul . . . but the metaphysical subject, the limit—not a part of the world." You'll probably be happy to know that Strawson, too, was puzzled by these words, which he called impressive but obscure.

Strawson was motivated by Wittgenstein's arcane statements to try to explain the idea of something that is both a subject of experiences (that is, is conscious) and is a part of the world (that is, is dependent on a body). He wanted to understand the relation between two questions: Why do we ascribe states of consciousness to our bodies, and why do we ascribe states of consciousness to anything at all? Descartes had raised the first when he said, "I am not lodged in my body like a pilot in a vessel," and Wittgenstein the second with his statement that "The thinking, presenting subject—there is no such thing."[23]

According to Strawson, because we can attribute our own states of consciousness to ourselves, others like us must also have similar states of consciousness. If we can figure out how to identify those who are like us, we can know to whom consciousness should be attributed—in other words, we can know who is a person. To do this, he distinguished between two kinds of statements: those that obviously can be applied to material bodies that also exhibit consciousness ("is in pain," "is thinking," "believes in God") and those that can be applied equally to material bodies that are conscious and that are not ("is heavy," "is tall," "is hard").

Like Locke and Strawson, many philosophers have taken the view that personhood is a characteristic of intelligent, conscious creatures, that consciousness is, in fact, the quality that defines personhood. But others demand more, in the form of a moral element. This was implicit in Locke, as well as in the writings of Kant. Daniel Dennett combined the thought of Locke and Kant, and other philosophers, proposing that there are two interrelated notions of a person, one moral and one metaphysical.[24] The metaphysical person is a thinking, feeling, intelligent, conscious agent, while the moral person is one who is accountable for his actions. Dennett asks whether being a person in the metaphysical sense automatically makes one a person in a moral sense, or does it merely make a moral capacity possible. He goes on to list several conditions of personhood. A being is a person if it is rational, verbal, conscious, and, in fact, self-conscious, capable of being acted toward in a certain way, and capable of reciprocating when acted toward in this way. Dennett's list also borrowed from John Rawls, who argued, "To recognize another as a person one must respond to him and act towards him in certain ways,"[25] and from Thomas Nagel, who affirmed that "extremely hostile behavior towards another is compatible with treating him as a person."[26] But Dennett, in the end, concludes that these are necessary but not sufficient conditions for defining a person—that there is, fundamentally, no way to set a passing grade for personhood that is not arbitrary.

The concept of the self, which is of utmost importance to us here, is closely related to the philosophical notion of a person. Within philosophy, there has, in fact, been a growing interest in the self,[27] an outcome of which has been the emergence of distinctions between different aspects of the self.[28] One much-discussed distinction is between the minimum and the narrative self.[29] The former is an immediate consciousness of one's self, and the latter a coherent self-consciousness that extends with past and future stories that we tell about ourselves.[30] The narrative self bears some relation to the postmodern notion that the self is socially constructed.[31] While social construction is often viewed as diametrically opposed to a scientific view of man,[32] the two are not necessarily at odds with each other since brains, in the end, are responsible for both the behaviors that collectively constitute the social milieu, and for the reception by each individual of the information conveyed by this milieu.

In focusing on consciousness as the leading metaphysical feature of who we are, philosophers interested in the question of personhood and the self leave out much of who we are—all the nonconscious aspects. And in dividing the world into material objects and conscious selves or persons, as Strawson did, nonhuman animals are placed in a kind of ontological limbo, since a nonhuman animal cannot be a person.

Although other animals are not conscious in the human sense, they are not simply objects, like rocks or chairs. They are living creatures with nervous systems that make it possible for their bodies to interact with and change the material world in ways that rocks and chairs cannot. The concept of a person, a conscious self, while useful as a way of evaluating issues related to being human, is thus less valuable as a general-purpose concept for understanding existence in the context of our animal ancestry. And because we must pursue many aspects of how the brain works through studies of nonhuman organisms, we need a conception of who we are that recognizes the evolutionary roots of the human body, including the brain.

Though not as widely discussed as conscious aspects of the self, nonconscious aspects are nevertheless important. They are essential to the Buddhist attempt to eliminate the conscious self,[33] to ideas about multiple selves by William James and others,[34] as well as to notions of a primitive, nonconceptual,[35] or ecological[36] self that exists outside of conscious awareness. Once we accept that the self of a human can have conscious and nonconscious aspects, it becomes easy to see how other animals can be thought of as having selves, so long as we are careful about which aspects of the self we are ascribing to each species in question.

The self, then, is a notion that can be conceived of along an evolutionary continuum. While only humans can have the unique aspects of the self made possible by the kind of brains that humans have, other animals have the kinds of selves made possible by their own brains. To the extent that many of the systems that function nonconsciously in the human brain function similarly in the brains of other animals, there is considerable overlap in the nonconscious aspects of the self between species. Obviously, the more similar the brains, the more the overlap.

The extent to which other animals have any kind of consciousness is, unfortunately, impossible to know. We can speculate, but because the human mind cannot become a cat, dog, bird, lizard, frog, or fish mind, we cannot know with certainty how such a question should be answered.[37] Descartes's greatest contribution was perhaps his conclusion that the only thing he could know with certainty was his own mind. So long as we are talking about other animals with brains like our own (that is, other humans), we can have some confidence that their mental states are like ours. But we cannot with any degree of certainty extrapolate from our own mental states to those of other species.

In spite of having gotten this far with some key concepts from philosophy, the fact is that philosophy will probably not give us the kind of foundation we need to pursue the relation between the self and the brain.[38] To state that a mind, or a person, or a self is all physical, or all mental, or partly physical and partly mental, or something else altogether (like the product of socially constructed relations between people),[39] lays out the territory in a way that is useful for analyzing broad categories of experience within and between species, but does not tell us much about how to pursue mechanisms in the brain. If we are going to figure out how it is that our brains make us who we are, we need a way of linking a fairly detailed conception of who we are to neural functions. Psychology may be more relevant to that purpose.

MIND SCIENCE

Psychology was actually a branch of philosophy until the late nineteenth century, when Wilhelm Wundt, a German physiologist, began doing experiments to understand the way the mind works rather than just speculating about it.[40] He and his followers, known as introspectionists, took the key steps required to convert psychology into an experimental science. Their main topic of investigation was conscious experience, which they explored by examining their own experiences, attempting to break them into essential, irreducible elements.

But early in the twentieth century, some psychologists began to argue that this was no way to conduct scientific research, since one's conscious experiences can only be known personally, and cannot be verified by others.[41] This idea caught on and eventually spawned behaviorism, which was based on the premise that a scientifically valid psychology had to focus on observable events (behavioral responses) rather than internal states.[42] Some of its adherents were methodological behaviorists, which meant they didn't necessarily reject the existence of consciousness, but simply believed it couldn't be studied. Radical behaviorists, in contrast, actually denied that consciousness existed. For them, mental states were illusions created by tendencies to act in one way or another. Philosophers like Gilbert Ryle adopted radical behaviorism as a resolution to the mind-body problem,[43] eliminating the mind entirely, leaving only the physical body to be explained in physical terms. Ryle called mental states "ghosts in the machine," after the "deus ex machina" of Greek tragedy, a god that was lowered onto the stage from above to solve the problems of mortals.

Toward the middle of the century, it dawned on some scientists that the operations (computations) performed by computers were not unlike what a human does when solving a problem.[44] This notion was embraced by some farsighted psychologists like Jerry Bruner[45] and George Miller,[46] and the cognitive approach to psychology, which emphasized internal mechanisms that process information, was born.[47] This was an attractive alternative to mindless behaviorism, and eventually the cognitive movement dethroned behaviorism and brought the mind back to psychology.

The mind that returned, though, was not exactly the one that the behaviorists had disposed of. Behaviorists had objected to the emphasis of introspectionists on mental content (the experience of the color red, for example). Cognitive scientists, however, were studying mental *processes* rather than the content of consciousness. They were more concerned with how colors are detected and discriminated than in what it is like actually to experience them.

It is now widely recognized that we can have conscious access to the outcome of cognitive processes, but we are not usually aware of the processes that were involved in generating that content.[48] Our perceptions, memories, and thoughts generally work in happy ignorance of the processes that make them possible. For cognitive scientists, and in stark contrast to Descartes, mind and consciousness are not at all the same.

The cognitive movement had a tremendous impact on psychology, but its influence did not stop there. Information-processing concepts were also adopted by workers in linguistics, anthropology, and other social sciences, as well as mathematics and physics. And just as psychologists were conceiving of minds in terms of computer operations, computer scientists and mathematicians were pursuing the notion that computers might perform mindlike operations, an idea that led to the field of artificial intelligence (AI). Ultimately, cognitive science emerged as an interdisciplinary approach to understanding how the mind works. It came to be called "the new science of mind."[49]

The fact that cognitive processes are not dependent on consciousness (actually, consciousness depends on *unconscious* cognitive processes) means that the mental vs. physical dilemma does not have to be overcome in order to study the brain mechanisms of cognition. Indeed, many of the processes studied by cognitive scientists are also topics of research pursued by so-called cognitive neuroscientists. Led by breakthroughs in understanding the psychology of cognition, cognitive neuroscientists have been very successful in relating perception, attention, memory, and thinking to underlying mechanisms in the brain.[50]

Cognitive psychology, and its sister, cognitive neuroscience, would thus seem to be taking us ever closer toward psychological and neurobiological understandings of the self. However, this is not exactly the case. Though we understand how specific cognitive processes work psychologically and neurologically, cognitive approaches fall short when it comes to explaining the self.

First of all, by its very definition, cognitive science is a science of only a part of the mind—the cognitive part—and not a science of the whole mind.[51] Traditionally, as we'll see in chapter 7, the mind has been viewed as a trilogy, consisting of cognition, affect (emotion), and conation (motivation).[52] The fact that emotion and motivation are not studied by cognitive science makes sense if cognitive science is regarded as a science of cognition, but is troubling if the field is supposed to be the science of mind. A mind without feelings and strivings (the kind of mind traditionally studied in cognitive science) might be able to solve certain problems given it by a cognitive psychologist, but it doesn't stack up well as the mental foundation of a self. The kind of mind modeled by cognitive science can, for example, play chess very well, and can even be programmed to cheat. But it is not plagued with guilt when it cheats, or distracted by love, anger, or fear. Neither is it self-motivated by a competitive streak, or by envy or compassion. If we are to understand how the mind, through the brain, makes us who we are, we need to consider the *whole* mind, not just the parts that subserve thinking.

A second shortcoming of cognitive science is that it has not grappled successfully with how various cognitive processes interact to form the mind. Considerable progress has been made in understanding how perception, memory, and thinking work, but not about how they work together. And in light of the tripartite nature of the mind, an understanding of the self is going to require that we not only figure out how various cognitive processes interact, but also that we include emotions and motivations in the mix and figure out how they interact with one another, as well as how they interact with cognitive processes. Our hopes, fears, and desires influence how we think, perceive, and remember. A science of mind needs to account for and understand these complex processes.

And, finally, cognitive science deals with the way the mind typically works in most of us, rather than the way it works uniquely in any one of us. While we all have basically the same mental processes mediated by the same brain mechanisms, the way these processes and mechanisms operate is determined by our particular genetic background and life experiences.

It would be hard to overstate the importance of cognitive science. It has been extremely successful as a research program, and has revolutionized the way we conceive of the mind. So when I single out the shortcomings of the field, I do so not to condemn it, but instead to simply point out that it's incomplete when it comes to understanding what makes us who we are.

THE PERSONALITY CONTEST

Psychology, as we know it today, is an imperfect marriage between two distinct approaches to the workings of the mind that emerged in the late nineteenth century.[53] One is the experimental approach, which emphasizes the way specific mental processes, like perception or memory, typically work. This is the approach that gave rise to cognitive psychology. The other approach is more concerned with how well-adjusted people are and how they might change their behavior to improve their psychological well-being. It focuses on individuals and their idiosyncratic traits, habits, feelings, and thoughts, rather than on the way things work in most people most of the time. The various forms of psychotherapy in use today are outgrowths of this approach, which has also been a fountainhead for theories of personality. This is the kind of psychology that is portrayed in films and novels and is what people usually have in mind when they think of what a psychologist is.

Ideas about personality are probably as ancient as ideas. Around 400 B.C., Hippocrates, for example, proposed that one's health and character were determined by the interaction among four bodily humors (blood, phlegm, black bile, and yellow bile), each of which, in turn, reflected four cosmic elements (earth, water, air, and fire).[54] Six hundred years later, Galen expanded the theory, proposing that excess in one or another humor gave people distinctive personalities. (Excess blood led to a sanguine, enthusiastic personality; too much black bile made one melancholic; abundance of yellow bile led to irritable or choleric temperament; and overproduction of phlegm gave rise to a slow, apathetic, or phlegmatic person.)[55]

Although views of personality, temperament, character, and the self continued to be developed over the centuries, modern approaches essentially began with Freud's psychoanalytic theory.[56] Subsequent theories have for the most part been variations on or reactions to Freud, and fall into several broad categories.[57] These include neo-Freudian psychodynamic theories, organismic or self theories, trait theories, behavioral or learning theories, and cognitive theories.

Personality theorists clearly have had valuable insights into the workings of the human mind, and have guided therapists in their efforts to help people adjust to life's challenges. But the various theories are often directly contradictory.[58] There were feuds within psychoanalysis even in Freud's day (Jung, for example, broke away from strict Freudianism). Later, the neo-Freudians had disputes as well. For example, some maintained Freud's emphasis on repressed sexual urges as the root of anxiety, while others replaced sexuality with social and/or cultural factors as the core psychoanalytic concept. Psychoanalytic theory today is probably best viewed as a family of theories rather than a single well-defined view of how the mind works and how it breaks down in psychopathology.

But differences within psychoanalytic theory pale compared to differences between psychoanalytic and other personality theories. Some theories focus on psychopathology, while others are more concerned with the nature of the well-adjusted person. Unconscious motivation plays a key role in some theories, while others go in the opposite direction and focus almost exclusively on conscious strivings. Behavior is motivated in multiple ways in many theories, whereas in others, a single motive is emphasized (e.g., sexual gratification or self-actualization). Social considerations are important in some but are less crucial in others. Biological factors, especially genes, are believed to underlie stable personality traits over one's life span for some theorists, but others emphasize the role of learning and situational (especially social) factors in determining behavioral and mental states.

One possible explanation for the diversity of personality theories is that the topic is simply so difficult that no one has quite figured it out yet. Alternatively, there may have not been a clear winner in this personality contest not because the various theories proposed to date are all wrong, but because many are at least partly correct. If this is true—and I believe this is the case—then the best way to construct a view of the self might be not to pit the various theories against one another but rather to synthesize across them.

A VIEW OF THE SELF

So far, I've used the terms *personality* and *the self* rather loosely. Now it's time to get more specific. From here on, when I use the term *the self* I am referring to the totality of the living organism. This notion subsumes the idea of personality[59] and is similar to what William James had in mind when he described the self as the sum total of who one is (see the opening paragraph of this chapter). But in order for this view of the self to be useful to our pursuit of how the brain makes the self possible, we need to refine it considerably.

In modern personality theory, as in philosophy, the notion of the self typically refers to the conscious self, in the sense of having self-knowledge, a self-concept, and self-esteem; of being self-aware, self-critical; of feeling self-important; and of striving toward self-actualization. Carl Rogers, a pioneer psychologist of the self, summed up this view, defining *the self* as "the organized, consistent conceptual gestalt composed of perceptions of the characteristics of the 'I' or 'me.'"[60] For Rogers, these perceptions are "available to awareness, though not necessarily in awareness." Modern self psychologists like Hazel Markus have a similar focus on self consciousness.[61] These psychologists do not deny that some aspects of mental life occur unconsciously, but instead minimize the importance of the unconscious components of the mind in favor of the notion of a self as an active agent in the control of mental states and behavior.

In spite of this long tradition of emphasis on the self as a conscious entity, the self that we are aware of, or can be aware of, is not the entirety of what the term *the self* refers to. The psychologist Ruth Munroe, for example, argues for a more fundamental view.[62] She points out that "a sense of self which *develops* in the course of living is too far confused with the truly necessary organismic self." Munroe is questioning whether the "sense of self" that develops over time is the whole self. In other words, she is arguing that the self that we are aware of and strive to improve, the self that we have a sense of, the self that many personality theorists have been enthralled with, is too narrow a view of what the self really is.

The existence of a self is a fundamental concomitant of being an animal. All animals, in other words, have a self, regardless of whether they have the capacity for self-awareness. As a result, the self consists of more than what self-aware organisms are consciously aware of. Indeed, recent research in social psychology has emphasized that many important aspects of human social behavior, including decision-making as well as the way we react to members of racial and ethnic groups, are mediated unconsciously.[63] These differences within organisms (conscious vs. unconscious aspects) and between organisms (creatures with and without consciousness) are not captured by an undifferentiated notion of the self, but can be accounted for by distinguishing between explicit and implicit aspects of the self.

Things we consciously know about who we are make up the *explicit* aspects of the self. These are what we refer to by the term *self-aware* and constitute what we call our self-concept; they are what the self psychologists are interested in. The *implicit* aspects of the self, by contrast, are all other aspects of who we are that are not immediately available to consciousness, either because they are by their nature inaccessible, or because they are accessible but not being accessed at the moment. All animals have implicit selves, but only animals that have the capacity for conscious self-awareness have explicit selves (this is why the existence of a personality in a pet does not necessarily mean that the pet is conscious in the human sense).

This view of the self contrasts with the idea of a person as elaborated by philosophers like Strawson, Dennett, and others. Only humans can be persons, but all animals can have selves, especially when we allow for the distinction between implicit and explicit selves. One might want to broaden the notion of a person to account for explicit and implicit aspects. This would solve one problem (the fact that there's more to a person than what that person is conscious of) but would leave another unaddressed (the relation between persons and other animals).

That explicit and implicit aspects of the self exist is not a particularly novel idea. It is closely related to Freud's partition of the mind into conscious, preconscious (accessible but not currently accessed), and unconscious (inaccessible) levels. However, Freud's terms carry much theoretical luggage that I want to leave behind.

The terms *implicit* and *explicit* are themselves not completely neutral. They are borrowed from the study of memory, where it is now widely recognized that the brain system involved in forming explicit, consciously accessible memories is distinct from a variety of other systems that are capable of learning and storing information implicitly, which is to say without conscious awareness.[64] Actually, since most brain systems are plastic, and work outside of consciousness, they can be thought of as implicit memory systems or, better yet, as systems that are able to store specific kinds of information implicitly. To the extent that our life's experiences contribute to who we are, implicit and explicit memory storage constitute key mechanisms through which the self is formed and maintained. Those aspects of the self that are learned and stored in explicit systems constitute the explicit aspects of the self. To be self-aware is to retrieve from long-term memory our understanding of who we are and place it in the forefront of thought. In contrast, those aspects of the self that are learned and stored in implicit systems make up the implicit aspects of the self. We use this information about our selves all the time, even though we may not be consciously aware of it. The way we characteristically walk and talk and even the way we think and feel all reflect the workings of systems that function on the basis of past experience, but their operation takes place outside of awareness. I will have much to say in later chapters about the workings of explicit and implicit memory functions of the brain.

The self is a unit, in the sense that organisms go to great pains to keep themselves alive and well. Physical damage to one's appearance is not taken lightly (remember the harm that face removal from the Egyptian tomb was meant to achieve), nor are insults to one's character. Both implicit and explicit systems are utilized to accomplish this unity in humans. But self-preservation is a universal motive, independent of whether the organism is aware that it is working toward this goal. A cockroach can scamper away when a human foot approaches without being explicitly aware of being in danger, the same way that a single-cell bacterium can detect and move away from harmful molecules in its chemical world.

The self is not static. It is added to and subtracted from by genetic maturation, learning, forgetting, stress, aging, and disease. This is true of both implicit and explicit aspects of the self, which may be influenced similarly or differently at any one point. For example, a mild compliment may only be registered and stored in explicit memory, but glowing praise, registered explicitly, might lead to the arousal of emotion systems that then also store aspects of the experience implicitly. On the other hand, stress is known to impair explicit memory while at the same time enhancing the implicit memory functions of emotion systems.[65]

As important as learning is, not all aspects of the self are learned. Some are due to our genetic heritage. All of the capacities that we have as *Homo sapiens,* including our capacities to learn and remember, are made possible by the genetic makeup of our species. What we place in our individual memory systems is a function of our unique experience, but the existence and basic mode of operation of these systems are due to our species's genes. At the same time, we each have a family genetic history that is a variation on the theme of being a human, and a personal set of genes that is a variation on our family's, and these variations also influence who we are.

The most well-articulated view of the role of genes in shaping behavioral and mental characteristics comes from biological trait theories of personality, which propose that one's enduring qualities are due to one's genetic background.[66] Considerable evidence has been amassed to support the view that some traits, such as the extent to which one is extroverted (gregarious) vs. introverted (shy, fearful, withdrawn), are highly influenced by one's genetic history. Nevertheless, there are two important caveats to genetic theory of personality.

First, genes have been found to account for only about 50 percent of a particular personality trait.[67] What this statement means is that genes account at most for half of a given trait, *not* that half of all of personality is accounted for by genes. For some traits, genetic influence is far less and is often not measurable. Introversion is probably the trait with the strongest genetic influence.[68] Although many extremely shy, introverted children tend to become anxious, depressed adults,[69] some do just fine. Is this because the genetic influence in the latter group was temporary, or because the genetic tendency was squelched? The fact that when extreme introversion is caught early, it can be reversed to some extent by a supportive family environment suggests that genes do not fully dictate psychological destiny.[70] Life's experiences, in the form of learning and memory, shape how one's genotype is expressed. Even the most ardent proponents of genetic determination of behavior admit that genes and environment interact to shape trait expression. It's a matter of how much, not whether, both contribute.

The second caveat to the genetic account of personality stability comes from research showing that people are not always true to their so-called personality traits. One may be shy at work or in social groups, for example, but domineering at home. In fact, when psychologists have examined the consistency of behavior across situations, the results have not supported the view that people act consistently in different situations. Observations such as these suggest to Walter Mischel that behavioral and mental states are not dictated by constitutional factors but instead are situationally determined. Mischel argues that the ability to predict behavior depends upon knowing about a person's thoughts, motivations, and emotions relative to a particular set of circumstances.[71] He describes these as "if . . . then relations." "If" you are in situation A, "then" you do X, but "if" in situation B, "then" you do Y. According to Mischel, people don't possess stable personality traits over time, but stable "if . . . then" profiles.

As with most polarized arguments in psychology, there is truth in both the situational and the trait views. The stronger the genetic contribution to a particular characteristic, the more likely it will be expressed uniformly in different situations. At the same time, situations vary in the extent to which they dictate the way we act. A red traffic light will cause most people to stop, regardless of whether they are generally aggressive or timid, whereas a yellow light allows more latitude for tendencies like aggression or timidity to be expressed.[72] We'll visit questions about genetics and personality again in chapter 4, when we explore how the brain is built.

In proposing that the self exists, I run the risk of reifying something that is, ultimately, not real. Bob Dylan, for example, said, "I change during the course of a day. I wake and I'm one person, and when I go to sleep I know for certain that I'm someone else. I don't know who I am most of the time. It doesn't even matter to me."[73] And, according to Philip Roth, "All I can tell you with certainty is that I, for one, have no self, and that I am unwilling or unable to perpetrate upon myself the joke of a self."[74] Mark Epstein, who has tried to integrate psychoanalysis and Buddhism, points out that the ego's image of itself (its object image) is always lacking as an account of the subject (the self), implying that much of the self is, in essence, implicit.[75] While the whole self is not usually encountered by the individual who possesses it (who *is* it), or by others, it nevertheless exists.

What then is it? In my view, the self is the totality of what an organism is physically, biologically, psychologically, socially, and culturally. Though it is a unit, it is not unitary. It includes things that we know and things that we do not know, things that others know about us that we do not realize. It includes features that we express and hide, and some that we simply don't call upon. It includes what we would like to be as well as what we hope we never become.

The fact that all aspects of the self are not usually manifest simultaneously, and that different aspects can even be contradictory, may seem to present a hopelessly complex problem. However, this simply means that different components of the self reflect the operation of different brain systems, which can be but are not always in sync. While explicit memory is mediated by a single system, there are a variety of different brain systems that store information implicitly, allowing for many aspects of the self to coexist. As William James said, "Neither threats nor pleadings can move a man unless they touch some one of his potential or actual selves."[76] In *Orlando,* Virginia Woolf pointed out, "A biography is considered complete if it merely accounts for six or seven selves, whereas a person may well have as many thousand."[77] Or as the painter Paul Klee expressed it, the self is a "dramatic ensemble."[78]

THE SELF AND THE BRAIN

Theories of the self and personality are not usually framed in ways that are compatible with our understanding of brain function.[79] How, then, can we relate the complex constellation I've called the self to the systems and synapses of the brain?[80] The goal of the rest of the book is to answer this question. However, a brief preview is in order.

The self can be understood in terms of brain systems involved in learning and storing information, in explicit and implicit systems, about things that are significant in people's lives. The processing by these systems always occurs in a physical and social context (a situation) and is performed by networks that function the way they do because of both genetic inheritance and past experiences. Put this way, in order to understand the self, we need to explain how brain systems underlying thinking, emotion, and motivation (the mental trilogy) develop under the influence of nature and nurture, and how these systems make it possible for us to attend to, perceive, learn about, and store and retrieve experiences. We especially need to explain how different systems interact with and influence one another. Without these interactions, and the mental integration they engender, each of us would simply be a collection of isolated mental functions rather than a coherent person.

The point, though, is not simply to state that learned and innate interactions between cognitive, emotional, and motivational processes make us who we are, but instead to explain *how* these interactions work. And the explanation that I will pursue in the remainder of this book involves neural, especially synaptic, mechanisms. I believe, in short, that an answer to the question of how our brains make us who we are can be found in synaptic processes that allow cooperative interactions to take place between the various brain systems that are involved in particular states and experiences, and for these interactions to be linked over time. It is probably not at all obvious what this statement means at this point in the book. Before it will begin to make sense, we need to cover more ground.

CHAPTER THREE

THE MOST UNACCOUNTABLE OF MACHINERY

MY OWN BRAIN IS TO ME THE MOST UNACCOUNTABLE OF MACHINERY—ALWAYS BUZZING, HUMMING, SOARING ROARING DIVING, AND THEN BURIED IN MUD. AND WHY? WHAT'S THIS PASSION FOR?

—Virginia Woolf

Most of us are as mystified by our brains as Virginia Woolf, though perhaps less eloquent in our ignorance. Still, everyone has heard a few things about the wrinkled blob in the noggin—for instance, that we use only 10 percent of it. But who came up with this number? And why would we even have the rest if it weren't useful? Evolution doesn't usually make organs in such a way that they mostly go unused, just in case someone figures out one day what to do with the extra material. It's hard to imagine how 90 percent of the brain, lacking in value for most of us most of the time, could have ever come into existence. Researchers have been looking into what the brain does for many years now, and from what they have discovered, it doesn't seem that most of it is, in fact, resting idly.

People also tend to carry around with them one or both of two additional erroneous beliefs about the brain. The first is that functions of the brain, like perception, memory, or emotion, are located in specific areas. The other is that chemicals floating around in the brain determine our mental states. Unlike the 10 percent myth, these are actually part truths that, taken out of context, are patently false. We know, at least in a general sense, how the brain works, and it's not by islands of brain tissue or by isolated chemicals operating independently. Particular areas are important, but not on their own: they participate in functions by way of their synaptic connections with other areas. Chemicals are also important, but mainly because of their work at synapses within functional systems.

This chapter will give an account (albeit an abbreviated one) of this "most unaccountable of machinery," describing some basic facts that are necessary to understand the brain's synaptic systems. Although the discussion will have to get a bit technical along the way, this information is essential to my attempt to relate the self to synapses. Because I've kept things simple, those already in the know may wish to skip ahead. However, the novice will get a crash course on what neurons are, how synapses connect them together, and why synaptic connections are the key to brain function.

BRAINS: SO DIFFERENT, YET ALL THE SAME

We mammals belong to the group of animals called vertebrates, a subphylum we share with other backboned creatures, including birds, reptiles, amphibians, and fish. Mammals and birds separately descended from reptiles millions of years ago. In spite of this common ancestry, the brains of reptiles, birds, and mammals look very different. Beneath these dissimilarities, though, there's a common plan that's rigorously adhered to.

Every vertebrate brain can be divided into three broad zones: the hindbrain, midbrain, and forebrain. In the early years of the twentieth century, neuroscientists discovered that damage to each zone had a different predictable consequence.[1] For example, in studies of cats, it was found that purposeful, voluntary behavior and problem-solving ability were impaired when the forebrain was damaged. Nevertheless, even with massive injuries to the forebrain, some semblance of normal coordinated behavior remained. Such compromised animals could orient toward a noise or withdraw their paw from heat, and could walk, eat, and groom. They could even display full-blown emotional responses, especially those typically expressed in anger or fear, if the hypothalamus, a small region situated at the base of the forebrain, was spared. When larger lesions were made that removed all of the forebrain, including the hypothalamus, only rudimentary responses remained. These animals, when challenged with intense stimulation, could hiss, bare their teeth, unsheathe their claws, or swipe a paw, but could not manage to put all of these behaviors together into a coordinated defense or attack response. When the midbrain was damaged, the animal was essentially comatose—

alive physically, but not behaviorally or psychologically. And when the hindbrain was destroyed, life itself ceased.

From these crude experiments, it was concluded that the hindbrain controls very basic functions, those necessary for staying alive; the midbrain is involved in maintaining wakefulness and coarse, isolated behavioral reactions; and the forebrain coordinates complex behavioral and mental processes. It should not be surprising, given these effects of brain damage, that the forebrain (necessary for thinking and problem-solving) is the region that differs the most between mammals and other vertebrates and the hindbrain (necessary for life) the least. Nevertheless, all three levels are represented in all vertebrates, and even the evolutionarily advanced forebrain is structured according to a common underlying organizational plan that is applicable to every vertebrate species.

For example, the human forebrain consists of several subdivisions,[2] one of which is the wrinkled outer layer, the neocortex. This is the part of the forebrain that makes possible many of our higher mental functions. The designation *neo* reflects the fact that this brain region was, for many years, believed to be evolutionarily new, having emerged when mammals evolved from reptiles.[3] Other vertebrates were thought to have a primordial or older cortex but not a mammalian or neocortex. This view began to change, though, in the late 1960s and early 1970s, when new techniques for studying the brain became available.[4] Based on the patterns of chemical staining and nerve connections discovered with these techniques, the organization of the brain came to be better appreciated, and researchers were able to use this information to find the equivalent (or at least the semblance) of a neocortex in both birds and reptiles, suggesting both that it wasn't so new after all and that it certainly wasn't unique to mammals. The reason this cortex had not been found in these animals earlier was because of its unusual location, buried beneath other brain areas, instead of resting on top, as it does in mammals.

While at the level of overall brain structure a similar organizational plan applies to many different animals, it is not the case that all brains are the same. A given brain area can vary enormously in size and complexity between different species, allowing some animals to do things that others cannot. In amphibians, for example, an area in the midbrain called the tectum is especially well developed, making it possible for most frogs to thrust their tongue into the flight path of an insect and capture it,[5] a feat most people can't accomplish. Bats and rats can hear things that we cannot, and bees use a magnetic sense, which we do not have, to guide their movements.[6] Different species have been subjected to different evolutionary pressures, and their brains reflect their unique histories.[7]

The most obvious difference between the mammalian and other vertebrate brains is the extent to which the cortex has expanded. Although, as we have seen, reptiles and birds are now known to have some neocortex, the mammalian neocortex is far more elaborate than the equivalent areas in these other species.[8] And within mammals, there are distinctions as well: the neocortex is bigger and more differentiated in primates than in rodents, and in humans more so than in monkeys. These changes in cortical size and complexity are, however, superimposed on a basic neocortical plan. For example, in all mammals, processes related to sensation (vision, audition, touch) are represented in the rear and processes involved in controlling movement in the front of the cortex.

Within a given species, the similarities of cortical organization are striking. Early anatomists discovered that the major patterns of cortical wrinkles, which appear to be randomly arranged to the uninitiated eye, are amazingly consistent from person to person, and can be used as landmarks to identify various regions of the neocortex.[9] What's remarkable is that these purely structural parcels, defined by the wrinkles, turn out to correspond to functional divisions, areas that participate (by way of their synaptic connections with other cortical and/or subcortical areas) in different aspects of mental life and behavior.[10] For example, the area of the cortex involved in controlling precise movements of various body parts is located just in front of the central sulcus, one of the major wrinkles in the cortex, while touch, hearing, and visual areas are defined by their own wrinkles, as are areas involved in language comprehension and speaking. On careful examination, some variation in the organization of cortical or other brain areas is evident in different people, but the basic overall architectural plan of the brain is pretty much the same in any two individuals.

In spite of the tremendous similarity of our brains, we all act differently, have unique abilities, and have distinct preferences, desires, hopes, dreams, and fears. The key to individuality, therefore, is not to be found in the overall organization of the brain, but rather in the fine-tuning of the underlying networks. To understand the defining qualities of each person, we need to go beyond the superficial organization of the brain (its division into broad regions and areas within these) and turn to the microscopic structure and function of neural systems, and especially to the cells and synapses that constitute them.

THE CELL WAR

All organs and tissues of the body are composed of cells. But unlike the cells in other body parts, brain cells, or neurons, directly communicate with one another. There's nothing magical about the process—neurons are simply built in a way that allows them to exchange information with one another in ways that other cells cannot.[11] Common patterns of communication between neurons ensure that all human brains work in basically the same way, whereas subtle differences in these patterns of communication give rise to the distinctive qualities that we each have.

The existence of cells in the brain and other parts of the body is taken for granted today, but this knowledge was only made possible by the further development of the microscope in the nineteenth century. Around 1837, Matthias Schleiden, a German botanist, first proposed that plants were made up of discrete units, or cells. The following year, his friend Theodor Schwann extended the notion to animals, and thereby brought botany and zoology together in a single theory, the so-called cell theory,[12] which argued that all living things are composed of cells.

Whether cell theory was applicable to the brain was a topic that was fiercely debated for decades. When early brain anatomists examined brain tissue under a microscope, they did see structures resembling cells. But unlike cells in other organs, brain cells had fine fibers extending out of them (fig. 3.1). Some scientists concluded that this meant that the brain was unique—not composed of discrete cells but instead made up of an entangled mesh or reticulum of continuously connected elements. Others, though, argued that the fundamentals of cell theory applied equally to the brain.

Two of the major figures in the debate were Santiago Ramón y Cajal of Spain and the Italian anatomist Camillo Golgi.[13] Golgi, working in his kitchen, invented methods for staining the brain that allowed better visualization of its microscopic anatomy. He favored the reticular theory. Ironically, on the basis of the methods pioneered by Golgi, Cajal argued forcefully for the application of the cell theory to the brain, and won many converts. One of these was Wilhelm Waldeyer, who in 1891 published a paper in which he suggested that brain cells be called neurons. In this paper, he also coined the phrase *the neuron doctrine* to account for the application of the cell theory to the brain. Cajal apparently considered the doctrine his, at least in spirit if not name, and was not happy to have had his thunder stolen by Waldeyer.[14] But the loss in stature, if any, was temporary. Every graduate student in

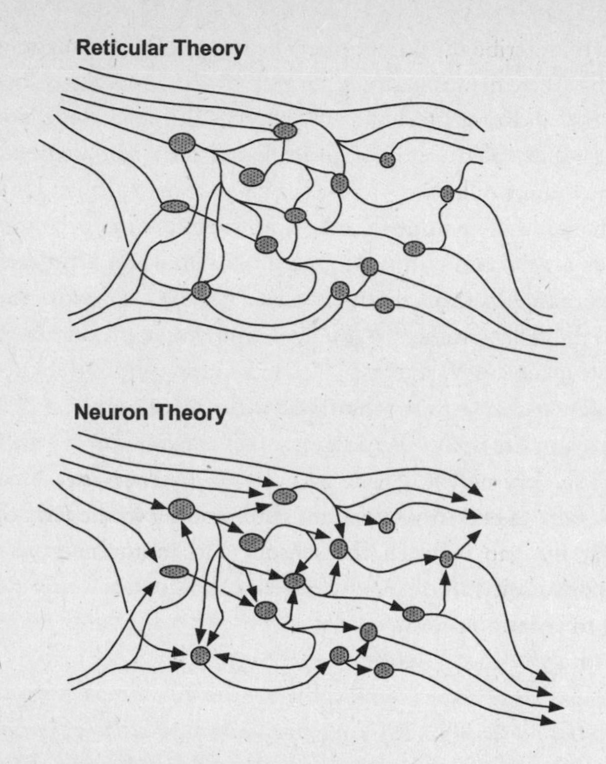

FIGURE 3.1 RETICULAR VERSUS NEURON THEORY

In the late nineteenth century, scientists fiercely debated the question of whether the brain was made up of a reticulum of continuously connected elements or, instead, of individual cells, neurons, that communicated with one another. By the beginning of the twentieth century, the so-called neuron doctrine had emerged as the prevailing view.

neuroscience today knows who Cajal was, whereas few have ever heard of Waldeyer.

One of the early and largely unrecognized soldiers in the neuron war was the young Sigmund Freud. After completing his medical training in Vienna, Freud accepted a position as a famulus, or research scholar, and studied the nervous system of fish and crayfish.[15] As early as 1883, long before the neuron doctrine was codified, he promoted the idea that nerve cells are physically separated from one another.[16] This concept later figured prominently in one of his earliest forays into psychological theory. In *Project for a Scientific Psychology,* written in 1895 but unpublished for many decades,[17] Freud stated that "the nervous system consists of distinct and similarly constructed neurones . . . which terminate upon one another." He introduced the term *con-*

tact barriers to describe the points where neurons abut, and suggested that interactions between neurons across contact barriers make possible memory, consciousness, and other facets of the mind. Although these notions were amazingly sophisticated for their time, Freud felt that progress in understanding the brain would be too slow for his taste and so abandoned a neural theory of the mind in favor of a purely psychological one.[18] The rest is history.

Two years after Freud wrote his *Project,* Sir Charles Sherrington proposed a different term for the connections between neurons.[19] Sherrington had been working on the reflex problem.[20] A reflex is the simplest kind of neural circuit that controls behavior. When your physician taps you on the knee, your leg jerks because the tap elicits sensations that are transmitted along *sensory* nerves that originate in your knee and travel to your spinal cord. The messages in the sensory nerves trigger activity in *motor* nerves that come out of the spinal cord and end in your leg muscles, leading to the jerk. Sherrington realized that the gap between the sensory and motor neurons had to be bridged somehow if information carried by the sensory nerves was to be transferred to the motor nerves. He was probably unaware of Freud's contact barriers, and chose to call the gaps synapses, derived from the Greek word meaning to clasp, connect, or join.[21] The notion of synapses as points of communication between cells is one to which we still adhere, and which is essential to our efforts to understand who we are in terms of brain mechanisms.

In 1906, Cajal and Golgi shared the Nobel Prize for their groundbreaking research on brain anatomy. Although the neuron doctrine had gained considerable support by then, Golgi clung bitterly to the reticular theory at the award ceremony.[22] Still, definitive proof that the nervous system is composed of cells did not come until many years later. With the invention of the electron microscope in the 1950s, scientists could finally examine the brain in sufficient resolution to see that the tiny fibers extending out of a neuron do not typically make direct physical contact with neighboring cells.[23] Indeed, they are separated by tiny spaces, synaptic spaces, across which the brain does its business.

WHAT MAKES NEURONS SPECIAL?

By knowing the function performed by a few cells of most organs in the body, whether the liver, kidney, or gall bladder, you can deduce the organ's overall function.[24] This is not true of the brain, however, where cells participate in myriad activities, from seeing and hearing to thinking and feeling, from awareness of self to the incomprehension of infinity. The architecture of a neuron helps us begin to understand why the brain is so multifunctional, while organs like the pancreas and spleen are not.

Neurons have two major parts. The first is the cell body (fig. 3.2), which is involved in important housekeeping functions, such as storing genetic material and making proteins and other molecules that are necessary for the cell's survival. The cell body does much the same work in neurons as it does in other cells. The major structural difference between neurons and other cells lies in the special appendages that neurons have—the nerves. These fibers, which extend out of the cell body, are what caused all the confusion in the nineteenth century about whether the brain was, like other organs, composed of discrete cells.

Nerve fibers are sort of like telephone wires. They allow neurons in one part of the brain to communicate with neurons in another. By way of these connections, communities of cells that work together to achieve a particular goal can be formed across space and time in the brain. This capacity underlies all of the brain's activities and is absent in other organs.

There are two varieties of nerve fibers, axons and dendrites (fig. 3.2). Axons are output channels, and dendrites are input channels. An axon carries messages to other cells. It can end nearby, allowing communication with its close neuronal neighbors, or it can stretch over very long distances, as much as several feet. If you are standing still and decide to take a step, the movement of your leg on the basis of your decision involves axons that originate in cell bodies located in the movement control regions in the frontal cortex (just behind your forehead) and that travel uninterrupted to the base of the spinal column (in the region of your lower back).

The end of the axon, called the terminal, is the point at which the sending neuron communicates with receiving neurons. Although terminals most of-

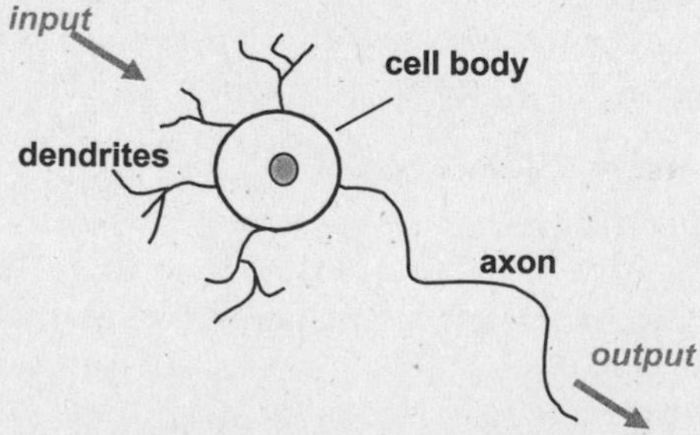

FIGURE 3.2 COMPONENTS OF A NEURON

All neurons contain three basic parts: a cell body and fibrous appendages called dendrites and axons.

ten form connections with dendrites, they can also contact cell bodies or other axons.[25] Dendrites, too, sometimes communicate between one another.[26] In order for the long axons descending from your frontal cortex to your spinal cord to cause your leg to move, the terminal has to contact dendrites of the receiving cells in the spinal cord. The axons of these receiving cells then extend out and terminate at muscles in your leg. The arrival of signals at the muscle leads to contraction, and thus movement.[27]

Many dendrites have little knobs called spines extending from them (fig. 3.3). These are readily seen when brain tissue is stained with the methods

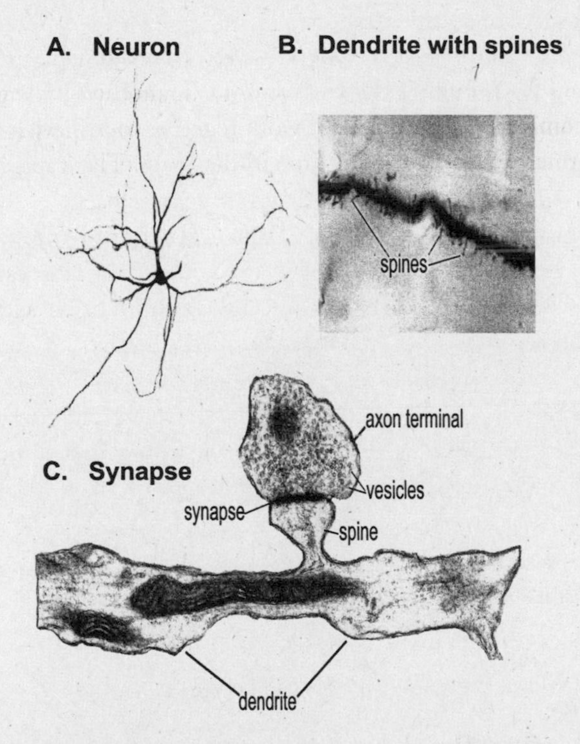

FIGURE 3.3 WHAT A NEURON LOOKS LIKE

Upper left: A single neuron and many of its dendrites. This neuron had been filled with a dye so that its shape can be seen. *Upper right:* A high magnification of a small piece of a dendrite showing the protrusion of the small spines from the dendritic shaft. Spines are often where axons from other neurons terminate and form synapses. *Bottom:* A highly magnified electron-microscopic picture of an axon terminal with vesicles forming a synapse with the spine of a dendrite. When an electrical charge travels down the axon to the terminal, neurotransmitter is released from the vesicles and drifts across the small synaptic space between the terminal and the spine. The neurotransmitter then binds to receptors on the spine and initiates electrical events in the receiving neuron.

developed by Golgi. Spines are especially important as receivers of messages from axons, and play a key role in brain development, as well as in learning and memory, as we will see later.

Most neurons have only one axon. However, each axon branches many times before it ends, allowing a single neuron to spawn many terminals. The result is that the messages sent out from one cell can affect many others. This is called divergence (fig. 3.4). At the same time, each neuron can receive inputs from numerous others. This is called convergence (fig. 3.4).

The point at which the sending and receiving elements of neurons meet is our star, the synapse. Because information usually flows across the synapse starting from the axon terminal, this side is said to be presynaptic, and the receiving side, often occupied by a dendritic spine, postsynaptic (fig. 3.5). As Sherrington noted, because a synapse is a space between the sending and receiving cells, something has to cross the synaptic space in order for the two cells to communicate.

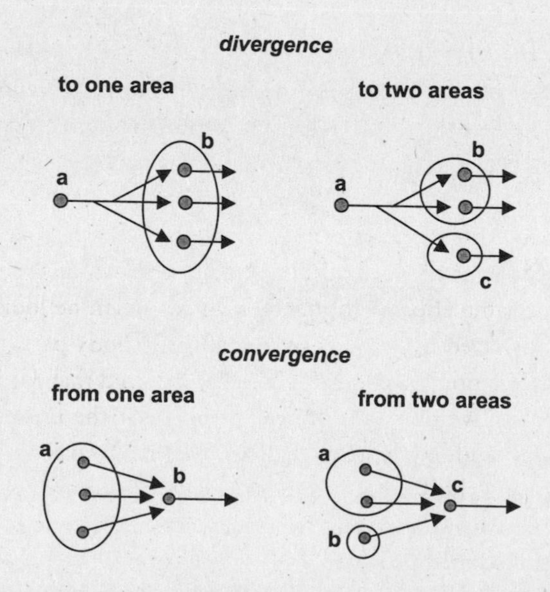

FIGURE 3.4 DIVERGENCE AND CONVERGENCE

Divergence exists when a neuron gives rise to axons that branch and terminate on multiple targets, whereas convergence exists when a single neuron receives inputs from multiple sources.

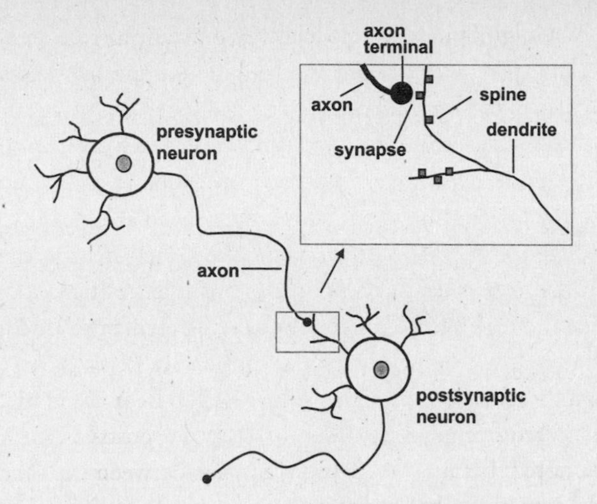

FIGURE 3.5 PRESYNAPTIC AND POSTSYNAPTIC NEURONS

This figure shows two neurons, one presynaptic to the other. The axon of the presynaptic neuron terminates at the dendrites of the postsynaptic neuron. Often, such terminations form synapses on the spines (small protrusions) on dendrites.

GALVANIZED FROG LEGS

The question of how information is exchanged between neurons across the synaptic space is tied in closely with that of how information is transferred along a nerve fiber of a single neuron. Before we consider synaptic transmission, we therefore need to consider nerve conduction.

In the 1770s, Anton Mesmer, a Viennese physician, had been using iron magnets to treat a variety of physical and mental maladies, until he found he could have the same effect without the magnets when he looked into a patient's eyes and waved his hands over the afflicted body part.[28] This was the birth of mesmerization, or hypnosis. Mesmer believed that some mysterious, magnetically sensitive fluid was present throughout the universe, including the human body, and that he could help his patients by using his own animal magnetism to alter this fluid.[29] At the time, little was known about the physiology of the nervous system, and any theory, including one as wacky as animal magnetism, seemed possible.

A few years later, the Italian Luigi Galvani noticed that the amputated leg of a frog hung from an iron trellis with a brass hook twitched during a lightning storm.[30] He also found he could make a frog leg kick at any time he wished if he touched the nerves within the wound with one metal and the foot with another. This was, in effect, the first battery. Galvani, in the tradition of Mesmer, concluded that the metals were conducting vital spirits from the frog. So-called animal electricity was an occult rather than a scientific phenomenon.[31]

Several decades later, Carlo Matteucci, another Italian, made the first measurements of genuine electrical activity in nerves.[32] In Germany, Johannes Müller and Emil Du Bois-Reymond, realizing the importance of this observation, began a research program that rescued electrical conduction in nerves from the world of mysticism and turned it into a thriving scientific research field.[33]

At the time, the assumption was that nerves conducted electricity like wires. But one of Du Bois-Reymond's students, Hermann von Helmholtz, did an experiment that suggested otherwise. He calculated the speed of electrical conduction in frog nerve fibers by measuring how much time elapsed before a given muscle twitched when nerves of different lengths were electrically stimulated. Although conduction time was fast—about 40 meters per second (roughly 40 mph)[34]—it wasn't as fast as electricity, which can under certain conditions flow through a wire at about the speed of light.

From these simple but informative experiments, it became clear that nerves do conduct electricity, but in a special way. Electricity does not flow passively through a nerve as it does through a wire. Rather, impulses conducted through nerves are *biologically* propagated, moved along by electrochemical reactions, a process that takes a lot longer than passive physical conduction.

The biologically propagated impulse in a nerve is called an action potential. This dramatic electrical event is normally initiated at the point where the axon emerges from the cell body. Once triggered, it travels like a rolling wave down the axon toward the terminal. The propagation occurs as a kind of neurodomino effect—an electrical change in one part of the axon membrane produces a similar change in adjacent parts, and so on, all the way down to the terminal. Action potentials can be triggered artificially by electrical stimulation, which makes them easy to study, but normally they occur in a cell when orders come from synaptic inputs.

Work by many pioneering neuroscientists established the basic principles of electrical propagation in axons, which became the foundation for much of what we now know about the working of neurons. A good deal of this research was performed using the giant axons of squids, the sheer size of which made it easier to investigate electrical conduction. Especially noteworthy were the studies performed in the 1940s by Alan Hodgkin and Andrew Huxley in England. Building on Ohm's law of electricity (which states that voltage is equal to current times resistance), they characterized in precise mathematical form the basic features of electrical transmission in axons. The Hodgkin-Huxley equations are still used today to calculate current, voltage, and resistance in axons.

SYNAPTIC CHATTER

The existence of electrical conduction in nerves suggested to late-nineteenth-century scientists that electrical impulses played a critical role in the normal functions performed by the brain. A key related question was whether electrical propagation was sufficient to explain how the brain worked. Sherrington's studies of reflexes determined that it was not.

Electrical impulses in sensory and motor nerves clearly seemed involved in reflexes: when sensory nerves detect a tap on your knee, they conduct electrical impulses that, in turn, lead to electrical impulses in motor nerves, and to the jerk. But how does the sensory nerve communicate with the motor neuron? Sherrington demonstrated that while electrical stimulation of a sensory nerve elicited an electrical response in the motor nerve, stimulation in the motor nerve did not evoke a response in the sensory nerve (fig. 3.6). He concluded that the junction between cells, the synapse, had a valvelike property—it only transmitted in one direction, from sensory to motor nerves.[35] This was particularly significant ammunition against the reticular theory, for if neurons were continuously connected and communicated only by electrical conduction, then motor nerves should have as sizable an effect on sensory nerves as the other way around. Neurons must therefore communicate with one another by some means other than mere electrical conduction.

Subsequent research revealed that the one-way conduction between neurons is due to the fact that synaptic transmission involves the release of chemicals from storage sites in the presynaptic axon terminal. These molecules are released when action potentials propagated from the cell body reach the terminal. The released chemicals then drift across the liquid-filled synaptic space[36] and come in contact with spines or other portions of the postsynaptic cell. Because the chemical storage sites usually are present in the presynaptic terminal and not in the postsynaptic dendrite, transmission only occurs in one direction. These chemicals are called neurotransmitters, since they allow neurons to communicate across the synaptic gap—they transmit between neurons.

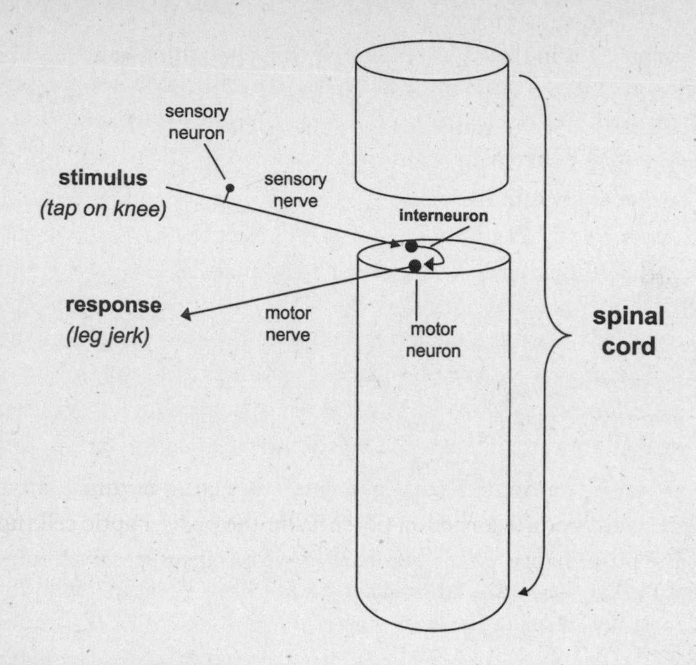

FIGURE 3.6 A SPINAL REFLEX

The basic elements of a spinal reflex include sensory neurons that receive messages about an external stimulus, motor neurons that initiate muscle movements, and interneurons in the spinal cord that link the sensory and motor neurons. Sherrington's studies of spinal reflexes led him to the conclusion that synaptic transmission is a one-way street.

The chemical nature of neuronal transmission was suspected from studies in the early 1900s showing that the effects of electrical stimulation of nerves could be mimicked or blocked by certain chemical agents. But it was an ingenious experiment by Otto Loewi in the 1920s that provided the ultimate proof.[37] He removed the hearts from two frogs, leaving the nerves connected to one heart but not the other, and infused each with a saltwater solution (similar to normal body fluid). He then electrically stimulated the nerves on one, which changed the beat rate of the heart (the heart is postsynaptic to these nerves). When he removed the solution from the stimulated heart and injected it into the other, the heartbeat changed in the unstimulated heart, much as if it had been stimulated, indicating that some chemical that had been released in the stimulated heart was transferred in the solution to the other.

While Loewi's experiments involved the connection between a nerve and a muscle—the heart, in this case—essentially the same thing happens when the connection is between two neurons. That is, the arrival of the action potential in the presynaptic terminal leads to the release of neurotransmitter into the synaptic space.

The release of neurotransmitter molecules from the presynaptic terminal is a means, not an end. Its goal is to generate an electrical response in the postsynaptic cell. Although it is often the dendrites that are the postsynaptic beneficiaries of the chemical message, the electrical change produced in the dendrite has to be propagated to the cell body, and then to the axon, before an action potential can occur. This is so because the action potential is generated in the initial part of the axon where it connects with the cell body (fig. 3.7).

The arrival of transmitter from a single presynaptic terminal is typically not sufficient to produce an action potential in the postsynaptic cell (fig. 3.7). Only if the postsynaptic cell is bombarded with transmitter molecules from many presynaptic terminals at about the same time—within milliseconds—will an action potential result.[38]

A given postsynaptic cell is believed to receive relatively few synaptic contacts from any one presynaptic neuron. As a result, much of the convergence that drives a postsynaptic cell toward action potentials comes from the convergence of different presynaptic cells onto the postsynaptic neuron (that is, the near-simultaneous arrival of neurotransmitter from different presynaptic neurons). In order for the inputs to arrive in the postsynaptic cell body at about the same time, action potentials have to have been triggered in the various presynaptic cells at about the same time. The timing has to be adjusted for different lengths of axons, since, as Helmholtz demonstrated, the longer the axon, the longer it takes for the action potential to travel down it. Keeping time in the nervous system is a very complex job.

Once the postsynaptic cell generates an action potential, its role shifts from that of a receiver to a sender. It now becomes a presynaptic neuron that helps fire action potentials in other cells.

The full sequence of communication between neurons is thus usually electrical-chemical-electrical: *electrical* signals coming down axons get converted into *chemical* messages that help trigger *electrical* signals in the next cell. There are also synapses through which communication between presynaptic and postsynaptic sites is purely electrical,[39] but chemical transmission is the more prevalent form. Thus, much of what the brain does involves electrical-to-chemical-to-electrical coding of experience. As hard as it may be to imag-

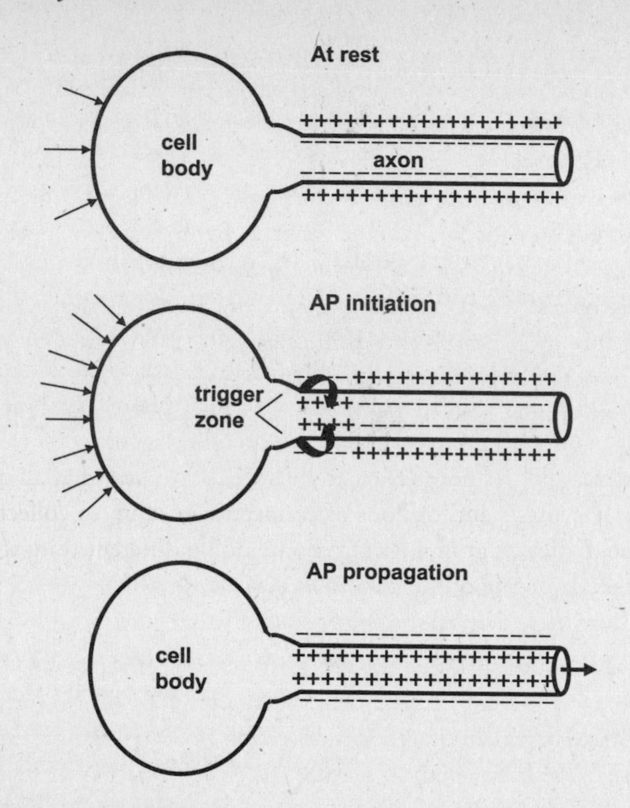

FIGURE 3.7 ACTION POTENTIALS

When a neuron is activated by other neurons, an action potential is initiated. This electrical storm begins at the trigger zone (the region where the axon joins with the cell body) and travels down the axon. *Top:* Neurons are "at rest" when they are not receiving sufficient inputs to alter their electrical properties. When at rest, the electrical charge of the inside of the axon is negative with respect to the outside (see the + and – signs along the axon). *Middle and bottom:* When enough inputs (arrows on left) converge at about the same time, an action potential is generated and propagated down the axon toward the terminal. The propagation process involves a wave of electrical change (inside becomes more positive at that spot) that moves step-by-step down the axon. When the terminal is reached, neurotransmitter is released into the synapse. Based on figure 2.6 in Guyton 1972.

ine, electrochemical conversations between neurons make possible all of the wondrous (and sometimes dreadful) accomplishments of human minds. Your very understanding that the brain works this way is itself an electrochemical event.

FROM CELLS TO CIRCUITS AND SYSTEMS

Every human brain has billions of neurons that together make trillions of synaptic connections among one another. Chemicals are oozing and sparks flying constantly, during wakefulness and during sleep, during thoughtfulness and during boredom. At any one moment, billions of synapses are active.

Imagine a large cocktail party at which hundreds of people are standing around and chatting with one another. If you were to place a microphone in the chandelier at the center of the room high above the crowd, you probably wouldn't be able to make out what was being said, for the many unrelated conversations would blend together in the microphone. You'd learn more by listening in on small groups than by eavesdropping on the whole room at once. In the same vein, it's not particularly instructive to ask what all of the brain's billions of neurons and trillions of connections are up to collectively at any one time. Different groupings of cells are doing different things, so attempting to take a reading of them all together doesn't tell you much. It would be more informative to examine the operation of specific circuits or systems.

A *circuit* is a group of neurons that are linked together by synaptic connections. A *system* is a complex circuit that performs some specific function, like seeing or hearing, or detecting and responding to danger. Seeing, for example, involves the detection of light by circuits in the retina, which sends signals, by way of the optic nerve, to the visual thalamus, where the visual information is processed by circuits that relay their output to the visual cortex, where additional circuits engage in further processing, ultimately creating visual perceptions. The visual system, like other brain systems, can thus be thought of as a series of hierarchically arranged circuits linked together by synaptic connections to perform some function.

Synaptic interactions between two types of neurons, called projection neurons and interneurons, are key to understanding how circuits and systems function.[40] Projection neurons have relatively long axons that extend out of the area in which their cell bodies are located. In a hierarchical circuit, their main job is to turn on the next projection cell in the hierarchy (fig. 3.8). They do this by releasing a chemical transmitter that increases the likelihood that the postsynaptic cell, the next projection cell, will fire its own action potential. Projection cells tend to activate or excite postsynaptic cells.

Interneurons, also called local circuit cells, send their short axons to nearby neurons, often projection neurons, and are involved in information processing within a given level of a hierarchical circuit (fig. 3.8). One of their main

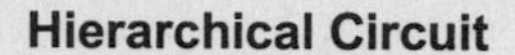

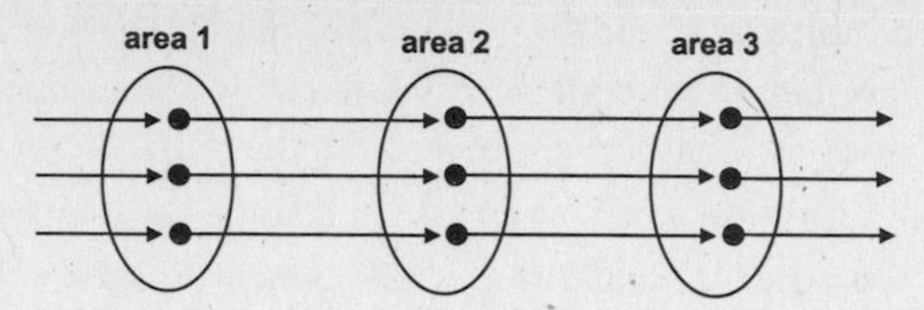

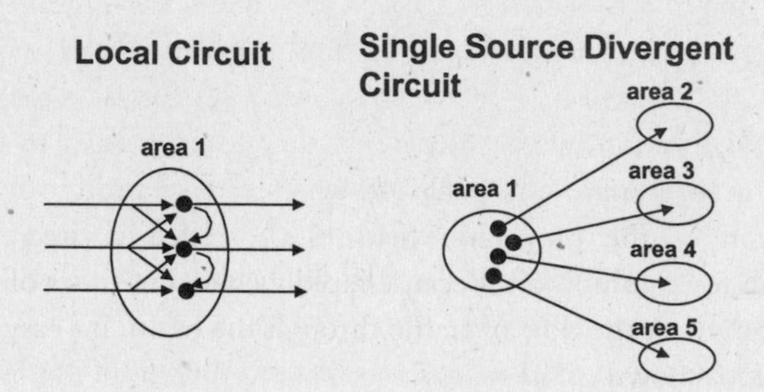

FIGURE 3.8 THREE TYPES OF CIRCUITS

Information is transmitted from area to area in sequence in hierarchical circuits. At each level of the hierarchy, though, processing is regulated by other kinds of circuits. Local circuit connections alter the processing at each hierarchical stage and also determine the ease with which activity in one area can influence the next. Single source divergent projections are typically made up of neurons located in one brain region that possess a particular chemical (typically, a neuromodulator like serotonin or dopamine—see text). These chemicals are then released at widespread areas and can influence processing by other circuits. Transfer of information from one level of a hierarchical circuit to another typically involves excitatory connections that are regulated by inhibitory local circuits, and both hierarchical and local circuit transmission is modulated by single source divergent connections. The terminology for these three circuit types is based on Bloom and Lazerson 1985.

jobs is to regulate the flow of synaptic traffic by controlling the activity of projection neurons. Inhibitory interneurons release a transmitter from their terminals that decreases the likelihood that the postsynaptic cell will fire an action potential. These neurons play an important role in counterbalancing the excitatory activity of projection cells.

Projection cells tend to be idle in the absence of inputs from other projection cells. Inhibitory interneurons, though, are often tonically active, which means they are firing all the time. Part of the reason why projection cells are inactive when not being stimulated is that they receive tonic inhibition from interneurons. As a result, when excitatory inputs try to turn on a projection cell, preexisting inhibition of the projection cell has to be overcome. The balance between excitatory and inhibitory inputs to a neuron determines whether it will fire.

The amount of inhibition affecting a cell can change from moment to moment, depending on other factors. For example, when projection cells in one area of a hierarchical circuit send enough convergent inputs at about the same time to activate projection cells in the next area, the level of inhibition in the second area usually goes up as well. This happens because the excitatory inputs to an area often activate interneurons as well as projection neurons. The momentary increase in excitatory inputs to interneurons leads to a momentary increase in their inhibitory behavior, which in turn produces a momentary inhibition of the projection neurons. So-called elicited inhibition contrasts with tonic inhibition. Because rapidly changing states of excitation and inhibition direct the flow of traffic through the brain, it's easy to understand how a breakdown in the flow of impulses could lead to neural gridlock.

Consider an example that will help illustrate how elicited and tonic inhibition regulate excitation. Imagine a circuit consisting of two projection neurons (A and B) linked together in a series (fig. 3.9). When A is active, B fires. If the job of the circuit were to make B fire action potentials as often as possible as long as A is active, these two neurons would be sufficient to do the job. But suppose its job instead is to take a barrage of action potentials in A and turn them into fewer action potentials in B, something that actually occurs quite often in the brain. This could be achieved by giving neuron B an inhibitory playmate (I). This local circuit neuron, like B, receives the output of A and then connects with B. So when A fires, it turns on B and I, and each produces an output. The output of B helps turn on the next cell in the circuit, while the output of I turns B off. As a result, B now produces fewer action potentials when it is fired by A.

Now suppose that the interneuron I is constantly inhibiting the projection cell B. With this tonic inhibition added in, it is going to be much harder for the input from A to trigger the projection cell. If we put more excitatory neurons in with A to drive B, and time arrival just so, the tonic inhibition can be overcome. The cell can now be continuously activated. But being stuck in fast-forward is not good for neurons, which can be damaged or even destroyed by unchecked excitation. Each burst of excitation thus needs to be countered with another round of inhibition. That's where elicited inhibition, like that described above, comes in. When an excitatory surge overcomes

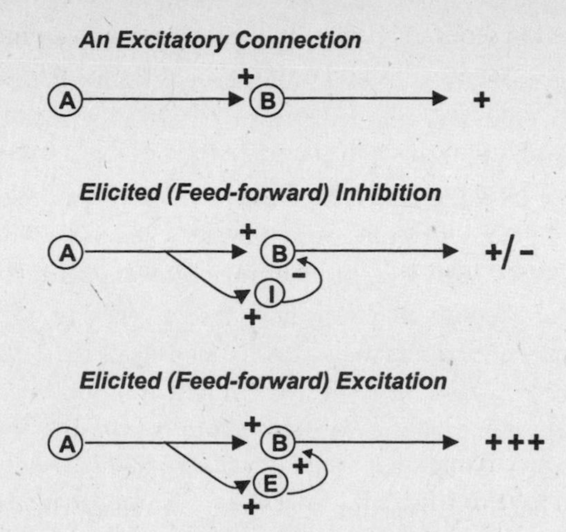

FIGURE 3.9 EXCITATION AND INHIBITION IN CIRCUITS
Excitation and inhibition are illustrated by way of an excitatory connection from A to B that is regulated by the inhibitory connection from I to B. The + and – signs to the right indicate the effect of the particular combination of connections. *Excitatory connection:* Activity in A leads to activity in B (+ on the right). *Feed-forward inhibition:* Activity in A leads to activity in B. A also activates I, which in turn inhibits B. B is thus first turned on by A (+ on right) but then turned off by I (– on right). The excitation of I by A thus gates or regulates the excitation of B by A. *Feed-forward excitation:* Activity in A leads to activity in B. Also activates E, which in turn further excites B. The excitation of B by A is thus amplified by E (+ + + on right).

tonic inhibition, elicited inhibition can rein in the excitation, resetting the circuit, preparing it for new inputs. There are many variations on this theme of tonic and elicited inhibition, but the scenario just described gives an idea of how inhibition in general works.

Inhibition is a very useful device in neural circuits. It adds tremendously to the specificity of information processing, filtering out random excitatory inputs, preventing them from triggering activity. Only if the excitatory inputs arrive simultaneously can they overcome the inhibition and elicit activity. And once activity is elicited, inhibition is important for keeping the excitation in check and resetting the circuit.

Although many local circuit cells are inhibitory, some are excitatory. Just as inhibitory interneurons can be thought of as filters, excitatory interneurons can be viewed as amplifiers. Again imagine a circuit consisting of neurons A and B connected in series. As before, B is associated with an interneuron, but in this case it's an excitatory interneuron (E), and the axon of A branches and contacts both B and E (fig. 3.9). When A turns on B, the interneuron E is also activated, and its output causes further excitation of B. As a result, the output of B is amplified by an excitatory interneuron just as it was reduced by the inhibitory interneuron. But, as we've seen, all this excitation ultimately has to be regulated, both to maintain normal functions and to prevent injury.

THE CHEMICAL BROTHERS

The job of a projection neuron, as we now know, is to turn on the next projection cell in the circuit. This means that action potentials in the axons of projection cells have to trigger the release of chemicals that cross the synapse and contribute to the firing of an action potential in the postsynaptic cell. Projection cells thus need to use a chemical neurotransmitter that has two properties. The transmitter first must be able to act quickly at postsynaptic sites—otherwise, our perceptions and other mental states could not keep up with rapidly changing events. And it must also be able to change the electrical state of the postsynaptic cell in such a way that the occurrence of an action potential is more likely to occur. Both requirements (speed and excitation) are fulfilled by the amino acid neurotransmitter glutamate, which is the main transmitter in projection neurons throughout the brain.

Glutamate actually has two roles in body function. In addition to serving as a neurotransmitter in the brain, it also plays a major part in basic life-sustaining metabolic processes that go on continuously throughout the body. For example, it is a building block in the construction of peptides and proteins, which are basic ingredients of living tissues. And, in the brain, it helps detoxify ammonia, which is a natural by-product of certain chemical reactions. Although glutamate is now known to be a ubiquitous excitatory transmitter in the brain, its role in transmission was for a long time hard to dissociate from its so-called metabolic functions.[41]

In contrast, inhibitory neurons, especially inhibitory interneurons, often release the amino acid GABA (short for gamma-aminobutyric acid) from the terminals of their short axons.[42] In contrast to glutamate, this inhibitory transmitter reduces the likelihood of an action potential being generated in the postsynaptic cell. By sending axons to nearby projection neurons, GABA interneurons thereby regulate the flow of traffic through a given area.

GABA actually was identified as a neurotransmitter long before glutamate.

Because it was well established that glutamate was one of the essential chemical components involved in the synthesis of GABA, its metabolic role in GABA production hampered the discovery that glutamate was itself a neurotransmitter.

Glutamate and GABA are together responsible for much of the neurotransmission business in the brain. If you understand the work done by these two chemicals, you will understand quite a lot about how synapses function. These and all other transmitters work by attaching to molecules called receptors on the postsynaptic cell. Receptors selectively recognize and bind (literally, hold on to) transmitter molecules. Glutamate receptors recognize and bind glutamate, but ignore GABA (fig. 3.10); GABA receptors are just as selective (fig. 3.10). How, then, does the binding of glutamate and GABA molecules to their receptor molecules lead to excitation and inhibition?

All cells in the body are completely enclosed by a membrane, which defines the boundary of an individual cell. The membrane is like a formfitting bag, a spandex suit, in which the cell is contained. In neurons, it covers the axons and dendrites as well as the cell body. The space outside the membrane between neurons is called the extracellular space. The fact that the extracellular space is filled with liquid has two important consequences.

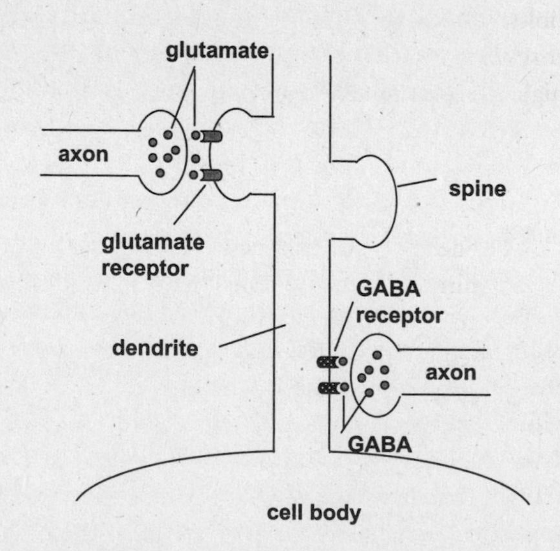

FIGURE 3.10 GLUTAMATE AND GABA SYNAPSES

The two major neurotransmitters are glutamate and GABA. These are released from different presynaptic neurons and bind to distinct postsynaptic receptors.

First of all, this liquid is a medium that allows transmitter molecules to cross the extracellular space between presynaptic and postsynaptic sites—the synapse. Transmitters do this by diffusing out from the terminal. The distance they have to travel is very small (it's measured in tiny units called angstroms, one of which equals one ten-millionth of a millimeter), making the postsynaptic site a close and easy target.

The second point is that extracellular liquid contains all sorts of chemicals, many with electric charges, that influence cellular function. The cell membrane keeps chemicals that are inside and outside the cell separate. At rest (when the cell is not being influenced by inputs), the chemical composition of the inside of the cell is more negatively charged than the fluid outside, due to the kinds of ions that are present on the other side of the cell membrane. Neuroscientists have measured the difference in electrical charge between the inside and outside of a nerve cell. In general, the inside of a neuron that is not being stimulated is about 60 millivolts (60 one-thousandths of a volt) more negative than the outside. In other words, the resting potential of the cell is about −60 mV.

For our purposes, the actual voltage is not that important. All we need to keep in mind is the fact that the membrane potential is fairly negative at rest. When a neuron is stimulated by excitatory inputs from other neurons, however, the membrane potential becomes more positive (see fig. 3.7). The reason for this is related to the way that glutamate works as a neurotransmitter.

Glutamate receptor molecules span the cell membrane, with part facing inside the cell and part facing outside. When glutamate (released from a presynaptic terminal) binds to the outside part of a postsynaptic receptor, a passage opens up through the receptor, allowing positively charged ions in the extracellular fluid to move inside the cell, which changes the chemical balance between outside and inside. If enough glutamate receptors are occupied on the postsynaptic cell at about the same time, and the voltage inside becomes sufficiently positive, then an action potential occurs.

In contrast, when GABA receptors are occupied, the inside of the cell becomes more negative (due to the influx of negative ions, especially chloride, through a passage in the GABA receptor). This makes it harder for glutamate released from other terminals to change the concentration of the positive ions in the postsynaptic cell sufficiently to trigger an action potential. Whether an action potential occurs, then, depends on the relation between glutamate (excitation) and GABA (inhibition). And since any one cell receives many excitatory and inhibitory inputs from many other cells, the likelihood of firing at any one moment depends on the net balance between excitation and inhibition across all of the inputs at that particular time.

Glutamate receptors tend to be located out on the dendrites, especially in the spines, whereas GABA receptors tend to be found on the cell body, or on the part of dendrites close to the cell body. In order for glutamate-mediated excitation to reach the cell body to help trigger an action potential, it has to get past the GABA guard. Excitation coming down a dendrite and headed for the cell body can be extinguished by GABA.

Without GABA inhibition, neurons would send out action potentials continuously under the influence of glutamate, and would eventually literally fire themselves to death. This effect has been demonstrated in experiments where the action of GABA is blocked artificially, or where powerful doses of glutamate-related compounds, too strong to be inhibited by natural levels of GABA, are administered. Overactivity of glutamate, and the resulting injury to neurons, actually plays an important role in stroke and other vascular disorders of the brain, as well as in epilepsy and possibly Alzheimer's disease. Some people have experienced mild versions of glutamate toxicity after eating Chinese food. Monosodium glutamate (MSG), sometimes used as an additive in this cuisine, can increase the amount of glutamate in the body to the point of causing headaches, ringing ears, and other physical symptoms. Regulation of GABA inhibition is one of the ways that psychoactive drugs work. For instance, the antianxiety drug Valium works by enhancing GABA's natural ability to regulate glutamate. Excitatory inputs that would normally elicit anxiety by firing action potentials in fear circuits are less able to do so in the presence of Valium and related drugs.

MOD SQUADS

Interactions between glutamate and GABA are key to understanding information processing by the brain, but these substances do not work alone or in isolation. For example, when receptors in the eye detect patterns of light, they send messages through the axons of the optic nerve to the brain. When the electrical signal reaches the axon terminal, glutamate is released. Whether the postsynaptic cell fires depends not only on the counterbalancing force of GABA inhibition, but also on other chemicals that are present at the time. These are called modulators.

Modulators are neurotransmitters in the sense that they provide a chemical link between the site from which they are released and the location of the receptors upon which they act. But in contrast to glutamate and GABA, they are less directly involved in the transfer of information from point to point in hierarchical circuits. The way a modulator is distinguished from a transmitter is different for different kinds of modulators, as we'll see soon. And sometimes, the distinction is murky. But one important difference is related to their speed. Glutamate and GABA are fast-acting:[43] they cause an electrical change in the postsynaptic cell within milliseconds of being released from the presynaptic terminal, and their effect is over in a matter of milliseconds.[44] Modulators, on the other hand, have slower and longer-lasting effects.

We'll consider three classes of modulators: peptides, amines, and hormones. Each can have excitatory or inhibitory effects, depending on the specifics of their participation in functional circuits.

Peptides represent a large class of slow-acting modulatory substances found throughout the brain. They are made up of many amino acids, and are larger molecules than simple amino acids like glutamate or GABA. Because peptides are often present in the same axon terminal as glutamate or GABA (but in their own separate storage compartments), they are released with the fast transmitter when an action potential comes down the axon (fig. 3.11). But peptides bind to distinct postsynaptic receptors and can, as a result, augment or reduce the effect of the fast transmitter with which they are released. However, since peptides are slow to affect the postsynaptic site, and their effects are long-lasting, they tend to have more of an effect on subsequent squirts of fast

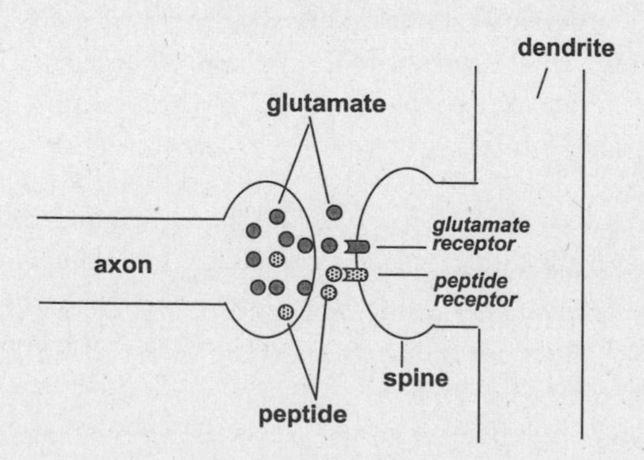

FIGURE 3.11 GLUTAMATE AND PEPTIDE RELEASED FROM THE SAME TERMINAL

Axon terminals sometimes release peptide transmitters along with glutamate (or GABA). In this case, different postsynaptic receptors bind the two kinds of molecules released from the same terminal.

transmitter. While glutamate and GABA can have slow effects as well as fast ones, depending on the receptors involved,[45] peptides typically only have slow modulatory actions. They can affect dramatically the ability of a cell to be fired by other inputs, but cannot do so with precise timing.

There are many, many peptides that participate in a wide variety of bodily functions. Our interest is in the neuroactive peptides, those that act in the nervous system. The best known of these are the opiates—endorphins and enkephalins. These are triggered by pain and stress and bind to their special receptors, altering pain sensations and mood. "Jogger's high" is said to be an opiate effect. Morphine generates its effects by binding to these receptors.

The monoamines, another class of modulators, include substances like serotonin, dopamine, epinephrine, and norepinephrine. Unlike most other transmitters and modulators, the cells that produce monoamines are found in only a few areas, mostly in the brain stem,[46] but the axons of these cells extend to widespread areas throughout the brain (figs. 3.8 and 3.12). In this way, a small number of highly localized neurons making monoamines can influence cells in many other locations. Monoamines achieve their effects by facilitating or inhibiting the actions of glutamate or GABA (and the peptides that are released with them). Because the axons are so widely distributed, monoamines have relatively nonspecific effects. They are thus not involved in precise representation of stimuli in specific circuits. Instead, monoamines produce global state changes in many brain areas simultaneously, such as the high degree of arousal occurring throughout the brain when we encounter a sudden danger or the low degree of arousal required when we are going to sleep.

Many drugs used in the treatment of psychiatric disorders work by altering monoamines. Prozac, for example, prevents the removal of serotonin from the synaptic space. Normally, as part of the process by which transmitter action is regulated, neurotransmitters are sucked back into the terminals that release them. By preventing the removal of serotonin, allowing more to stay around longer, Prozac amplifies its effects. One theory holds that there is a deficiency of serotonin in depressed or anxious brains, which Prozac helps correct.[47] The exact means by which the increase in serotonin levels relieves anxiety or depression is not known.

Antidepressant drugs (like monoamine oxidase inhibitors and tricyclic antidepressants) and antipsychotics (like chlorpromazine or phenothiazine) also work by altering monoamine levels. Amines are also targets of recreational drugs: cocaine and amphetamine affect norepinephrine and dopamine levels, while LSD acts on serotonin receptors.

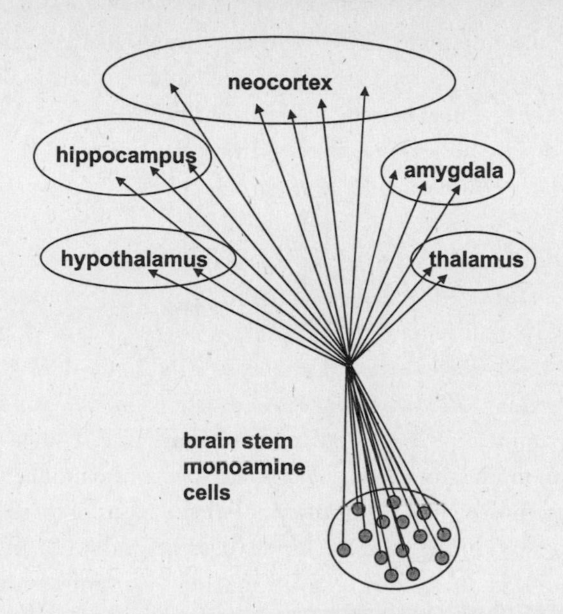

FIGURE 3.12 DIFFUSE PROJECTION OF BRAIN STEM MONOAMINE CELLS TO FOREBRAIN AREAS

Monoamine neuromodulators are made in discrete areas of the brain stem, but because of their diffuse connections, they can simultaneously modulate transmission in widespread areas of the brain.

Another monoamine is acetylcholine, which functions as a fast transmitter when it works with one receptor and as a modulator with a different receptor.[48] Disruption of acetylcholine in the neocortex is believed to play a role in Alzheimer's disease,[49] and many drugs that have been tested as treatments for Alzheimer's alter acetylcholine function.[50] Acetylcholine is also a very important transmitter in the body, involved with nerves such as those that control muscle movements and heart rhythm. Nerve gas works by disrupting acetylcholine transmission at muscles, especially muscles required for normal breathing. Many insecticides have similar effects in bugs.

Hormones are the last class of modulators we will consider (fig. 3.13). Typically, they are released from bodily organs (like the adrenal, pituitary, or sex glands) into the bloodstream where they travel to the brain. There they can, like other modulators, alter the efficacy of glutamate or GABA transmission by binding to specific receptors on cells. For example, cortisol, a steroid hormone released from the adrenal gland during stress, is known to alter informa-

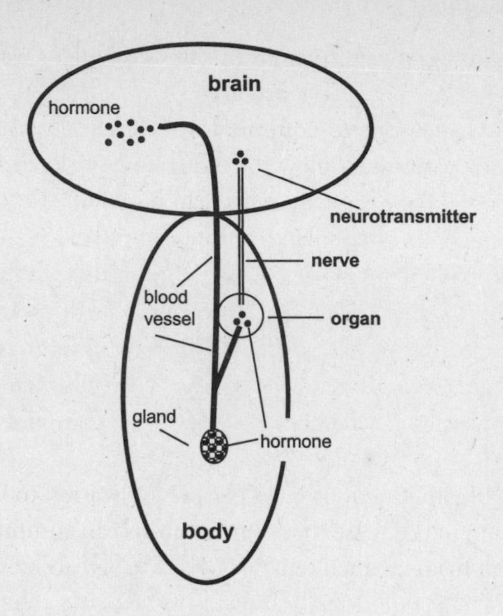

FIGURE 3.13 HOW HORMONES REACH THE BRAIN

Hormones released from glands in the body travel in the bloodstream and either can influence the brain directly or can influence the brain indirectly by acting in body organs that send nerves into the brain.

tion transmission in a variety of circuits involved in memory and emotional processes,[51] in part by altering the ability of GABA to inhibit glutamate.[52] Sex hormones, such as testosterone and estrogen, also can have profound effects on neural transmission and other brain functions. The mood-altering effects of monthly variation in estrogen levels in females are widely discussed, and estrogen replacement therapy during and after menopause is believed to counter some of the effects of aging on brain functions.[53] Because hormones reach the brain through the bloodstream, they can influence many regions simultaneously. However, since only certain areas, and only certain circuits in those areas, possess the relevant receptors, considerable specificity can be achieved by hormonal modulation.

GOLGI AND THE GAP

As important as chemical synaptic transmission is in the brain, another form, called electrical transmission, also occurs. Although the extent to which electrical synapses operate is not known, it is becoming more and more apparent that they are significant forces for us to deal with as we conceive brain function.

In order for two neurons to communicate electrically, their membranes have to fuse in such a way as to allow the direct flow of electricity from one to the other. These points of fusion are called gap junctions. Recent studies have shown that in some brain areas, like the hippocampus, a region important for the formation of explicit memories, GABA (inhibitory) cells are linked together, or electrically coupled, by gap junctions.[54] In this way, when GABA cells are activated, excitation can spread between them in such a way as to activate many of the interconnected cells at once. The cells then fire together, in synchrony, and thereby can regulate activity of projection cells throughout the region.

The existence of gap junctions gives partial vindication to Golgi's reticular theory of the brain, in the sense that some neurons can communicate directly by way of physical fusion. Much remains to be learned about them, and their contribution to synaptic transmission needs to be better integrated with our knowledge of chemical transmission.

CIRCUITS IN ACTION

The same basic transmitters, modulators, and hormones can be involved in very different functions. Our abilities to see, hear, remember, fear danger, and desire happiness all involve excitatory (glutamate) synaptic transmission regulated by inhibitory (GABA) synapses and modulated by peptides, amines, and hormones. What makes a sound different from a sight, a memory different from a perception, a fear different from a desire is not so much the chemistry involved but instead the specific circuits in which the chemicals act. As a way of illustrating how glutamate, GABA, and modulators work, let us consider their role in the detection of danger by the amygdala.

The amygdala detects danger by virtue of its position in a synaptically connected system. In its simplest form, this system can be described in terms of a three-level excitatory chain of cells that releases glutamate—projection cells in sensory systems activate projection cells in the amygdala, which activate projection cells in motor control areas (fig. 3.14). This scheme leaves much out, but we'll have the opportunity to embellish it later.

Amygdala cells receive inputs from the sensory world constantly, but they ignore the majority of them. In fact, they tend to be quiescent most of the time. They do get worked up, though, when the right kind of stimulus is present—one that signifies danger or some other biologically significant event.

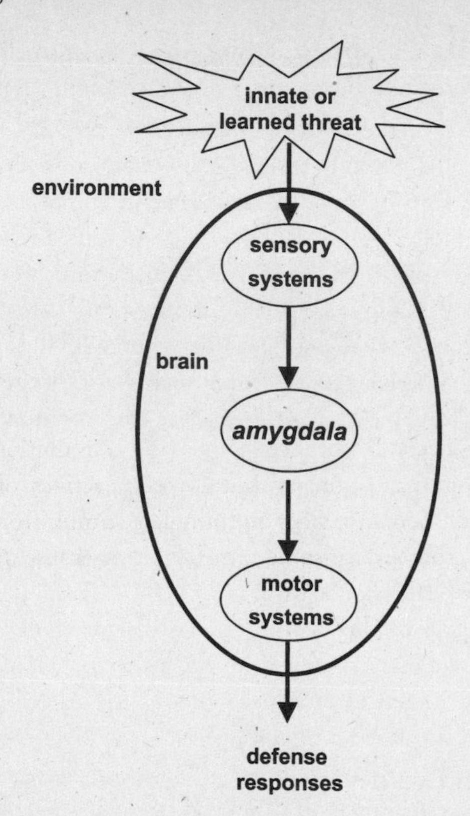

FIGURE 3.14 INPUT AND OUTPUT CONNECTION OF THE AMYGDALA IN FEAR

The amygdala is able to serve as an interface between threatening stimuli in the environment and defense responses because it is connected with sensory processing systems on the one hand and with motor control regions on the other.

This has been shown to be true in studies of both lower animals[55] and humans.[56] So what keeps a projection cell in the amygdala from firing in response to meaningless stimuli? The answer, as you've probably guessed, is GABA.[57]

As we've seen, the resting membrane potential of cells in many brain areas is about −60 mV. In the amygdala, however, some cells can be as negative as −80 mV,[58] due to sustained or tonic inhibition by GABA. With GABA receptors on amygdala projection cells occupied and passing chloride, the inside of the cells becomes more negative, which means it takes extra excitation to turn the amygdala on. As a result, not any old stimulus will do the trick. The stimulus has to have special qualities that allow it to overcome the tonic inhibition produced by GABA.

Stimuli that are inherently dangerous (the sight or smell of a predator) or unpleasant (intense stimuli, like loud noises or stimuli that cause pain) are able to overcome the tonic inhibition, as are stimuli that have emotional resonance acquired through past learning. Thus, an otherwise meaningless sound of modest intensity that previously occurred in association with pain has the same effect as a natural (innate) form of danger.[59] Both innate (hard-wired) and learned danger signals cause amygdala cells to fire rapidly for a sustained period, and are thus able to overcome the GABA guard.

Even after fear-arousing stimuli get past tonic inhibition and cause amygdala cells to fire, however, they are still subject to GABA control. The inputs to the amygdala activate GABA cells as well as projection neurons.[60] As a result, as the inputs become more active, the elicited inhibition in the amygdala builds up, which in turn begins to shut down the activity of amygdala cells.[61]

If the ability of GABA to keep meaningless stimuli from turning on the amygdala is compromised for some reason (either because the projection cells come to fire more easily or because the GABA cells fire less easily), stimuli that are not dangerous come to be responded to as though they were. This may occur in certain fear and anxiety disorders. By the same logic, things that make projection cells fire less readily or that make GABA cells fire more readily should reduce fear and anxiety. Indeed, one of the most popular medications ever invented for the treatment of anxiety is Valium, which works by facilitating GABA transmission. Although drugs taken orally reach many sites in the brain, it is likely that at least some of their effects on fear and anxiety are achieved by enhancing inhibition in the amygdala, and thereby making it harder for external or internal stimuli to elicit fear responses by activating amygdala circuits.

The amygdala also receives modulatory inputs of various types. For example, serotonin fibers terminate there, and when the amount of serotonin rises in the amygdala the activity of excitatory projection cells is inhibited.[62] The inhibition in this case is not due to the fact that serotonin directly affects projection cells, but rather that serotonin excites GABA cells, and thus increases the degree to which they inhibit projection neurons.

Drugs like Prozac work by increasing the amount of serotonin available at synapses. By enhancing serotonin transmission at GABA synapses in the amygdala, and thereby reducing the activity of projection neurons, Prozac may, like Valium, help control anxiety by reducing the ability of inputs to the amygdala to activate fear circuits.

The amygdala is also the target of many hormones. One of these is cortisol, which is released from the adrenal cortex during fear-arousing and otherwise stressful events.[63] The facilitation of GABA inhibition of amygdala projection cells by serotonin is modulated by cortisol.[64] Serotonin's ability to facilitate inhibition by exciting GABA cells thus depends on the binding of cortisol to receptors located on amygdala neurons. Cortisol is elevated in a variety of psychiatric disorders,[65] and cortisol increases the intensity of fear reactions.[66] Drugs like Prozac may reduce exaggerated fear and anxiety in psychiatric disorders by enhancing the ability of serotonin to facilitate GABA inhibition in the presence of elevated cortisol.

The fear system thus nicely illustrates the basic elements of neural transmission in the brain and its regulation by modulatory chemicals. We will build upon these points at various times in later chapters.

ARE SYNAPSES ENOUGH?

My emphasis on the importance of synapses in brain function is not intended to minimize the role of other factors. For example, the rate at which a cell fires spontaneously is a function of certain electrical and chemical characteristics of the cell.[67] These are called intrinsic properties to distinguish them from extrinsic influences from other cells mediated by synaptic transmission and modulation. A cell's intrinsic properties, which may have a strong genetic component, will greatly influence everything that cell does, including its participation in synaptic transmission. But because psychological and behavioral functions are mediated by aggregates of cells joined by synapses and working together rather than by individual neurons in isolation, the contribution of the intrinsic properties of a cell to mental life or behavior occurs only by way of the role of that cell in circuits. While synapses themselves don't account for everything the brain does, they do participate crucially in every act or thought that we have, and in every emotion we express and experience. Synapses are ultimately the key to the brain's many functions, and thus to the self.

CHAPTER FOUR

BUILDING THE BRAIN

EACH CHILD IS AN ADVENTURE INTO A BETTER LIFE—AN OPPORTUNITY TO CHANGE THE OLD PATTERN AND MAKE IT NEW.

—Hubert H. Humphrey

The brain's billions of neurons are intricately connected in ways that make possible the mundane (such as the regulation of breathing) and the marvelous (the belief in an idea). But how do cells in the developing embryo become neurons, and how do they end up in just the right places? How do the axons of all these cells find their way to their target areas? And once having reached them, how do the terminals figure out exactly which neurons to make synapses with? Because the various steps take time, and because different circuits go through these steps on different schedules, our behavioral and mental repertoire unfolds gradually, and unevenly, during childhood. Somehow, though, it all comes together, and a person, a self with all its aspects, emerges.

Brain development is the major battlefield of the nature-nurture conflict. In its simplest form, the debate is about whether mental and behavioral characteristics are determined more by genes or by environment. To the extent that mental and behavioral characteristics are functions of the brain, and synaptically connected circuits underlie brain functions, the nature-nurture debate essentially reduces to questions about how circuits are built during development.

No one today seriously proposes that the brain is a blank slate at birth, waiting to be written on by experience, or, alternatively, that it is a genetically predetermined, unchangeable repository of tendencies to act, think, and feel a particular way. Instead, it is widely believed that brain circuits come about through a combination of genetic and nongenetic influences, and the debate now hinges less on the dichotomy than on the manner in which nature and nurture contribute to brain construction.

The dichotomy, in fact, begins to dissolve when it is realized that, regarding questions of mind and behavior, nature and nurture are really two ways of doing the same thing—wiring up synapses—and both are needed to get the job done. We commonly think of experiences as leaving their mark on the brain through the record of memory, and, as we'll see in later chapters, memory is a product of synapses. It is less common, but no less appropriate, to think of genes as also influencing us in the form of memory. In this case, though, the synaptic memory comes about as a result of ancestral rather than personal history. The shaping of synaptic connections in early life by genes and experience is the topic of this chapter.

FIRST STEPS

The earliest events in embryonic brain development are largely controlled by genes, their products, and the local chemical environment in which they exist.[1] The job of genes is to make proteins, which regulate many aspects of brain development. Some proteins are enzymes that trigger chemical reactions, others induce additional genes to make additional proteins, some form barriers that guide and restrict the many cell movements that take place, and still others provide adhesive surfaces on cells to which other cells cling while making their way to their final destinations. When we speak of genetic influences on brain development, we are essentially describing the effects of proteins and their chemical spin-offs.

The young embryo, lacking sensory systems, is largely isolated from direct perceptual contact with the external environment. But even in the earliest stages of development, genes do not operate completely independently of the outside world. The chemical environment of the embryo is, by necessity, in direct contact with the body chemistry of the mother. The embryo cannot make on its own the amino acids that are used to assemble the proteins that are required for brain and body development. They have to be obtained from the mother, who gets them from the food she eats. The mother's diet can also be the source of less desirable substances—toxins and chemical additives in foods, for example—as can the air she breathes, drugs she takes (prescription and recreational), and cigarettes she may smoke. The mother's level of stress will affect her hormonal state, which can influence the embryo, as can antibodies she makes to fight off infections. Although the major features of the brain are dictated by a genetic plan (which guarantees that all human brains look and work much the same), this plan requires certain conditions in the internal chemical environment in which neurons will grow. If this gene–internal environment interaction is perturbed, so, too, will be the normal development of the brain. Nature and nurture interact from the start.

Brain development begins in the ectoderm, which, together with the mesoderm and endoderm, make up the three major parts of the embryo. These give rise to the various regions and organs of the body. A thickening of a part of the ectoderm forms the neural plate, which folds to form the neural tube, out of which the brain begins to be constructed.

In the neural tube, nonneural precursor cells divide and give rise to neurons. In humans, the vast majority of neurons are made in the months just prior to birth. At the peak production point, about 250,000 neurons are being generated per minute. This process is controlled by hormones that diffuse up into the neural tube from underlying tissues and turn on genes that make proteins, which, in turn, regulate neuron production by precursor cells. (The hormones themselves are gene products made in nonneural parts of the embryo.) If production of neurons is interfered with during this early period (say, by consumption of alcohol or drugs), birth defects, such as spina bifida or anencephaly, can result. However, not all birth defects are due to drug or alcohol consumption, and drug or alcohol consumption does not always lead to birth defects.

According to conventional wisdom, no new brain cells are formed in adult mammals—the brain cells you acquire in early life are pretty much the ones you die with, minus the ones you lose for various reasons over the course of years. However, recent studies by Fernando Nottebohm and Bruce McEwen of Rockefeller University, Elizabeth Gould of Princeton, and others have challenged the conventional view, demonstrating that neurons continue to be generated in some areas of the adult brain.[2] At this point, it is not known how widespread the birth of new neurons is, and it may be fairly rare. However, this finding offers at least the glimmer of a hope that brain scientists might someday be able to develop therapies fostering the growth of new neurons for largely untreatable neurological disorders, such as spinal cord injuries, stroke, and Alzheimer's disease. It's important to make a distinction here between neurogenesis—the birth of new neurons—and synaptogenesis—the creation of new synapses between existing neurons. Synaptogenesis is a common phenomenon, one that probably occurs up until the moment of death. As we will see in this chapter, synapses change dramatically during early life, and in the next two chapters, we'll see how synapses are changed every time our brain records an experience.

Soon after their birth, neurons begin to segregate, with cells destined to be part of the hindbrain, midbrain, and forebrain taking up different territories in the growing neural tube. Segregation is under the direct control of a special set of genes called homeotic genes, which make proteins that control the placement of cells, providing boundaries that guide and restrict cell movement, and adhesive surfaces that allow cells to cluster together. Some scientists believe that autism might be due to a mutation of homeotic genes that leads to faulty brain construction and connection.[3] The function of homeotic genes was discovered in studies of fruit flies, and is being elaborated on in studies of flies and worms, especially the worm *C. elegans.* It may be difficult to believe that we have much in common with flies and worms, but the fact is that these genes, as well as a number of others, have been preserved through many levels of evolutionary history.[4]

Segregated cells eventually differentiate, which means that they take on different shapes and sizes, and end up with different transmitters and modulators. This is how, for example, projection and interneurons come to differ. Differentiation is under genetic control but is not strictly a genetically determined process. Studies have shown that if cells from one brain region are transplanted to a different region in another animal prior to differentiation, they take on the characteristics expected of cells in the region of the host rather than that of the donor.[5] Chemical factors in the local environment appear to determine the ultimate type of cell that will be expressed. However, the switch can only occur in very young cells, indicating that once a cell's type is determined, its fate is sealed. This suggests that cell type is not rigidly dictated by genes and is strongly influenced by the environment, albeit in this case the local chemical environment that surrounds the cell rather than the environment that is external to the organism as a whole. The local cues involved, though, are proteins that have themselves been genetically coded, so the nongenetic contributions to cell differentiation are not too far removed from genetic ones.

As development proceeds, the neural tube expands and bends, eventually assuming a shape that begins to resemble the brain. While segregation initially places cells destined to be in the same brain regions together in the neural tube, as the tube grows, the cells have to migrate out from their segregated resting place to reach their final destinations in the growing brain. Migration, for example, allows the many regions that make up the adult neocortex to receive from the neural tube the correct numbers and types of cells that will be needed for those areas to perform their functions.

Although the exact manner in which migrating cells find their destinations is not completely understood, work on the development of the cortex by Pasko Rakic of Yale suggests that the process involves the building of scaffolds or chemical trails that migrating cells follow (fig. 4.1).[6] These trails are formed by glial cells (nonneural cells found in the brain that contribute in important ways to development and other aspects of brain function). The glia are themselves guided by local chemical cues, made by genes and their by-products, that serve as molecular signposts, creating barriers that restrict movement and

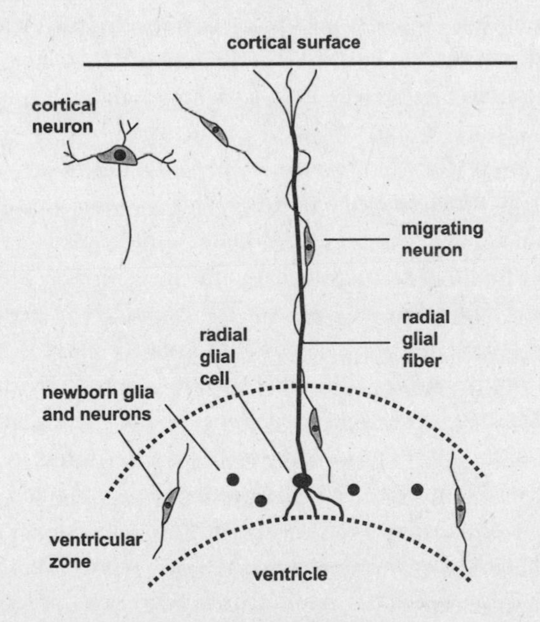

FIGURE 4.1 CELL MIGRATION ACROSS THE GLIAL TRAIL

New neurons and glial cells are born in the ventricular region. From there, neurons have to find their way to their destination in order to form brain regions. Radial glial cells seem to aid in this migratory process. They give rise to fibers that extend toward the brain's surface (the external surface of the cortex is shown). By climbing the glial trail, neurons find their homes. Thus, neurons that are born next to each other in the ventricular zone end up near each other in the cortex. This is believed to contribute to the orderly construction of brain areas. Based on figure 1 in Rakic 1995.

providing adhesive surfaces. As the emerging brain expands in size, the traveling glia grow fibers that extend out from the proliferative zone (the place where the precursor cells are making neurons and glia) toward the brain surface (where the neurons need to go to form the cortex). By crawling along the glial trail, the young neurons find their way to their target.

The glial trail makes it possible for some of the neurons living next to each other in the segregated neural tube to end up next to each other in the cortex. This accounts in part for how cortical areas come to have the cells they do—those adjacent to each other in the tube travel together along the same glial fibers and tend to end up in the same cortical location. However, the manner in which cortical or other brain areas are assembled is not fully understood. One possibility is that the function of a cell (whether it will be involved in sight vs. touch sensations, for example) is specified genetically, so that it is destined to end up in specific areas devoted to a specific function before it leaves the neural tube. Another possibility is that cells learn what to do from the context in which they end up. In support of the second possibility is the fact that if a piece of visual cortex is removed and transplanted into the location of the somatosensory (touch) cortex, it will accept somatosensory axons from a lower station in the brain (the somatosensory thalamus) and then function as somatosensory cortex.[7] This suggests that cells adapt to the local situation (the chemical environment and the axons that are coming into it) in the process of taking on their function. Support for the first possibility (that the functional fate of a cell is genetically determined) comes from a recent study showing that under certain conditions where axons are prevented from growing into the cortex, the cells in the cortex still come to possess the chemical molecules that characterize that region.[8] This suggests that the chemical signature of the cell is determined by the cell itself, likely by its genes. However, from the point of view of brain function, the job performed by a piece of cortex is often more dependent on its synaptic connections than on the molecules present. As a result, even if the chemical signature of a region is fixed by genes, the function of the region will in all likelihood be more determined by its synaptic connections. Cells traveling along the glial trail thus seem to be ambivalent functionally, at least to some extent.

Once neurons reach their destinations, they begin to sprout axons (fig. 4.2). These fledgling fibers then have to find their way to their targets, the neurons with which they will form synapses. The targets can be located nearby or in faraway areas, or both. Their so-called pathfinding[9] depends on growth cones, structures shaped like Japanese fans that extend from the ends of growing axons. The tips of the growth cones have chemical affinities with specific proteins in the surrounding space—they only attach to certain proteins, and are repelled by others. When one of the tips of a growth cone finds the right substance, it attaches. If the other tips are not attached, the growth cone is twisted in the direction of the attachment, and the axon follows. Recent studies have shown that the same protein can be both an attractant and a repellent for an axon, depending on the chemical state of the axon.[10] Slowly, step by step, the growth cone crawls along the chemical trail, pulling the axon forward, until it reaches its destination. Some signal, the nature of which is still not fully understood, then causes the growth to stop. The first axon along the trail, the pioneer, leads the way, making a path that other nerve fibers from the same area can follow. Pathfinding by fibers is not unlike migration along the glial trail by cells.

When axons reach their target, and receive the signal to stop growing,

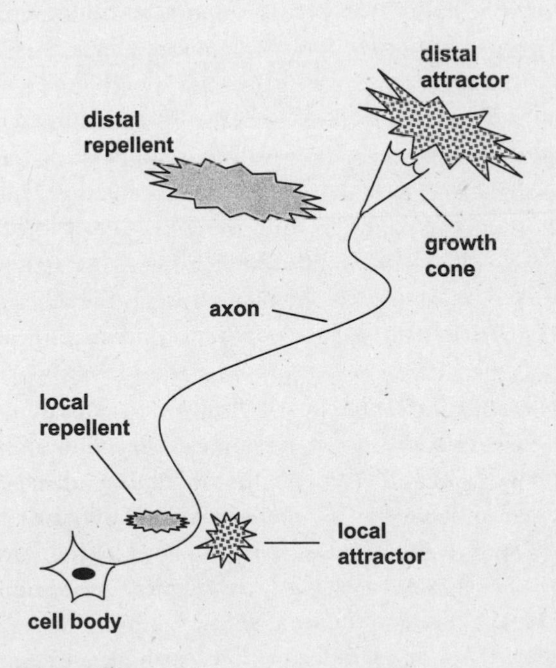

FIGURE 4.2 PATHFINDING

Once cells have migrated to their ultimate destinations (fig. 4.1), they begin to generate axons that must then find their targets. The end of the axon forms a growth cone that leads the search, being pulled by local and distal attractor molecules and pushed away by repellent molecules. Based on Jessell and Sanes 2000.

growth cones give way to axonal terminals, which begin to form synaptic connections with postsynaptic cells. The initial connections, like the other early events in brain development already discussed, are largely established by genetic and other intrinsic factors. But the brain is not completely built from the inside out. The final steps in its construction, especially the steps through which connections are fine-tuned in a way that enables them to do their jobs, require more.

PRO-CHOICE

The human genome contains instructions for body development, including, as we've seen, the development of the overall structure of the brain. Somewhere between 50 and 70 percent of all genes in the human body are believed to be in the brain. Nevertheless, we probably don't have enough genes to account for the wiring of each of the brain's trillions of synaptic connections.[11]

Pretty much everyone agrees that the transition from the uncommitted, immature initial connections of the young brain to the mature and very specific connectivity that characterizes that of the adult requires neural activity, that is, transmission across synapses. The exact role of neural activity, though, is heatedly debated, with the main issue of contention being whether activity, especially activity initiated by environmental stimulation, helps create the mature connections or just selects from the initial set of intrinsically established connections those that will be retained. This instruction vs. selection debate cuts deep into the heart of human nature: Is the self sculpted from a preexisting set of synaptic choices, or does experience instruct and add to the synaptic basis of the self as we go through early life? As we will see, since environmentally triggered neural activity is involved in both instructing and selecting connectivity, this is not so much a debate about genes vs. environmental experience as one about the precise contribution of experience.

Selectionist ideas originated in evolutionary (Darwinian) biology, were adopted and adapted by the field of immunology, and were then applied to brain function. Niels Jerne, a Nobel Prize recipient in immunology, introduced selectionist thinking to the study of the brain in the late 1960s.[12] Jerne pointed out that the history of biology is filled with instances of instructional ideas giving way to selectionist ones. In his own field of immunology, for example, it was once thought that foreign antigens enter cells and instruct them to make antibody molecules. That model is now known to be incorrect. To do this, a cell would have to be able to identify foreign antigens. Since the recognition of antigens is the business of antibodies, the only way to make an antibody to a novel antigen would be to already have the antibody. Jerne saw the fallacy in this logic, and subsequent research showed, as he predicted, that foreign antigens select precursor molecules from a preexisting pool that can be assembled into a large variety of antibodies.

Jerne later applied his antibody logic to the topic of learning. He objected to the instructional implications of John Locke's seventeenth-century empiricist hypothesis that the mind is a blank slate filled in by experience, and instead sided with the Greek Sophists, who believed that learning, as such, is impossible. He suggested that the idea of learning from experience (instruction) be replaced with the concept that experience just selects from preexisting latent knowledge. Paraphrasing Socrates, Jerne noted that "learning consists of being reminded of what is already in the brain."

A few years after publication of Jerne's hypothesis, Jean-Pierre Changeux, a prominent French neuroscientist, used selectionist principles to explain certain findings he had obtained regarding the role of neural activity in the development of synaptic connections between nerves and muscles. He concluded that neural activity "does not create novel connections but, rather, contributes to the elimination of pre-existing ones."[13]

Perhaps the most vocal contemporary practitioner of neural selectionism is Gerald Edelman, who like Jerne received a Nobel Prize for his work on the immune system. In *Neural Darwinism,*[14] Edelman argued that synapses in the brain, like animals in their environments, compete to stay alive. Synapses that are used compete successfully and survive, while those that are not used perish. According to Edelman, "The pattern of neural circuitry . . . is neither established nor rearranged instructively in response to external influences."[15] External influences, instead, select synapses by initiating and reinforcing certain patterns of neural activity that involve them.

Selectionists assume that genetic and nongenetic factors interact at each step of brain development. Selection operates on preexisting connections set up by genes (which make proteins that help guide axons to the right areas) working in concert with nongenetic factors (chemicals from the mother, for example). But genes and the chemical environment are not wholly responsible for establishing the initial connections. Selectionists also assume that there's a good deal of randomness involved—terminals and dendrites that happen to be in the same vicinity take the opportunity to form synaptic connections, independent of the overall guidance plan specified by genes. As a result, in spite of a general genetically programmed plan, the preexisting connections upon which selection ultimately operates also have a unique, individualistic nature, from which experience then does the selecting. Because each person's experiences are different, different patterns of connectivity are selected. Genes thus dictate that we will all have a human kind of brain with roughly the same kinds of circuits, but random individual differences will exist, and the connectivity of the circuits, selected by synaptic activity, will shape the individual brain.[16]

SYNAPTIC MATH

Three of the main tenets of neural selectionism are described by the terms *exuberance* (more synapses are made than are kept), *use* (the synapses that are kept are the ones that are active), and *subtraction* (connections not used are eliminated). Building the nervous system in this fashion is thought to provide a means of coping with the paucity of information available to the brain from other sources, such as the outside world, during early development.[17] If the selectionists are right, the connections we end up with as adults are those that were not subtracted—the self, they would say, is not constructed, it is selected from preexisting possibilities.

Pioneering studies by Viktor Hamburger and Rita Levi-Montalcini in the 1930s showed that subtractive or regressive events do occur in the nervous system,[18] a view that is now widely accepted.[19] During development, cells are made and then are killed off; chemicals come and go; functions change (fig. 4.3). Our main concern here is with synaptic regression, the pruning back of exuberant, unused projections during early development.

Evidence for synaptic regression was provided in a widely cited study by Pasko Rakic and his colleagues,[20] who found that the number of synapses in several areas of the primate cortex rises and then falls during the first year of life. However, a follow-up study involving a more detailed analysis of one cortical area indicated that the decrease didn't occur until puberty.[21] Given that cognitive development is close to completion by puberty, a decrease in synapses that doesn't occur until puberty can't account for the maturation of the mind.[22] Although few such investigations have been performed in humans, one widely cited study suggests that across the cortex the greatest number of synapses are present at around twenty-four months of age (obviously long before puberty), but with different areas reaching their peak at different times.[23]

Certainly, much more work is needed to determine the extent to which synapse elimination occurs, when it occurs in specific neural circuits, whether there are significant species differences, and how such data should be interpreted.[24] The key issue is not whether synapses are eliminated during devel-

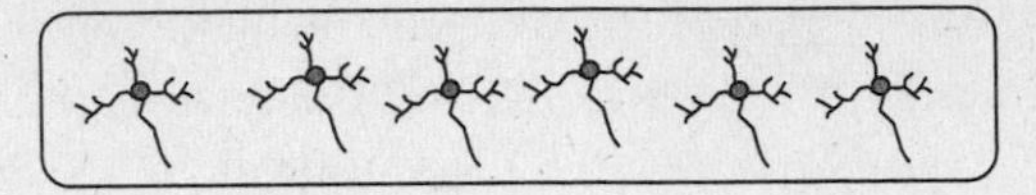

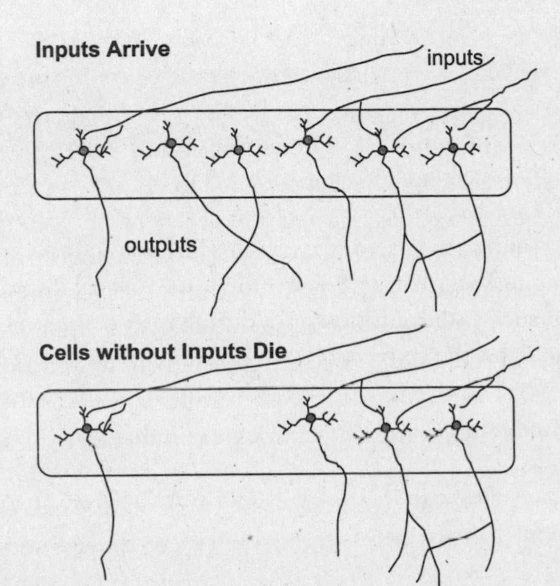

FIGURE 4.3 SYNAPTIC ACTIVITY PREVENTS CELL DEATH
After migration, cells receive inputs and start growing axons (outputs). Those cells that receive inputs are more likely to survive, whereas those that do not are more likely to die. Based on figure 20.4 in Oppenheim 1998.

opment (they clearly are)[25] but instead whether the decreases that do occur provide conclusive support for the strong version of selectionism—the version that claims that activity *only* prevents the elimination of synapses.[26] This takes us to the second part—the "use it or lose it" part—of the selectionist argument.

USE IT OR LOSE IT

That use prevents synaptic demise was suggested by a set of classic studies started in the 1960s by David Hubel and Torsten Wiesel on visual system development.[27] Like most other sensory systems, the visual system uses receptors near the body surface (in this case, cells in the part of the eye called the retina) to take in information from the environment. Retinal cells give rise to axons that terminate in several postsynaptic targets, the major one of interest here being the visual thalamus. Thalamic cells, in turn, send their axons to areas of the visual cortex, where, to state it most simplistically, our visual perceptions emerge.[28]

Neocortical areas, by definition, have six layers. In the visual cortex, the axons from the thalamus end mostly in the middle layers and are distributed to the rest of the visual cortex from there. (Axons from regions other than the thalamus end in other layers.) In studies of mature cats, Hubel and Wiesel found that cells in these middle layers tended mainly to fire action potentials when stimulated visually through one eye or the other, with fewer cells responding to stimulation of both eyes. The eye-specific cells tended to cluster together in separate patches, leading to areas with cells responsive to only one eye. But when they performed the same study on very young cats, most cells responded to stimulation of both eyes. This led them to surmise that initially, cortical cells must receive synaptic inputs from each eye, and then, during later development, inputs from one eye were eliminated, leaving only the inputs from the other.

To test whether their hunch was correct, Hubel and Wiesel closed one eye of their experimental subjects early in life before the eye segregation took place in the visual cortex. The eye was opened when the animals became adults. By then, relatively few if any cells responded to visual stimulation of the previously closed eye. Nevertheless, the cells that would have normally responded to the closed eye were still functional—but now they only responded to the eye that had been left open. The open eye, in other words, retained access not only to its own cells but also to the cells normally devoted to the other eye. Closure of an eye in an adult cat had no such effects, suggesting that the cortex only has this capacity to reorganize during early development. Hubel and Wiesel concluded that the development of synaptic connections is a competitive process, one in which the connections that are used are kept and those that go unused are eliminated.

These findings, which have been confirmed and elaborated on by many investigators over the years,[29] are certainly consistent with the selectionist principle that adult synapses are picked from a preexisting set by activity. However, since this early study was conducted, the tools available for exploring how the brain works have advanced considerably, allowing neuroscientists to now ask more sophisticated questions. Michael Stryker and his colleagues, for example, studied the issue using a more exacting measure of the effects of synaptic activity.[30] Rather than measuring the overall clustering of eye-specific cells in a region, they asked what happens to individual axon terminals and the synapses they make during development. Individual cells in the thalamus were injected with a chemical tracer at different times.[31] The tracer then spread throughout the neuron, all the way to the axon terminal and its presynaptic endings in the cortex.[32] They found that as development proceeded, many axons did indeed retract (consistent with a selectionist view), but that the remaining axons also increased in complexity (consistent with an instructional view). Activity thus leads to an increase in synaptic complexity rather than merely a stabilization of the preexisting pattern—activity therefore is capable of instructing the formation of new synaptic connections.[33]

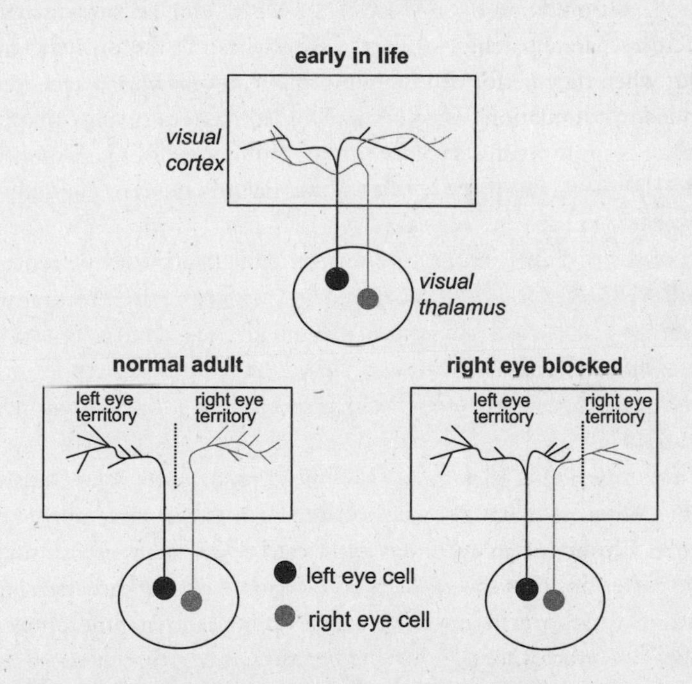

FIGURE 4.4 SYNAPTIC ACTIVITY DETERMINES CONNECTIVITY

Early in life, thalamic cells connected to the left and right eyes give rise to axons that end in overlapping areas of the visual cortex. By the time animals mature, the axons from the two eyes have segregated into areas devoted to one eye or the other. The axons also increase in complexity. When one eye is closed or its activity blocked, it vacates the territory and the axons from the other eye invade. The retraction of axons from an area supports the selectionist view of development, whereas the increase in synaptic complexity supports the instructional view, suggesting both are relevant. Based on Goodman and Shatz 1993, and Quartz and Sejnowski 1997.

Activity also appears to help define the demarcation between areas of the cortex. For example, the axons from the visual thalamus normally spread into areas of the auditory cortex, and vice versa, in early life.[34] As development proceeds, the stray connections are pruned back, since the cells in a given area are more strongly connected with one system or the other, and the stronger inputs compete more successfully for the attention of the postsynaptic cells. This can be seen with congenitally deaf adults, in whom the region of the cortex responsive to visual stimulation includes the area that would typically be occupied by the auditory cortex.[35] It is the absence of activity in the auditory area that has allowed the visual fibers to compete for those synapses.

While we would tend to assume that synaptic activity in a sensory system, like the visual system, is due exclusively to environmental stimulation, studies by Carla Shatz and others suggest that a broader view of such activity is needed.[36] In contrast to cats and some other animals, primates have eye-specific patches present in the cortex at birth.[37] Several weeks before birth, cells in the retina are spontaneously firing action potentials in these animals.[38] Because many cells in a given eye fire simultaneously, a wave of synchronized activity is transmitted to the brain from each eye. However, the waves coming out of the two eyes are not in sync and, in fact, typically occur at different times. This suggests that endogenous activity generated in the two eyes independent of external visual stimulation might be sufficient to provide the competitive activity in the cortex to create eye-specific clusters of cells. Indeed, if this endogenous activity is blocked chemically, the clusters fail to develop.[39]

It's important to make a clarification at this point. New connections formed by activity are not created as entirely new entities but rather are added to intrinsically determined preexisting connections. Added connections are therefore more like new buds on a branch rather than new branches. Activity thus does not produce wholesale rewiring of the brain. After all, most of the connections in my brain and yours are the same. Activity adds those little adjustments that make you and me different.

In reviewing the evidence on the role of activity in the establishment of the adult pattern of synapses, two leaders in the field, Larry Katz and Carla Shatz, recently noted that "neural activity is likely providing cues that drive the formation of new synapses and axon branches, as well as cues that act to select and stabilize existing ones." While it may seem that such an "everybody wins" summary statement implies that nothing has been accomplished in the field, this would be the wrong conclusion. It has taken a tremendous amount of work to demonstrate that activity both instructs the formation of new connections and selects used ones from a preexisting pool. While activity may not be necessary for the establishment of all circuits,[40] the fact that it plays a key role in constructing at least some of them suggests that a purely selectionist view of brain development is not viable. At the same time, the fact that activity instructs the growth of connections does not rule out its role in selecting and eliminating synapses. Activity is involved in both processes.[41]

It is probably best to think of instruction and selection as two complementary means by which circuits can be constructed rather than as mutually exclusive theories of brain development. Unfortunately, the only way to determine the precise contribution of each to the construction of a particular circuit is to do the experiments. Since only a few circuits have been studied in this degree of detail, developmental brain researchers are not likely to be bored any time soon.

WIRING BY FIRING

New connections made by neural activity are just as subject to the "use it or lose it" rule as old ones that have been selected from the initial pool of synapses. When new connections are made, only some survive—those that go on to be used. But how exactly does "use" prevent loss? What occurs between a postsynaptic cell and its presynaptic terminal to ensure a connection's survival? Or, stated more positively, how does activity strengthen connections?

In 1949, Donald Hebb, a Canadian psychologist, proposed that if two neurons are active at the same time, and one is presynaptic to the other, then the connection between them will be strengthened.[42] In Hebb's words, "When an axon of cell A is near enough to excite cell B or repeatedly and consistently takes part in firing it, some growth process or metabolic changes take place in one or both cells such that A's efficiency, as one of the cells firing B, is increased." The essence of Hebb's idea is captured by the slogan "Cells that fire together wire together."[43]

Although Hebb originally proposed his fire-and-wire theory to account for the nature of learning and memory, it has been used to explain other aspects of synaptic function, especially the construction of synapses during development.[44] Consider again the establishment of visual cortex connectivity. As we've seen, in primates, the weeks before birth are an important period for visual system development, as waves of spontaneous activity from the two eyes set up patterns of activity that result in the preferential activation of certain cortical cells by one eye or the other. Because cells in the retina of one eye are more likely to fire spontaneously at the same time, and are much less likely to fire at the same time as cells in the other eye, chances are that when a postsynaptic cortical cell is activated by presynaptic inputs from one eye, presynaptic inputs from other cells in the same eye will be arriving more or less simultaneously. According to Hebb's rule, this concurrent activity in presynaptic and postsynaptic cells then leads to a strengthening of the connections from that eye to the postsynaptic cell.[45]

Quite a lot has been discovered about the neurotransmitter mechanisms that underlie Hebbian plasticity, especially in the context of learning and memory. We will examine this work in detail in chapter 6. For now, we can say that the best-understood form of Hebbian plasticity involves the release of glutamate from presynaptic terminals, and the binding of the released transmitter to two kinds of postsynaptic receptors. One of these receptors records that the postsynaptic cell was active, and the other records which presynaptic terminals contacting the postsynaptic cell were active at the same time. In this way, the postsynaptic cell is able to detect the co-occurrence of presynaptic and postsynaptic activity, and thereby help link the postsynaptic cell with all of the presynaptic terminals that were active when the postsynaptic cell fired an action potential.

Hebb suggested that in order for the increased efficiency of the connection between the presynaptic and postsynaptic cell to be maintained, something like new growth needs to occur. Although there was little evidence of it at the time, activity-induced growth has now been observed in many different cir-

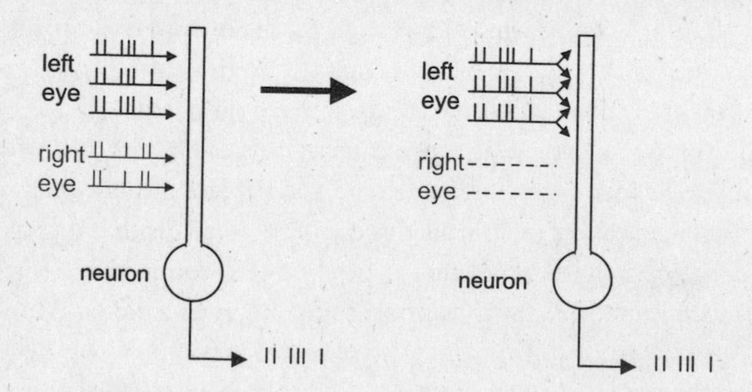

FIGURE 4.5 WIRING BY FIRING

Hebbian plasticity during development. The output firing pattern matches the input pattern from the left but not the right eye (shown on left). According to Hebb's learning rule (see text), the coordinated pattern of presynaptic and postsynaptic activity from the left eye leads to a strengthening of its synapses and a weakening of the right eye synapses (shown on right). Based on figure 22.8 in Purves et al. 1996.

cumstances. For example, Stryker's results described earlier showed that active axons branch and sprout new connections. Also, studies of Eric Kandel and colleagues at Columbia have demonstrated that physiological plasticity is accompanied by axon branching and new synapse formation both during development and following learning.[46] Once this occurs, a given action potential coming down the axon will be more effective in firing the postsynaptic cell because it activates more synapses on that cell.

The manner in which activity promotes growth during early development is being actively researched. One of the key players is a class of postsynaptic receptors for the excitatory transmitter glutamate (see chapter 3). These special receptors, known as NMDA receptors, appear to be particularly important in that they are able to detect the match between activity in the presynaptic and postsynaptic neuron.[47] When these receptors are blocked or otherwise interfered with, normal development is disrupted.[48] We will have much to say about the role of NMDA receptors in plasticity in chapter 6. For now, it is mainly important to know that mechanisms exist that can indeed allow Hebb's magic to be worked.

Another important set of molecules are neurotrophins, special tonics that promote the survival and growth of neurons.[49] When an action potential occurs in a postsynaptic cell, neurotrophins are released from the cell and diffuse backward across the synapse, where they are taken up by presynaptic terminals. Under the influence of neurotrophins, the terminals begin to branch and sprout new synaptic connections. Since only those presynaptic cells that were just active (that just released transmitter) take up the molecules, only they sprout new connections. Activity thus induces growth, and the growth that occurs is restricted to the active terminals.

In addition to this role in the active construction of circuits, neurotrophins are also involved in synapse selection. The natural fate of many cells during development is an early exit. So-called programmed cell death is one of the regressive events that help shape the final pattern of connectivity (fig. 4.3).[50] Cell death is prevented if a presynaptic terminal receives a life-sustaining shot of neurotrophins from its postsynaptic partner. The survival rate of neurons is in this way regulated by the limited availability of neurotrophins. Only those cells that compete successfully for neurotrophins (those that are active) survive. In the presence of neurotrophins, the surviving terminals (those that were active) also begin to sprout new connections. Selection can be a step along the path toward activity-instructed growth—in other words, selection and instruction are partners in synaptic development.

WHAT ABOUT INNATENESS?

The two main ways circuits are known to get built—selection and instruction—both assume that the synaptic connections that ultimately determine what a circuit will do and how it will do it are wired up epigenetically, that is, by interactions between genes and environment (including both the internal and external environment). Where does this leave innateness, the idea that certain functions mediated by the brain might develop mainly under genetic control?

The topic of innateness often comes up in the context of psychological (mental and behavioral) development. To the extent that psychological characteristics are mediated by synapses in the brain, the psychological and synaptic levels are intimately related. The question raised in the preceding paragraph can therefore be rephrased: Where does the epigenetic nature of synaptic development leave the synaptic basis of innateness? In order to consider this question meaningfully, we need to make a distinction between the two different meanings of innateness. One is the idea that genes make us all the same, giving us traits that we share with other members of our species. The other is that genes make us unique.

INNATENESS I: SPECIES SPECIFICITY

For a long time, innateness was the business of ethologists, biologists like Konrad Lorenz and Niko Tinbergen who studied behavior.[51] A wide range of behaviors in a variety of different animals—from nest-building, maternal care, and foraging to predation and defense—were said to be innate and thus were believed to emerge during early life independent of learning and other environmental influences. Learning was deemed to be important in shaping the individual, but many of the characteristics that make members of a species act alike were thought to be free from environmental influence, or at least more dependent on nature than nurture.

The classic method used by ethologists to test innateness was to show that a behavior was present at birth or shortly thereafter, or, more generally, that it emerged in the absence of an opportunity for the organism to have learned from environmental stimulation. But in the 1950s, Daniel Lehrman, an American animal psychologist, convincingly argued that even under the most stringent isolation conditions, the organism is never completely isolated from nongenetic influences, such as things that happened prenatally or shortly after birth.[52] As a result, the epigenetic nature of behavior development came to be accepted by ethologists, who shifted away from terms like *innate* and *instinctual* and toward *species-typical* in their efforts to understand the development of the unique behavioral repertoires of different species.[53]

Ironically, just as ethology was coming to accept epigenetic contributions to species-typical behavior, psychology (especially American psychology) turned toward an emphasis on innateness. Today, the innateness of certain mental capacities in humans and other species is emphasized by many psychologists. Below, we'll examine where this trend originated and where it is going.

THE DEMISE OF UNIVERSAL LEARNING. Through the mid-twentieth century, American psychology paid little attention to special capacities of individual species.[54] Under the influence of behaviorists like John Watson, B. F. Skinner, and Clark Hull, behavior came to be understood as dependent on what organisms learned from their experiences. For behaviorists, learning was a universal capacity that worked more or less the same regardless of which animal was doing the learning and what was being learned.[55] If one wanted to understand how humans learn language, math, telephone numbers, tennis, or what someone looks like, studies of the way that key-pecking in a pigeon or bar-pressing in a rat was reinforced by food were just as valid as anything else.

Things began to change in the 1950s, when the linguist Noam Chomsky challenged the notion of a universal learning capacity.[56] He proposed that natural language is unique to humans, and that language acquisition is different from other learning capacities that humans have. His ideas about rules and representations in our use of language helped spark the cognitive revolution in psychology,[57] and his conception of a universal grammar encoded in the human genome has ever since been a cornerstone of cognitive nativism, which assumes that certain psychological capacities are innate.[58]

The plain vanilla view of learning was also challenged when animal psychologists in the 1960s began to realize that the behaviorists' "law of learning" didn't apply to all forms of learning. For example, John Garcia's discovery that some stimuli (those that make you sick to your stomach) can be associated with tastes, but not with sounds or sights, was inconsistent with the behaviorist notion that any given stimulus could be associated with any other one, though it makes perfect sense from a biological point of view.[59] This and other findings suggested that there might be biological constraints on learning, factors that tailor the workings of the learning process to the specific learning task and to the particular learning requirements of the species.[60] A universal learning mechanism was clearly not sufficient, and the study of animal behavior by American psychologists became more ethological.

THE RISE OF EVOLUTIONARY NATIVISM. Eventually, biological constraints came to play an important role in ideas not just about animal behavior but also about human thought. For example, the psychologist Steven Pinker has recently blended the Chomsky nativist tradition, which views language as a special capacity of humans, with ideas about evolution, asserting that humans have a language instinct,[61] as well as a variety of other innate mental organs (capacities).[62] These are believed to have resulted from Darwinian natural selection and "owe their basic design to our genetic program."[63] That language may be innate is supported by studies of pidgin language, a simple, crude form of verbal communication among people who speak different languages but live together and have to communicate with one another. Pidgin lacks grammar when it is initially pieced together, but by the second generation, it begins to take on a grammatical structure, suggesting that the underlying innateness of grammar ultimately emerges.[64] The innateness of language, though not universally agreed-upon,[65] is supported additionally by certain genetic conditions that have been linked to specific chromosomes, including the spared language abilities in spite of other mental difficulties in children with Williams syndrome, and the language impairment in spite of otherwise intact mental abilities in specific language impairment (SLI).[66] Although the evidence for innateness is strongest for language[67] and emotional expressions,[68] innateness has been postulated to account for many mental capacities.[69] For example, it has been proposed that humans have innate knowledge about numbers, physics, and even about how other human minds work.[70]

The recent groundswell of interest in the natural selection of mental functions has a name—evolutionary psychology.[71] The aim of this discipline is to try to determine what factors might have driven the evolution, by natural selection, of our various mental capacities. Taking a clue from Kipling, critics like Stephen Jay Gould call the explanations proposed by evolutionary psychologists "Just So Stories": they may sound good, but whether they are true or not can't be known.[72] After all, there is no fossil record of the mind, and evolutionary explanations about the mind are basically untestable. Gould suggests, in contrast, that language and perhaps other mental functions that are being touted as innate are not evolutionary adaptations, traits that were selected for, but rather are exaptations, features that enhance our fitness but that weren't originally designed to do what we use them for.[73] For example, birds probably developed feathers as a means of controlling body temperature, and only later used them to fly. As Gould notes, "Once you build a complex machine, it can perform so many unanticipated tasks. Build a computer 'for' processing checks at the plant, and it can also . . . whip anyone's ass (or at least tie them perpetually) in tic-tac-toe."[74] Another critic of the evolutionary explanation of language is the psychologist David Premack, a pioneer in the study of chimp language. As he notes, "Human language is an embarrassment for evolutionary theory because it is vastly more powerful than one can account for in terms of selective fitness."[75] Evolutionary psychologists have responded to such attacks,[76] pointing out, among other things, that there are bad evolutionary explanations (those that are indeed "Just So Stories") but also good ones (those that are better grounded). The problem is there's a great deal of room (maybe too much) for interpretation in the decision about what constitutes a good as opposed to a bad evolutionary psychological explanation. A more general critique of evolutionary psychology can be found in *Alas, Poor Darwin,* a recent collection of essays put together by Hilary and Steven Rose (Steven is a neuroscientist who studies memory and a vocal opponent of nativist psychology).[77] Regardless of how this ongoing debate ends, however, for now nativists are enjoying a good ride (though not a free one) in psychology.

Even if evolutionary psychology turns out to be dead wrong (and though I don't think that's the case, I am concerned about whether its explanations can be put to rigorous scientific tests), it wouldn't mean that nativism is also completely wrong. It's important to distinguish between the question "Is the function X innate?" and the question "Why did the innate function X evolve?" The former is a question about genetics, which can potentially be answered, while the latter concerns historical facts that are not easily verified scientifically. As Gould's exaptation argument points out, something can be innate (passed on genetically) even if the role it plays today is not the one for which it evolved.

LEARNING DOMAINS. Classic nativism, dating back to the Greeks, assumed the existence of a priori knowledge, of innate ideas, such as ideas about God, numbers, and other so-called universal truths.[78] Modern nativists do not argue so much that we have a priori knowledge per se but rather that we have innate predispositions to acquire (learn about) specific kinds of environmental information.[79] For example, although nativists view language as an innate capacity in humans, they do not assume that we automatically know words and rules: to acquire a specific language we have to be exposed to people who speak that language. However, the information obtained from listening to others is, on its own, insufficient to fully account for language acquisition (especially the use of grammar)—it is this "poverty of the stimulus" argument that is key to the nativist assumption that language is innate.[80] Nativists further assume that the innate cognitive apparatus that makes language acquisition possible does not help us learn other things, which are managed by other independent innate mental modules.[81] The capacities underlying innate mental modules have been described as domain-specific or information-specific learning biases, "specialized structures tuned to learn a particular kind of knowledge."[82] The nativists' modular view of learning is obviously inconsistent with the behaviorist notion of a universal learning capacity present in all species and used for all varieties of learning.

We now know, and will see in detail in later chapters, that the brain does indeed learn different things using different systems, and that different animals can have different learning capacities, which is consistent with the nativist view of innate learning modules and inconsistent with the behaviorist notion of a universal learning function.[83] It seems possible, maybe even likely (though not yet proven), that the overall design of these systems is to a large degree programmed genetically. This would account for why members of a species share capacities, and why some capacities differ between species. At the same time, while the various learning systems may well have unique genetic histories, at the level of cells and molecules it appears, as we'll see in the next two chapters, that relatively few solutions are used to achieve learning both within and across species. This implies that deep down there is a kind of universality to learning. We've encountered one of these mechanisms already, Hebbian plasticity, and will have more to say about it and related forms of plasticity in subsequent chapters.

Most systems in the brain can learn from experience, which means that the transmission properties of their synapses can be altered by experience. In this sense, learning itself is not the specific function that circuits were built (by evolution) to perform. Learning is instead a capacity of synapses, something that contributes to the way circuits work. The defense circuit involving the amygdala, for example, exists to detect and respond to danger. Learning is a very useful feature in a circuit like this, allowing it to respond to stimuli that are not in themselves dangerous but that have come to be associated with danger in the past—the sound made by a predator prior to an attack, for example. Without the capacity to learn from experience, the fear system would still be able to do its job (that is, to detect and respond to danger) but would be limited in that it could only respond to preprogrammed dangers.

SELECTIONIST NATIVISM VS. INSTRUCTIONAL CONSTRUCTIONISM. How, then, might an innate circuit, one used for learning a certain kind of knowledge, be wired during development? As we've seen, selection and instruction are the two main choices. Psychological or cognitive nativists reject on principle the idea that the environment "instructs" the mind, and therefore reject environmental instruction of circuits as well. As a result, they tend to be selectionists. Massimo Piattelli-Palmarini, for example, builds upon Jerne, Changeux, and Edelman in denouncing the empiricist implications of instruction, and argues that "there is no such thing as learning." He is not questioning "learning" in the general sense of the term but in the behaviorist sense of being stamped in by the environment. He goes on: "There is no known process, either in biology or in cognition, that literally amounts to learning in the traditional 'instructive' sense, that is, to a transfer of information *from* the environment *to* the organism. . . . all the mechanisms of acquisition . . . are due to a process of internal selection." In a similar vein, Jacques Mehler has stated that knowledge comes from "unlearning," implying that we come to know things by a process of elimination.

Nativist psychologists assume that all human capacities are present in the exuberantly wired brain, waiting to be picked, or parameterized (to use Piattelli-Palmarini's term). However, as we've seen, synaptic selection alone can't fully explain brain development—instructed growth also plays an important role.

Nativists might at this point counter that the evidence for activity-induced growth is strongest for sensory systems, leaving open the possibility that the brain systems involved in higher cognitive processes (the mental modules of interest to nativists) might be built by selection. However, until it can be shown that higher cognitive functions are built by selection alone (that activity does not induce growth in areas involved in higher cognition), we have to assume that the basic principles discovered so far for sensory systems apply to other systems—that activity-induced growth is a viable possibility in cognitive systems.

Not all cognitive scientists are nativists—some, in fact, strongly oppose nativism. For example, in 1997, two influential publications attacked nativism head-on. One was *Rethinking Innateness,* a book by Jeffrey Elman and colleagues;[84] the other was a journal article by Steven Quartz and Terry Sejnowski titled "The neural basis of cognitive development: a constructivist manifesto."[85] Both publications argued for a *constructivist* approach to neural development, one that places instructed growth in the spotlight and emphasizes cortical plasticity rather than cortical specificity. While constructivists are especially opposed to the idea of innate knowledge (innate mental content, like words or concepts), they are generally displeased with most notions of innateness, including the innateness of capacities (like the capacity to learn a language).

A major tenet of constructivism is the idea that the wiring of cognition in the neocortex can be achieved on the basis of extrinsic inputs—that neocortical circuits can be constructed by extracting structure from the sensory environment. Constructivists accept the principle that genes build the overall structure, and that regressive events like synaptic selection take place, but they argue that epigenetic prewiring and selection alone cannot explain cognition. In addition to citing evidence for activity-induced growth of synapses, they refer to studies showing that damage to areas involved in language processing in the brain of a child leads to only minor deficit, if any, whereas the same lesion in an adult causes severe problems in using language. Thus, in early life, before cortical circuits are finalized, areas normally not involved in language can take on language functions. Similarly, they cite the transplantation studies discussed earlier in which pieces of cortex are removed in young animals and put in a different location in another animal's brain. (Recall that the relocated bit of cortex then acquires the mental properties of the host.) Findings like these lead constructivists to conclude that circuits are not genetically programmed but acquire their functions from the neural activity that the environment elicits in them. At the same time, constructivists accept that there are constraints on how synapse construction works in the cortex. For example, cortical areas develop in conjunction with subcortical circuits, which are believed to be more subject to innate assembly than cortical ones.

The constructivist position accounts for a good deal of the data on cortical plasticity and is consistent with the notion that learning may be accomplished in fairly universal ways at the cellular level. Although I have considerable sympathy for much of the constructivist argument, I feel that it, too, has several limitations. First, it assumes that mental development can be understood more or less exclusively from the point of view of the cortex; noncortical contributions, though acknowledged, are viewed as mostly hardwired, and not given much consideration. However, because cortical areas depend on subcortical areas for the relay of environmental information, hardwiring in a subcortical circuit will strongly influence what happens in a related cortical area, especially if the subcortical circuit matures earlier, as is often the case. The information ultimately extracted by the cortex is thus as much due to the organization of the subcortical circuit as to the structure of the environment.[86] Second, in arguing against the innateness of some capacities, especially language, the constructivists may be throwing the baby out with the bathwater for the sake of consistency. Subcortical circuits are clearly more likely to be hardwired than cortical ones, but we shouldn't reject outright the possibility that some cortical circuits may also be hardwired. (For example, it has been proposed that the perception of facial expressions of emotion is performed by a species-specific face perception module in the primate, including human, cortex.)[87] Third, in proposing that the cortex is a general-purpose cognitive learning device, not enough credit is given to the multiplicity of systems (both cortical and subcortical ones) that engage in learning. This throwback to universal learning just won't do. Fourth, in emphasizing the ability of a piece of cortex to adapt to new circumstances (due to deprivation or damage or surgical transplantation), constructivists overlook the fact that the adaptation that results from such unusual conditions may reveal abnormal rather than normal function: just because the color-processing part of the primate cortex can take on a new function when the color-processing area of the thalamus is damaged or removed does not contradict the possibility that color vision is an innate specialization of primates.[88] Plasticity, again, is a feature of brain systems, not their evolved function.

A BOTTOM LINE. Clearly, more research rather than more speculation is in order at this point. However, armed with the information we've considered, let's return to the question posed earlier: Where does the epigenetic nature of synapse development leave innateness? It seems that there is no good reason to reject the possibility that the overall function of some circuits is more or less innately determined by genetic coding, especially if we are talking about innate predispositions to acquire certain kinds of information rather than about innate knowledge per se. At the same time, there seems to be no reason to assume that the epigenetic expression of an innate function has to involve *only* selection from preexisting synapses in finalizing the circuit: selection and instruction probably work together to shape the final connections of most circuits. The innateness of species-typical characteristics is thus not at all incompatible with the epigenetic nature of brain development (that is, the fact that genetic and nongenetic factors interact), so long as we do *not* insist that specific knowledge is innate or that nongenetic factors (like environmentally driven neural activity in neural systems) influence only the selection of preexisting synapses. Further, we have to remember that while innateness (genetic coding of an existing function) can mean adaptation (natural selection of that function), this is not always the case.

INNATENESS II: INDIVIDUALITY

One logical assumption might be that if a capacity is innate at the level of the species, individual differences in that capacity should also be innate. Suppose, for example, that after a thorough analysis of the problem, scientists were to come to the conclusion that humans have a specific capacity called intelligence, and that this capacity is innately coded in the human genome (by the way, neither of these propositions is generally accepted).[89] Would we then be justified in assuming that differences in intelligence between individuals or between groups of people are due to genetic variation between individuals? Not necessarily!

The complex relation between the innateness of individual and species characteristics is nicely illustrated by a study performed on two groups of garter snakes living in California.[90] The snakes are members of the same species, but some live in the marshy coastal areas and others in the drier inland regions. The coastal snakes eat slugs; the inland ones don't. Steven Arnold was interested in whether this difference in food preference was due to genetic or environmental causes.[91] He took eggs from snakes living in both regions, and when the babies hatched, he immediately put them in separate cages, isolated both from littermates and mother and from their natural habitat. A few days later, he offered slugs to snakes from the two groups. The coastal babies ate the slugs, while the inland ones ignored them. He then placed essence of slug on cotton swabs and measured how often the snakes flicked their tongues at the swabs. (Tongue-flicking is how snakes take in smells.) Isolated coastal snakes flicked their tongues vigorously, while the inland ones showed little interest. Even within each group, however, there was some variation from individual to individual. Arnold therefore examined the heritability of the within-group trait by comparing tongue-flicking scores of siblings to see if they were more like one another than unrelated snakes. If related individuals were more alike than unrelated ones, then genes would be implicated. It turned out that family relations had little to do with the way the snakes responded. Slug preference *within* the population is thus not due to genetic causes. Arnold then crossbred coastal and inland snakes and found that this led to more variability in the behavior of the offspring than in the "pure" groups. Mixing the genes of the two populations reduced the genetic influence on the behavior, suggesting that the differences in the behavior *between* the groups is genetic in origin. Together, these findings lead to the conclusion that genes help in the explanation of the behavioral difference *between* groups but not *within* groups, at least in this case.

The snake study suggests that even if our capacity for some mental function like intelligence or language or emotion is innate, we are not necessarily able to explain differences between people in these capacities in terms of genes. Genes may play an essential role in placing a function in the brain of every human, and at the same time make a relatively small contribution to differences in the way that function is wired in individuals. The only way to know for certain is to do the proper experiments. Indeed, for some behaviors in some species, crossbreeding and heritability studies do show a significant relation in the behavior of siblings, including those reared separately, which suggests (but, as we'll see below, does not conclusively prove) a genetic contribution to behavior at the level of individuals and their families.[92]

Clearly, the most compelling case for a role of genes in individual human behavior comes from studies of identical twins separated at birth. Because their common postnatal environmental influences are minimal, this situation is believed to be the best test for a role of genes in shaping personality traits of an individual. Some remarkable observations have been made. For example, one pair of identical twins separated shortly after birth ended up in very different situations—one grew up as a Catholic in Nazi Germany, and the other lived as a Jew on a Caribbean island.[93] In spite of being raised under vastly different environmental conditions, the pair ended up with peculiar shared traits, including reading magazines back to front, and keeping rubber bands on their wrists.

Though interesting, it's hard to extrapolate from these idiosyncratic findings. However, large studies of many pairs of twins have been conducted to assess their similarity in personality traits and intelligence.[94] This was done by computing heritability scores, which estimate how much of a given trait is due to environment and how much to genes, for the different measures. Such studies have shown that while identical twins, living together or apart, are more similar than fraternal twins or nontwin siblings, they are by no means indistinguishable. In fact, the correlation between pairs on any particular measure was at most only about .50 for identical twins living together, and was a bit less for those living apart, suggesting that genes account for at most half of a given trait.

It has been argued that the contribution of genes is overestimated in twin studies in at least two ways.[95] First, an assessment of the genetic component of heritability does not distinguish the direct effects of genes from indirect effects: a child's inherited timidity may lead his parents to reassure him constantly, his peers to pick on him, and his teachers to ignore him, each of which might in turn reinforce his timidity or bring about other behaviors.

Second, heritability scores do not accurately contrast genes with environment, for things that are similar in the environments in which separated twins live inevitably get mixed in with the genetic contribution. There are many opportunities for twins to have shared experiences of one kind or another. The renowned developmental psychologist Howard Gardner notes the following: "For at least nine crucial months, the twins share the same environment—the womb of the birth mother. . . . Then, too, they may not have been separated right at birth. (And under what extraordinary circumstances does such separation occur?) They may or may not have been raised for a while by family members. The children are not randomly placed; in nearly all cases they are raised within the same culture and very often in the same community, with similar social settings. Also, infants who look the same and behave the same may elicit similar responses from adults."[96]

In the earliest days of life, as we've seen, the main environmental factors that could affect the developing twins are chemicals passed from the mother's bloodstream to the fetus—hardly trivial influences. Also, many connections form in the brain prenatally, allowing the twin fetuses to begin to receive information from the external environment in utero. For example, by the thirtieth week of pregnancy, the heart rate of the fetus changes when sounds occur in the environment.[97] The fetal brain can even discriminate environmental events and can learn and retain information about environmental stimuli.[98] And after birth, as Gardner noted, they are often not separated immediately, and frequently end up in similar environments, where tendencies to look or act alike could help convert subtle into more powerful similarities. The case of the twins described above shows that this is not always true, but heritability studies live and die on group averages, not on interesting observations of individuals. When identical twins, separated at birth, end up sharing characteristics as adults, we need to wonder whether this is due to their common genetic heritage, to common influences within the womb, or to subtle environmental similarities that shape the development of their synaptic connections. At the same time, we also need to place the similarities in the context of the many ways in which twins differ from one another.[99]

In such quantitative analyses of groups, the goal has been to determine how the variability of a trait relates to genetic similarity (identical vs. fraternal twins, fraternal twins vs. siblings, children vs. parents) in the general population, rather than in individuals. But in the last few years, research on genes has exploded, and new approaches are taking over. The older behavioral genetics approach had as its aim distinguishing the contribution of the environment from that of genes (that is, of all genes taken together). Complex mathematical analyses were used to extract subtle differences in the contributions of genetic and environmental factors to variation in a trait in large groups. This approach is giving way to efforts to identify *specific* genes that might be involved in specific traits. The entire genome has been mapped in some animals, including humans, which opens the door for some very sophisticated studies of the role of particular genes rather than of genes in general. For example, genetically altered mice can be created that have a single gene missing. Depending on what gene is knocked out, dramatic differences in behavior can result. Elimination of the gene for a certain enzyme used in regulating the effects of calcium or other molecules inside brain cells, for example, can interfere with Hebbian plasticity at synapses and thereby disrupt long-term memory, as we will see in chapter 6.

But as important as genes are, they are usually contributors to traits rather than dictators. And seldom is any brain function as complex as those involved in behavioral or mental processes under the control of only one gene. The fact that disruption of one gene can lead to a memory deficit does not mean that memory is due to that gene alone. All effects of genes are expressed epigenetically, by way of interactions in the internal chemical environment between the proteins made by multiple genes, or by way of external environmental stimulation that elicits synaptic activity that then induces genes to make proteins. The proteins made by genes can in turn modulate neural activity at synapses. Assembly of the individual is a big job, and genes have lots of partners in the process.

TIMES FOR LEARNING?

Many psychological theories of development propose that the child matures in stages.[100] Freud divided them into the anal, oral, phallic, and genital. Piaget talked about sensorimotor, preoperational, concrete operation, and formal operation stages, while other theorists have proposed different terms. Underlying the unfolding of any of these developmental stages in behavioral and mental processes must be neural changes. And, as discussed earlier, there's plenty of evidence that the brain does go through all sorts of changes during development—synapse numbers are rising and falling, cells are being born and dying, dendrites are growing, brain energy use (metabolism) increases. The problem is that it is not clear how the changes at the neural level relate to those at the psychological level.

A movement called brain-based education attempts to use information about brain development and function to guide education policy and practice.[101] One of its basic principles is that there are special times for learning—critical periods, windows of opportunity—times during which information has to be obtained or it won't be stored.

A good deal of evidence exists for critical or sensitive periods. As we've seen, if a kitten's eye is closed during a narrow time span in early life, it won't develop proper binocular vision because binocular cells in the cortex don't mature normally. The effect is reversible if the eye is opened within the window of opportunity, but not afterward. This is why misalignment of a child's eyes has to be corrected in early life. Once the critical period is over—that is, once synapses in the cortex are wired—the window closes. Another example is birdsong.[102] If birds that sing aren't exposed to their species-typical song during the sensitive period (which begins in the weeks after hatching and continues until sexual maturity), they can't sing the song as adults. If they are exposed only immediately before the critical period (when their synapses are not ready) or afterward (when the synapses are already wired), they cannot learn to sing. This is somewhat similar to what happens in language development: language learning is more flexible in children than adults, as anyone who has tried to learn a second language after puberty knows.[103]

But not all forms of learning are subject to critical periods. And because different systems of the brain are involved in learning different things, there's no way to generalize from a few instances to all forms of learning. Whether math or music learning has a critical period for synaptic development is not known. However, as far as education is concerned, these may not be the most relevant questions, at least not at this point.

John Bruer, a leading thinker in educational practice, has cautioned that brain-based education is an idea whose time has not yet come, in spite of the fact that it is a thriving approach.[104] Bruer argues that we simply don't know enough about how changes in the brain (such as changes in synapse number or density, or changes in brain energy metabolism) relate to learning, or at least not enough on which to base something as important as education policy.

Instead, Bruer argues, educators should make the most of psychological findings in their efforts to improve the educational process. Once we know more about how specific capacities are used and when they appear, we might then gain clues about what to look for and where to look for it in the brain.

An excellent example of how psychology and brain science can interact in a way relevant to education comes from the work of Paula Tallal and Michael Merzenich.[105] Tallal, a cognitive psychologist who studies language, discovered that dyslexic children are deficient at detecting and discriminating rapidly changing sounds, and that the ability of infants to perceive such stimuli predicts later language skills. When she played the same sounds to dyslexic children at a slower rate, they perceived the sounds much more accurately. Tallal's work caught the attention of Merzenich, a neuroscientist who had been studying the ability of the neocortex to adapt to environmental stimulation, showing, for example, that giving monkeys extensive experience with a particular sound increases the area of cortex that processes that sound (this is described in detail in the next chapter). Tallal and Merzenich collaborated to create special video games that would expose children with dyslexia to certain carefully selected stimuli in an effort to improve their ability to process spoken and written language. By starting with slow stimulus-presentation rates and building up toward the rate at which speech and reading normally occur, they hoped to take advantage of the plastic nature of the cortex to actively shape its synaptic organization by training it to process stimuli more effectively. It is too early to tell how effective their program is, but regardless of its ultimate outcome, they are to be praised for their efforts to take basic research from the laboratory and use it to help children with reading disorders learn to read. Perhaps there are also lessons from their studies that might suggest how neuroscience research can also be used to improve the way children without learning problems are educated.

This work is at the heart of the debate between those who believe in specialized learning modules vs. more general learning capacities, since Tallal and Merzenich are using nonlinguistic stimuli to try to influence language learning. Those who believe that language is an innate module with its own learning rules argue that nonlanguage stimuli should have no effect. So not only will the outcome of this work have practical implications, it may also be very relevant to fundamental issues about brain organization.

John Bruer's most recent book attacks what he calls the myth of the first three years.[106] He argues that too much emphasis is given to the idea that specific kinds of learning have to take place very early in life, or the brain will never again be able to learn that kind of information. Learning is indeed a lifelong process. And as Alison Gopnik and colleagues argued in another recent book, every time the infant learns something, his or her brain is changed in a way that helps it learn something else.[107] In a review of this book, a prominent developmental expert, Mark Johnson, noted that the early years are crucial not because the window of opportunity closes but because what is learned at this time becomes the foundation for subsequent learning. Indeed, much of the self is learned by making new memories out of old ones. Just as learning is the process of creating memories, the memories created are dependent on things we've learned before.

MOVING ON

Learning and development are two sides of the same coin. We can't learn until we have synapses. And as soon as synapses start forming on the basis of intrinsic commands, they are subject to being influenced by our worldly experiences. Genes, environment, selection, instruction, learning—these all contribute to the building of the brain and the shaping of the developing self by wiring synapses. Although the extensive plasticity that is present in early life eventually stops, our synapses do not stop changing, but remain subtly changeable by experience. It's time to see precisely how this happens.

CHAPTER FIVE

ADVENTURES IN TIME

YOU HAVE TO BEGIN TO LOSE YOUR MEMORY, IF ONLY IN BITS AND PIECES, TO REALIZE THAT MEMORY IS WHAT MAKES OUR LIVES. LIFE WITHOUT MEMORY IS NO LIFE AT ALL. . . . OUR MEMORY IS OUR COHERENCE, OUR REASON, OUR FEELING, EVEN OUR ACTION. WITHOUT IT, WE ARE NOTHING.

—Luis Buñuel

Memory is a marvelous device, a means of transporting ourselves to earlier times. We can go back a moment, or most of a life. But as we all know, it's not perfect, and is certainly not literal. It's a reconstruction of facts and experiences on the basis of the way they were stored, not as they actually occurred.[1] And it's a reconstruction by a brain that is different from the one that formed the memory.[2] Sometimes, details are lost, but the gist is there. At other times, we just can't come up with what we're looking for, though we know we once had the information. Occasionally, we remember things that didn't actually happen. Fallible as it is, however, memory usually does a pretty good job for us. Adelle Davis's indispensable guide to sixties dietary habits told us, "You are what you eat." But Buñuel's lament is probably more accurate: we are our memories, and without them, we are nothing.

But what is memory? To most people, it is the ability to consciously recollect, to remember what happened days, weeks, or years ago. This is what psychologists call explicit or declarative memory. The information it summons is explicitly available for conscious recollection, and can be verbally stated or declared. This kind of memory is extremely flexible, allowing, for example, the sound of two cars crashing together to trigger a vision in your brain of an accident you once experienced. Through explicit memory you can recall a phone number, the way someone looks, what you had for lunch yesterday, or what you did on your last birthday. It is this memory that is so savagely attacked in Alzheimer's disease. Though extremely important in our lives, it is, in fact, only one kind of memory. My mother, an Alzheimer's victim, can't consciously recall much of anything that happens to her, but she can flawlessly execute the Cajun tune "J'ai Passé D'avant Ta Port" on her accordion. The kind of memory that allows her to do this is called implicit or nondeclarative memory. Such memories are reflected in the way we act more than in what we consciously know. Underlying this psychological difference is the fact that different brain systems are involved in implicit forms of memory, on the one hand, and conscious/explicit/declarative memory, on the other. To state this in the context of terms used in the last chapter, implicit memories are formed by systems that engage in domain-specific learning, whereas a domain-independent system forms explicit memories.

In chapter 2, I presented the outlines of a theory of the self that built on this view of memory. I proposed that the self is in part made and maintained by memory, and that both implicit and explicit forms are involved. In this chapter, we'll look at what is known about the neural circuits underlying explicit and implicit memory capacities. The synaptic mechanisms of memory are the topic of the next chapter.

IN SEARCH OF THE ENGRAM

In 1904, Richard Semon, a German scientist, coined the term *engram* to refer to the neural representation of a memory.[3] Two decades later, Karl Lashley, an American psychologist, had begun a search for the engram, the locus of memory in the brain, a quest that occupied him for much of his career.[4]

Lashley was one of the pioneers in the use of behavioral tasks, such as mazes, to investigate the relation between brain and behavior in animals. In his now-famous studies, he trained rats to "run a maze." Once the habit was learned, he then examined the effects of removal of different parts of the neocortex, which was believed even then to be where cognitive processes live. The animals were subsequently tested to see if they remembered the task. While these lesions sometimes disrupted memory, the impairment was not systematically related to the location of the damage. Instead, it seemed that the memory loss was related more to the size of the lesion than to its placement in the cortex.

The results suggested two principles to Lashley. The first was mass action:

the effects of brain damage on memory are due more to the amount of tissue removed than to the actual area removed. The second was equipotentiality: different areas contribute equally to memory, and when one is damaged, the others can take over. Lashley's bottom line: the brain doesn't have a specific system devoted to the formation and storage of memory. In his view, memories are stored in a widely distributed fashion in the cortex.

In retrospect, Lashley's approach was faulty. He used a behavioral task that could be solved in different ways, and made lesions more or less randomly. The reason that bigger lesions had bigger effects was not that all areas made equal contributions, but instead that different areas made unique contributions. As the lesions got bigger, more processes were disrupted, leaving the animal with fewer options for solving the problem. And in many instances the deficits were due less to the involvement of the cortical areas storing memory than to their role in basic sensory or motor processes necessary to perform the task. For example, a small lesion in the back of the cortex disrupted vision, but blind rats could solve the maze in other ways, using their senses of touch or smell. Bigger lesions left the animal with fewer sensory capacities, and the deficits got bigger.

Today, researchers go about things differently. They use knowledge about the organization of the brain to guide their attempts to find memory circuits. But hindsight is always visionary. The organization of the cortex was not well understood in Lashley's day, so he took the only approach available, one that now seems crude. Lashley didn't find the engram, but he made the search for it a legitimate enterprise.

THE SEA MONSTER SURFACES

In 1950, Lashley proposed somewhat facetiously that learning was not possible, given what he had discovered (or failed to) about the brain.[5] Unfortunately, this was the state of things at the time. But soon thereafter the tide turned when a young man named Henry underwent brain surgery in Hartford, Connecticut.[6]

Henry was one of several patients in whom parts of the temporal lobe were removed in an effort to control epilepsy. The temporal lobe is one of the four major lobules that make up the cerebral cortex, and is often the place in the brain attacked by epilepsy. Sometimes, temporal lobe seizures are so severe and uncontrollable that the only recourse is to remove the epileptic site. In Henry's case, temporal lobe areas were removed on both sides of his brain. After the surgery, his epilepsy was better, but the improvement came at a high cost: Henry lost his memory.

Henry, known in the memory literature as H. M., is probably the most famous case in neurological history, and he has been the subject of countless published studies on memory. Much of the initial work was carried out by Brenda Milner and her colleagues in Montreal.[7] They found that although H. M. could recall many of the events of his earlier life, especially those that occurred up to several years before the surgery, he was unable to form memories for experiences that occurred after the surgery. He could remember things for a few seconds (he had short-term memories), but he couldn't convert this information into long-term memories. He failed to recognize himself in a mirror, for example, but did recognize himself from old pictures. Though his IQ was in the normal range for his age, his view of who he was—his visual concept of his self—got stuck in the past, welded to his old surviving memories.[8]

The temporal lobe is a complicated structure with many subregions. An analysis of H. M.'s lesion, based on the surgical report, indicated that the main temporal lobe areas affected were the hippocampus, amygdala, and parts of the surrounding cortex.[9] By comparing H. M.'s lesion with those in other patients, it seemed that the hippocampus was the area damaged most consistently when memory deficits resulted.

The hippocampus, like many other brain areas, was named because its shape reminded early anatomists of something—in this case, a sea horse. The word comes from the Greek word *hipokampus,* which translates as "sea monster." In the early part of the century, the hippocampus was viewed as part of the rhinencephalon (which refers to smell or olfaction). It later emerged as one of the major structures of the so-called limbic system, and came to be thought of as playing a crucial role in emotional functions.[10] Paul MacLean, the originator of the limbic system concept, conceived of the hippocampus as the seat of the Freudian id, a place in the brain where ideas could be mixed together and confused. MacLean located the id in the hippocampus because he thought that the primitive architecture of this structure could blend things in a surreal, dreamlike fashion, and would be ill-suited to participate in cognitive functions. But, by the end of the 1950s, studies of H. M. and other patients had led to the conclusion that the hippocampus was crucially involved in one of the most important cognitive functions of the brain—memory.

CANS AND CAN'TS

At first, it was thought that H. M. and similar patients had a global amnesia, a complete loss of memory, and that the hippocampus was indeed a general-purpose memory machine. However, it was soon noticed that amnesics could learn to perform well on some tasks that depended on prior learning.[11] For example, Brenda Milner asked H. M. to copy a picture of a star viewed through a mirror. To do this, the movements had to be done in the direction opposite from the way they seemed they should be made. It's hard at first, but eventually most people can do it. H. M. was no exception. He learned the task, and he retained the learning. But if asked about the drawing, he had no conscious memory of having made it. Suzanne Corkin of MIT then found that H. M. also improved with practice in another manual skill learning task—one in which he was required to keep a stick held in his hand on a dot spinning on a turntable. As with the mirror drawing task, the more times he did it, the better he got. His ability to form memories about how to make precise movements (called motor skills) seemed intact.

Neal Cohen later examined whether the spared skill-learning ability in amnesia was restricted to motor skills or might extend to what are called cognitive skills, the ability to get better at doing well-practiced mental tasks.[12] He showed that amnesics could learn and retain the ability to read mirror images of words (*egral* is the mirror image of *large*) just as they could learn to draw mirror images. He also showed that the patients could learn some complicated rule-based strategies required to play certain games or solve puzzles. As in other examples, although they could learn to use rules to play games, they later had no recollection of playing the games themselves.

Studies of a phenomenon now called priming also turned out to be very influential.[13] Elizabeth Warrington and Larry Weiskrantz demonstrated that amnesic patients were able to learn to recognize stimuli (pictures or words) on the basis of incomplete depictions of the stimuli. Each stimulus came in five versions, from very incomplete to complete. Initially, they had to see the more complete stimuli to successfully figure out what each was. But with repeated trials, they were able to recognize the stimuli on the basis of the weaker cues. This kind of learning, now called priming, might therefore be independent of the hippocampus. Later studies by Larry Squire showed that the instructions given to subjects made a big difference in whether priming occurs. For example, in one version of priming, the subject is given a list of words. Later, if asked to recall the items that were on the list, amnesic patients perform poorly. However, if instead of being asked to recall the items, they are given fragments of words (*mot*) and asked to complete them, like normal subjects, they are more likely to come up with words that were in the original study list (*motel*) than with other words (*mother*) that were not in the list. Although the amnesic patients had little conscious memory of the study list items, prior exposure to the list had clearly been retained at some level, since it affected their performance.

Weiskrantz and Warrington also showed that the classical conditioning of eye blink responses is preserved in amnesia.[14] In this task, a tone is paired with an aversive stimulus (usually an air puff to the eye). After hundreds of trials, the tone elicits eyelid closure immediately before the onset of the air puff. This precisely timed response protects the delicate tissues of the eye from the air puff. Amnesic patients show normal eye blink conditioning, though they later have no memory of having seen the conditioning apparatus.

In the early 1980s, Larry Squire and Neal Cohen introduced the distinction between declarative vs. procedural memory to account for memory functions of the hippocampus vs. other brain systems.[15] Their proposal was that the hippocampus mediates conscious memory (memory that can be verbally declared) and that other brain systems mediate other forms of memory that are not dependent on conscious processes (fig. 5.1). They initially used the term *procedural memory* because at the time, many of the memory abilities that survived in amnesic patients involved the execution of skills that could be learned as rules or procedures. But as more examples emerged of things that hippocampal-damaged patients could learn and retain, it was clear that not all

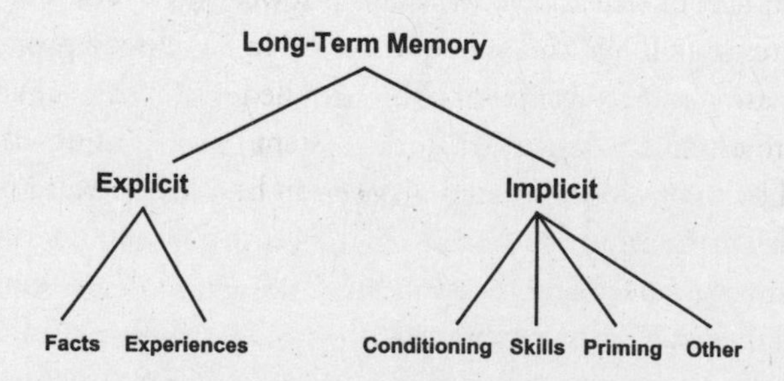

FIGURE 5.1 A TAXONOMY OF LONG-TERM MEMORY

Long-term memory is often divided into two broad classes, explicit and implicit, each of which has further divisions. Based on Squire 1987, and Squire and Kandel 1999.

of these involved procedures. Procedural memory was therefore later renamed with the more neutral designation: *nondeclarative memory*. The terms *explicit* and *implicit* memory, suggested by Dan Schacter, essentially refer to the same memory processes as declarative and nondeclarative memory.[16] Damage to my mother's hippocampus probably explains why she can't consciously remember the things that happen to her (her declarative or explicit memory system is gone), and lack of damage to the nondeclarative or implicit system that stores the skill of playing the accordion explains why she can remember how to play.

INS AND OUTS

The hippocampal contributions to memory, like the contributions of any brain area to any function, are made possible by the synaptic pathways. These deliver information to the hippocampus, process it within the structure, and transfer the outcome to target areas. Much is known about hippocampal anatomy, but we will only touch on some of the key points here.

As we've seen in previous chapters, information about the external world comes into the brain through sensory systems that relay signals to the neocortex, where sensory representations of objects and events are created. Outputs of each of the neocortical sensory systems then converge in the rhinal cortical areas, also known as the parahippocampal region,[17] which integrates information from the different sensory modalities before shipping it to the hippocampus proper (fig. 5.2). Studies by Wendy Suzuki, David Amaral, and others have elucidated many of the details of connectivity.[18]

The hippocampus and parahippocampal (rhinal) areas constitute what has come to be called the medial temporal lobe memory system, the system involved in explicit or declarative memory. (However, they are not part of the neocortical temporal lobe and are not involved in its sensory-processing functions. They are instead examples of the so-called old cortex [chap. 3]; this is why they are associated with the limbic system, which, as described earlier, was defined by the now-antiquated distinction between old and new cortex.)

Although within the hippocampus many complex circuits participate in the processing of incoming signals, there is a main-line circuit running through it called the trisynaptic circuit that is especially important. The trisynaptic circuit involves the relay of signals from the rhinal areas to an input area of the hippocampus (dentate gyrus) and from there to other areas (the CA3 and CA1 regions) and finally to the output region (the subiculum), which projects back to the rhinal cortex, closing the loop.

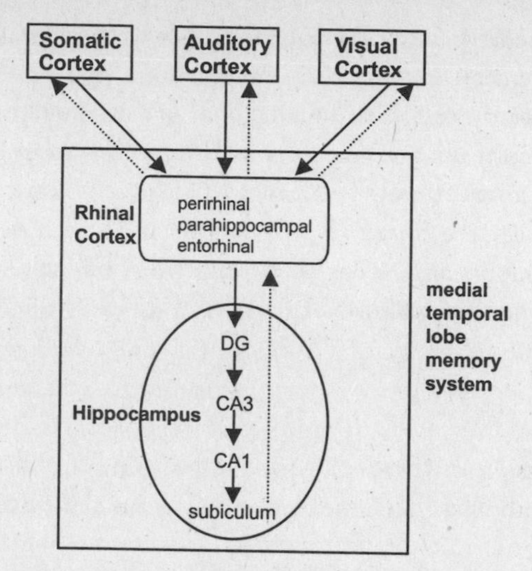

FIGURE 5.2 THE MEDIAL TEMPORAL LOBE MEMORY SYSTEM

Information about life's experiences is processed in the various sensory systems of the neocortex (auditory, visual, somatosensory [somatic], etc.). These areas in turn send their information to the rhinal cortical regions, where multimodal (multisensory) representations are formed. The rhinal regions then converge on the hippocampus. Within the hippocampus, information coming into the dentate gyrus (DG) is processed and sent to the CA3 region, which connects with the CA1 region, which in turn connects with the subiculum. Outputs of the subiculum are transferred to the rhinal areas, which then can send the information back into the hippocampus or back to the sensory neocortex for additional processing and storage.

The connections between the hippocampus and neocortex are all more or less reciprocal. As a result, the pathways taking information from the neocortex to the rhinal areas and then into the hippocampus are mirrored by pathways coming out of the hippocampus to the rhinal areas and ending in the same neocortical areas that originated the inputs (fig. 5.2). In this way, cortical areas involved in processing a stimulus can, as we will soon see, also participate in the long-term storage of memories about that stimulus.

It's important to reflect for a moment on the nature of these connections and their implications for what the hippocampus and rhinal areas do. The rhinal areas[19] serve as convergence zones,[20] brain regions that integrate information across sensory modalities and create representations that are independent of the original modality through which the information was processed (fig. 5.3). As a result, sights, sounds, and smells can be put together in the form of a global memory of a situation. Without this capacity, memories would be fragmented. Convergence zones also allow mental representations to go beyond perceptions and to become conceptions—they make possible abstract representations that are independent of the concrete stimulus. While the primate neocortex has several cortical convergence zones, fewer such areas exist in the neocortex of other mammals. This may be an important clue to differences between the cognitive capacities of primates and other animals.

Because the hippocampus receives inputs from several convergence zones in the rhinal region, it can be thought of as a superconvergence zone.[21] This no doubt accounts for why the hippocampus plays an essential role in our domain-independent memory capacity. It can form explicit memories about the implicit workings of many domain-specific systems, such as face- and language-processing systems, allowing us, for example, to form a memory that includes both what someone says and what he looks like.

NOMADIC MEMORIES

There are basically two kinds of amnesia: one is an inability to remember things that happened before the surgery or brain injury (called retrograde amnesia), and the other is an inability to form new memories (called anterograde

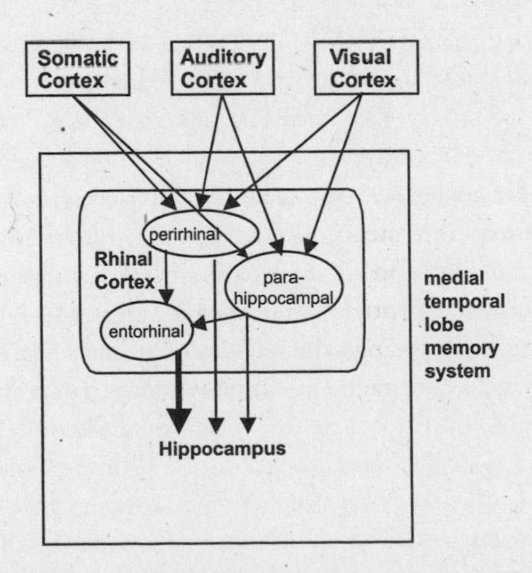

FIGURE 5.3 CONVERGENCE ZONES IN THE MEDIAL TEMPORAL LOBE

The rhinal cortical areas and hippocampus are convergence zones, regions that receive and integrate inputs from diverse regions.

amnesia). While we've focused so far on anterograde amnesia, both forms are present in patients with damage to the temporal lobe. Interestingly, the degree of retrograde amnesia is graded over time—memory is worse for events right before the surgery than for earlier ones. H. M. remembers his childhood, but not his present life.

The idea that amnesia affects recent more than remote memories is known as Ribot's law, after the French psychologist Théodule Ribot, who proposed that "the new perishes before the old."[22] Larry Squire performed a seminal study of Ribot's law on patients who received electroconvulsive therapy (ECT) for depression, a procedure that often produces memory disturbances as a side effect.[23] Squire tested patients before and after ECT treatment in 1974 on their memory of television shows. Before ECT, the patients had fairly reliable memory of shows from the early seventies and late sixties, with decreasing memory for earlier periods. This is quite normal, since we tend to remember recently learned things better than older ones. After ECT, though, their memory of shows from the early seventies was worse than before ECT, while memory for the earlier periods was unaffected.

The graded effect of retrograde amnesia is now believed to occur because the role of the hippocampus changes over time—the hippocampus is needed for memory storage initially, but its role decreases as time goes by. Why would the brain work this way? Why would memories be nomadic?

Jay McClelland, Randy O'Reilly, and Bruce McNaughton have proposed an answer, one inspired by results of a kind of computer simulation called connectionist modeling.[24] These simulations have been very useful in attempts to understand learning. For example, it is known that when a connectionist model tries to learn about relations between stimuli, it does better when new information is incorporated gradually into the memory store rather than when it is put in all at once. So-called interleaved learning prevents new information from interfering with old memories. As a result, if the model is trained to recognize characteristics of animals like birds ("can fly") and fish ("can swim"), and then encounters the fact that a penguin is a bird that "can swim" but "can't fly," the result is very different when interleaved learning is used than if rapid learning is used. With rapid learning, the new information tends to result in a shift of the knowledge base such that both fish and birds come to be treated as swimming animals. But with interleaved learning, where the representation is built up slowly over many repetitions, the network gradually refines the representation of a penguin as a bird that swims but doesn't fly. In other words, the new information doesn't interfere with the knowledge base and is instead gradually added to it.

Many researchers believe that explicit memories are stored in the cortical systems that were involved in the initial processing of the stimulus, and that the hippocampus is needed to direct the storage process. For example, the creation of a memory for a visual scene involves the transfer of the perception from the visual cortex to the parahippocampal cortical areas and then into the hippocampal circuits. The processed signal, the memory, is then fed back through the parahippocampal areas back to the visual cortex.

According to the interleaved learning hypothesis, then, the memory is initially stored via synaptic changes that take place in the hippocampus. When some aspect of the stimulus situation recurs, the hippocampus participates in the reinstatement of the pattern of cortical activation that occurred during the original experience. Each reinstatement changes cortical synapses a little. Because the reinstatements depend on the hippocampus, damage to the hippocampus affects recent memories, but not old ones that have already been consolidated in the cortex. Old memories are the result of accumulations of synaptic changes in the cortex as a result of multiple reinstatements of the memory. The slow rate of change of the cortex prevents the acquisition of new knowledge from interfering with old cortical memories. Eventually, the cortical representation comes to be self-sufficient. At that time, the memory becomes independent of the hippocampus.

Researchers like Jonathan Winson, Gyorgy Buzsaki, Bruce McNaughton, and Matt Wilson believe memory consolidation occurs during sleep,[25] and specifically that it is during sleep that the slow interleaving of information into cortical networks takes place. Recent studies support this notion.[26] For example, Wilson and McNaughton recorded the activity of neurons in the rat hippocampus. Using technically sophisticated procedures, they were able to identify precise patterns of cell activity in the hippocampus as rats explored a novel environment. Subsequently, when the rats went to sleep, the neural patterns seemed to be repeated in the hippocampus, as if the rats were dreaming about the places they had explored. This is an impressive finding. Although it has not yet been demonstrated that the hippocampal playback during sleep is actually read and used by the cortex, the existing data are consistent with the possibility.

Recently, the nomadic memory hypothesis has come under fire. Morris Moscovitch and Lynn Nadel have argued that the hippocampus always remains involved in memory storage.[27] However, with time, the memory trace comes to involve more and more brain regions, especially cortical regions, so that damage to any one area, like the hippocampus, fails to produce a deficit, since the other areas compensate.[28] While this issue will no doubt continue to be debated in the coming years, a recent study has provided some important new evidence.[29] Mice were trained in a spatial task that is usually impaired by hippocampal damage. Neural activity was then measured in the hippocampus at various times after training. Initially, neural activity was high in the hippocampus and was directly related to memory performance. (Mice with higher activity performed better.) With time, hippocampal activity decreased and came to be unrelated to memory performance, while cortical activity increased and came to be related to memory performance. Chalk one up for nomadic memory.

FACT AND EXPERIENCES: ARE THEY DIFFERENT?

So far, we have treated explicit or declarative memory as a single kind of memory capacity. However, in the early 1970s, the psychologist Endel Tulving proposed that long-term memories could be distinguished on the basis of what they were memories *about*.[30] He argued that episodic memories are about personal experiences (things that happened to you at a particular time and place), while semantic memories are about facts (things you come to know, but have not necessarily experienced). For example, you can know that deserts are hot and dry from personal experience, or by learning about deserts in school, without ever having been to one. Tulving suggested that episodic memory, which requires conscious recollection of the time and place of some personal experience, is particularly characteristic of humans, whereas semantic memory, being the simple storage of a fact rather than a personal experience, is within the capacity of many animals.

Recent studies by Faraneh Vargha-Khadem, Mortimer Mishkin, and colleagues support Tulving's distinction.[31] Their research involved children in whom the hippocampus was damaged early in life. Remarkably, these children managed to attend mainstream schools and learn the basic facts that schools impart, in spite of the fact that they had poor memory for their own experiences. Taken at face value, the results suggest that the hippocampus is involved in remembering personal experiences, but not in remembering facts. This was unexpected by most researchers, since both semantic and episodic memory are lost in patients with temporal lobe lesions. However, most such patients have had damage to the hippocampus and the surrounding cortical areas (parahippocampal region), while the children studied by Vargha-Khadem are believed to have mainly sustained hippocampal damage.

Not everyone accepts the conclusion that the hippocampus is selectively involved in episodic aspects of declarative memory. Two prominent memory researchers, Larry Squire and Stuart Zola, for example, have long promoted the view that the hippocampus and parahippocampal cortex (along with certain thalamic areas) constitute a single system that works as a declarative memory unit, making both semantic and episodic memory possible.[32] They mention several problems with Vargha-Khadem's conclusions. One is that episodic memory, while poor, is still present in the children, and that even a little episodic memory capacity may be sufficient to support a good deal of semantic knowledge. As a result, the deficit may have involved both episodic and semantic memory, but the episodic deficit may be more readily detected. Further, they argue that Vargha-Khadem and colleagues are not justified in saying, on the basis of brain imaging, that the damage involved only the hippocampus. This can be determined only from an autopsy in which a detailed analysis of the brain is conducted, and not merely by taking brain scans from living persons. According to Squire and Zola, even if there was more of a deficit in episodic than semantic memory, the conclusion that one involves the hippocampus and the other doesn't would still be suspect.

Obviously, it is very difficult to pursue the fine-grained analysis of questions about the involvement of particular brain regions situated next to each other in studies of patients with brain lesions, for brain damage has little respect for anatomical boundaries of interest to scientists. In order to understand the detailed anatomy of memory, even of human memory, it is necessary to attack the problem through studies of experimental animals, the topic to which we now turn.

IN SEARCH OF H. M.

Soon after H. M.'s amnesia was discovered, researchers began trying to achieve what Lashley was unable to do—namely, to create amnesia by damaging the brain in animals. If this could be accomplished, it would greatly facilitate the study of memory from a biological point of view. With the hippocampus identified as the experimental target based on H. M.'s misfortune, it seemed that success would come quickly. However, efforts to produce amnesia in animals by damaging the hippocampus were not resoundingly successful at first. As late as 1970, well over a decade after Brenda Milner first described H. M.'s amnesia, Peter Milner (Brenda's ex-husband and a neuroscientist at McGill University in Montreal) wrote, "Unfortunately for the experimental study of the hippocampal amnesia effect, efforts to reproduce it in animals have so far proved unsuccessful."[33]

A big part of the problem was that it was not understood initially that hippocampal lesions in humans affected only a certain kind of memory, explicit or declarative memory. In the animal studies, which were conducted under the influence of the behaviorist view that learning was learning was learning (chap. 4), any task that measured the effects of prior learning was a legitimate way of assessing the general capacity for memory. We now know that only some kinds of memory depend on the hippocampus, and so only tasks that measure that kind of memory will be affected when the hippocampus is damaged.

In humans, it turned out to be fairly clear how to define hippocampal-dependent memory. If conscious retrieval was required, the hippocampus tended to be involved. But this criterion isn't really suitable for animal studies. We have no way of knowing whether animals have the kinds of conscious experiences we have, so from a scientific point of view, consciousness is not a very good notion on which to base a comparative approach to memory.

So how can we study hippocampal contributions to memory in animals? As Peter Milner's quote above implied, animal studies are important for working out the details and mechanisms. Over the years, two approaches emerged. One focused on studies of recognition memory in primates under the assumption that the best way to figure out why H. M. had the particular memory disorder he had was to find a hippocampal-dependent form of memory in a species close to humans. The other was the result of a serendipitous observation in a rat.

MONKEY BUSINESS

The history of primate studies of recognition memory is long and complicated. I reviewed it in some detail in *The Emotional Brain,* and will give only a brief summary of the story here. Most of the work has involved a test of recognition memory called delayed nonmatch to sample developed by David Gaffan, an Oxford psychologist. In this task, a monkey sees one object and then, after a delay, is given two objects, one that was seen before and a new one. The correct choice, rewarded with a peanut or Froot Loop, is the new object (the one that does not match the sample). Many of the key findings using this task were obtained by Mort Mishkin and Betsy Murray at the National Institute of Mental Health and Larry Squire and Stuart Zola in San Diego, as well as Gaffan.[34] Initial studies showed that damage to the hippocampus produced a significant deficit if the delay between the first and the second presentation of the objects was long enough. An animal model of H. M. seemed to exist. Then, it was shown that damage to the amygdala and hippocampus together produced even more of a deficit. This model seemed even better, since H. M. had damage to both. It turned out, though, that the effects of the lesions were not due to damage to the amygdala or hippocampus, but to damage of the surrounding parahippocampal region, which was incidentally injured during the monkey brain surgery. This was still satisfactory from the point of view of modeling H. M., since he also had damage in these areas, but it raised problems for the view that the hippocampus is a key part of the memory system. However, studies of other patients who had damage restricted pretty much to the hippocampus (as determined by post-mortem examination of their brains) showed that severe explicit memory deficits could result from hippocampal pathology alone. This suggested that the hippocampus is involved in human explicit memory, but that delayed nonmatch to sample is not a good test of the hippocampal contribution. It may instead be a better measure of the contribution of the parahippocampal cortex to recognition memory.

It is now believed by some researchers that the hippocampus and parahippocampal areas are separate components of a temporal lobe memory system.[35] As we've already seen, sensory information comes into the hippocampus from the neocortex by way of the parahippocampal areas, and memories are established in the neocortex by way of the reverse connections. In sum it seems that the parahippocampal areas and hippocampus make unique contributions, and the delayed nonmatch to sample task reflects the parahippocampal more than the hippocampal contribution.

This work is relevant to the issue discussed earlier regarding the role of the temporal lobe in episodic vs. semantic memory. Perhaps delayed nonmatch to sample in the monkey is more like human semantic than episodic memory. If so, the distinction between the parahippocampal areas and hippocampus proposed by Vargha-Khadem would be supported. This would still leave open questions about what role the hippocampus itself plays. While more work will be needed to sort out exactly what the hippocampus does for memory, studies in rats have already begun to suggest some answers, as we now will see.

A PLACE FOR SPACE[36]

The second approach to animal studies of explicit memory began more or less accidentally. In the early 1970s, John O'Keefe was a researcher at McGill University in Montreal. Brenda Milner, who did much of the initial work on H. M., was just up the street. So was Wilder Pendfield, a neurosurgeon who found that conscious memories were elicited in patients when particular areas of the temporal lobe, like the hippocampus, were electrically stimulated in an effort to isolate the locus of epilepsy prior to surgery. Memory and the hippocampus were big topics in Montreal, and O'Keefe wanted to know more about how the latter made the former possible.

He had placed electrodes in the hippocampus of a rat in an effort to record the electrical activity in this structure during the neural coding of experience. Normally, projection cells (chap. 3) in the hippocampus fire fairly slowly, about once per second. But O'Keefe noticed that the firing rate in individual cells rose dramatically, up to hundreds of times per second, when the rat was in a certain location. As soon as the rat left that spot, the cell stopped firing. When it came back to the same spot, the cell started firing again. Because these cells seemed to be encoding the spatial location of the rat, O'Keefe called them place cells[37] (fig. 5.4).

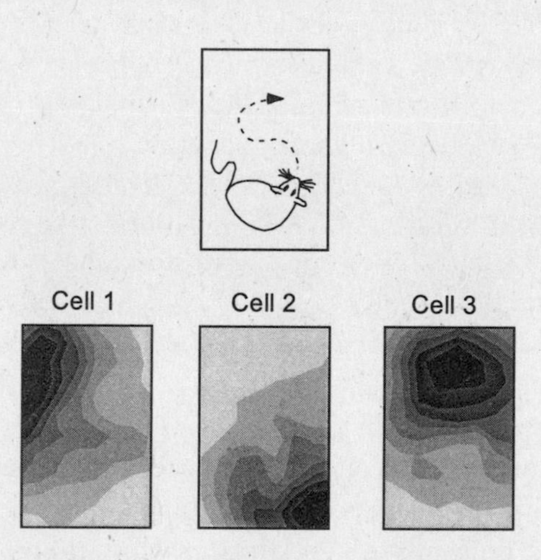

FIGURE 5.4 PLACE CELLS

"Place cells" in the hippocampus encode the organism's location as it navigates in its environment. For example, as a rat explores a rectangular chamber in search of small food pellets, different cells in the hippocampus are active at different locations. The responses of three such cells are shown. Dark shading corresponds to regions of the chamber where the cell was highly active, whereas light shading indicates lower activity. The region within the environment where a given cell is active is called its "place field." Across many hippocampal cells, the entire environment can be represented. For this reason, some researchers argue that the hippocampus creates a map of external space that is used in navigating in the environment. Others believe that spatial information is just one example of the kinds of complex relations encoded by the hippocampus (see text). Illustration provided by Marta Moita and Tad Blair.

Sometimes, one finding can help make sense of others. Recall that in 1970 Peter Milner pointed out that researchers had been unable to create amnesia in animals. He went on to note that some tasks were affected by deficits more consistently than others. Why this occurred was not well understood at the time. However, one of the few tasks that hippocampal lesions seemed to impair with any consistency was maze learning, a task that is often solved by spatial cues. Maybe O'Keefe's place cells were responsible for memory performance in mazes. Indeed, in the mid-seventies, David Olton of Johns Hopkins University devised a maze-learning task (the eight-arm radial maze) that could only be solved using spatial information, and damage to the hippocampus interfered with performance of this memory task.[38]

In the late seventies, O'Keefe, now in London, together with a former colleague from McGill, Lynn Nadel, wrote a famous book, *The Hippocampus as a Cognitive Map,* in which they proposed that the hippocampus is fundamentally a spatial cognition machine, and its contribution to other aspects of memory is secondary to its role in spatial processing.[39] The book was a tour de force that established spatial cognition as a major area of study in neuroscience. The study of the hippocampus in spatial memory in particular became a thriving area of investigation.

Many different research groups have studied the role of the hippocampus in spatial processing. In 1973, Jim Ranck published a seminal study of hippocampal cell activity that has inspired much work since.[40] Ranck, and especially his colleagues Bob Muller and John Kubie, have remained key players in the effort to understand place cell function.[41] Also noteworthy is the research of Bruce McNaughton, Carol Barnes, Matt Wilson, and their colleagues.[42] They have used technical ingenuity to push studies of place cells to the forefront of innovation. In one study, they recorded simultaneously from more than one hundred cells, showing how hippocampal neurons represent space and how the firing of these cells predicts where the rat will go next. As we saw above, they have demonstrated that the record of the rats' travels is encoded by place cells and played back during sleep, a finding that has been used to support the notion that the hippocampus slowly feeds new memories to the cortex during sleep. Another major contributor is Richard Morris, who developed a now widely used maze, the Morris water maze, in which rats use external cues to swim toward a submerged platform[43] (fig. 5.5). Damage to the hippocampus disrupts this form of spatial learning. Together, the studies of hippocampal firing and the effects of hippocampal lesions on maze learning have unequivocally shown that the hippocampus is involved in spatial memory. Might it be a place for space?

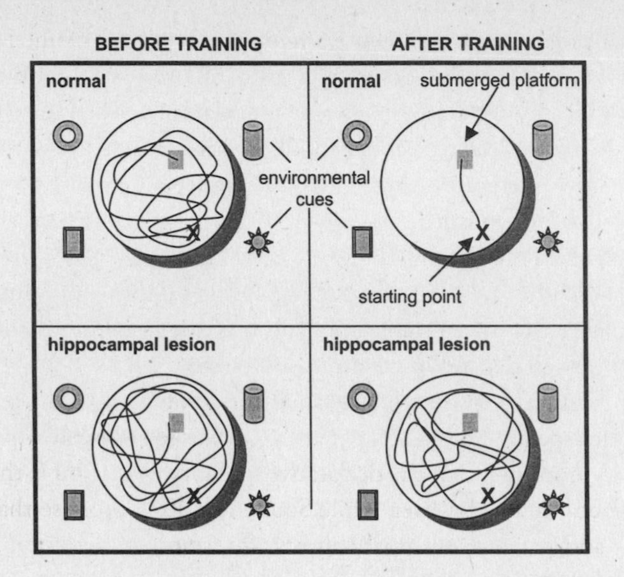

FIGURE 5.5 THE MORRIS WATER MAZE

In the Morris water maze, rats are placed at the starting point (X) and swim in an opaque solution until they find the submerged platform. Normal animals easily learn this task—they use stable environmental cues in the room to guide them straight to the submerged platform after they have undergone training. Rats with hippocampal lesions are unable to learn the task.

RELATIONS

Although most of the evidence from rats for a hippocampal role in memory has involved spatial learning, many tasks that monkeys and humans with hippocampal damage fail on have nothing obvious to do with spatial processing. Does this mean that the rat hippocampus and the primate hippocampus are fundamentally different? Not necessarily. Damage to the hippocampus in primates does produce deficits in spatial memory, along with deficits in nonspatial memory. Spatial processing might be a specific example, one that is especially important to rats, of a more general capacity of the hippocampus.

Howard Eichenbaum has long promoted the idea that spatial memory is one specific example of what the hippocampus does. Its general function, according to Eichenbaum and Neal Cohen, is declarative memory.[44] But this takes us back to the problem we encountered earlier. If declarative memory is conscious memory, how can we legitimately study it in animals? Eichenbaum and Cohen proposed that declarative memory should be defined not by its conscious aspects, but instead by the kind of processing requirements it makes. Specifically, declarative memories are relational.[45] Thus, activation of a declarative memory leads to the activation of other related memories. As a result, declarative memories can be activated independent of the context in which they were established, and by stimuli other than those that were initially involved in the learning. Eichenbaum and Cohen propose that the anatomical architecture of the hippocampus allows it to engage in relational processing, and that the various kinds of tasks that implicate the hippocampus in memory all depend on relational processing.

In Eichenbaum's view, the hippocampus is involved in both the semantic and episodic aspects of declarative memory. In this sense, he sides with Squire and Zola in proposing that the declarative memory in general is the business of the hippocampus. However, while Squire and Zola propose that the hippocampus and parahippocampal cortex form a unified declarative memory system, Eichenbaum suggests that only the hippocampus is involved in the flexible and relational processing that encompasses declarative memory, and that the parahippocampal areas are involved in the less flexible, nonrelational memory of individual component representations that go into declarative memory.

CONJUNCTIONS

The relational theory of hippocampal function is not universally accepted. The predominant challenge to it, other than the spatial theory, has been the view that the hippocampus binds stimuli together, blending the various elements of an experience (the way it looks, sounds, and feels) into a unified representation that is divorced from component elements. In this theory, the original elements, though, are lost, having been formed into conjunctions, while in the relational theory, the elements remain discrete and the hippocampus links them together. Consider an example. In the relational theory, your memory of a meal in a restaurant would involve the simultaneous binding of separate memories of the people you were with, the food you ate, and the general ambience of the restaurant, whereas the conjunction theory proposes that all of the elements are blended as a single memory of the situation.

Although the notion of conjunctions has been a part of many theories,[46] a recent proposal by Randy O'Reilly and Jerry Rudy seems particularly promising.[47] They built upon the earlier theory of McClelland, McNaughton, and

O'Reilly,[48] and on Rudy's earlier configural theory,[49] which posited that individual objects are 'configured' in such a way to account for all of the objects in a single representation. The theory failed to hold up because such global representations were difficult to substantiate. O'Reilly and Rudy argue that the configural theory was indeed partly wrong because much research subsequently showed that animals with hippocampal lesions could learn conjunctive relations. In the new theory, though, O'Reilly and Rudy propose that the data can be explained by assuming that the spared conjunctive learning was performed by the neocortex, and that the neocortex and hippocampus learn using fundamentally different rules. The hippocampus learns conjunctions naturally and rapidly. When the hippocampus is damaged, the neocortex can be forced, through extensive training, to also learn conjunctions, but normally it does not do so. As McClelland, McNaughton, and O'Reilly had proposed, the neocortex, in contrast to the hippocampus, learns by slow, interleaved learning. Ongoing research is testing this new version of the configural theory in relation to spatial and relational theories.

Clearly, animal models have generated many interesting ideas about exactly what explicit or declarative memory is and how it is supported by the anatomy of the temporal lobe. Although we still don't have all the answers, impressive progress has been made. We can expect that progress in understanding the role of the hippocampus in memory processes will continue to be made in animal models, and that this information will help, in turn, to understand the nature of explicit memory in humans.

THE WAY YOU DO THE THINGS YOU DO

Though we have firsthand knowledge of our explicit memories (which are about things we were once aware of), many aspects of our outward behavior and inner life are controlled by brain systems that store and use information implicitly, that is, without our awareness of their operation. Implicit memories are reflected more in the things we do, and the way we do them, than in the things we know.

We've already encountered some examples of implicit memory. Almost by definition, these are memories that can be formed in H. M. and other patients who have amnesia due to damage to the hippocampus and related brain regions. For example, learned motoric and cognitive skills, priming, and classically conditioned responses can each be learned and performed in the absence of the hippocampus. They must be mediated by brain systems other than the one involved in explicit memory.

While explicit memory is mediated by a particular system (contained within circuits of the medial temporal lobe), many different systems in the brain engage in implicit learning. While these are often described as "implicit memory systems," this is something of a misnomer. The systems that engage in implicit learning are not strictly speaking memory systems. They were designed to perform specific functions, like perceiving stimuli, controlling precise movements, maintaining balance, regulating circadian rhythm, detecting friend and foe, finding food, and so on; plasticity (the ability to change as a result of experience) is simply a feature of the neuronal infrastructure of these systems that facilitates their operation.

It's worth noting that John O'Keefe suggests the same basic scenario holds for the temporal lobe memory system. As we saw above, he believes that the hippocampus was built to process space, and that space processing demands a certain kind of plasticity that has been co-opted in the service of the more general capacity expressed as explicit memory in humans. In this sense, explicit memory is, in Stephen Jay Gould's terms, an exaptation (chap. 4).

Many of the neural systems that engage in implicit learning in humans have existed throughout much of the evolutionary history of mammals, and probably other vertebrates. They work unconsciously not because of some grand design to hide aspects of mental life from our sentient self but simply because their operation is not directly accessible by the conscious brain.

Systems that function implicitly contribute in important ways to our most characteristic traits. Each of us has his or her own style of walking, talking, and thinking. We hold our bodies in a certain manner when we are standing or sitting. We notice things that some others ignore, and ignore things that some people notice. The way we smile and the kinds of vocal inflections we use also help define who we are. The extent to which we are calm and collected, or emotionally reactive, when things go awry is also revealing, as are the logical paths and illogical leaps of thought we have. These and many other aspects of mind and behavior are expressed so automatically, so implicitly, that they may seem unchangeable, perhaps innate. But we should not overlook the crucial role of experience, which is to say of learning and memory, in establishing and maintaining them.

In recent years, considerable progress has been made in elucidating the neural circuits underlying certain examples of implicit learning and memory. While observations on H. M. and other humans led researchers to the hippocampus and its role in explicit memory, it has been studies of nonhuman species that have revealed the neural circuits of implicit memory. Studies in humans have, in many instances, verified that the same circuits are at work in our brains, but the basic findings have for the most part been discovered through studies of other animals.

THE RIGHT STUFF

Although the forms of implicit learning and memory that are now understood in most biological detail in mammals are each examples of classical conditioning, few studies before 1970 employed classical conditioning as a tool to explore the neural basis of learning and memory. Earlier research was inspired by the behaviorist tradition in psychology, which emphasized operant or instrumental conditioning over classical conditioning. In classical conditioning, the subject learns an association between two stimuli—a bell and food, for example. The responses that result are automatic (e.g., salivation or heart rate changes). In contrast, in instrumental conditioning, the association is between a stimulus (reward) and response—food occurs when a rat presses a bar or turns the corner in a maze. The food reinforces the response, and, as a result, the subject repeats the response to get the food. The subject is passively involved in classical conditioning (the food comes when the stimulus occurs, regardless of what the subject is doing) but in operant conditioning, the subject is actively involved since the reward does not come unless the subject responds in a certain way. Instrumental conditioning was viewed as a more suitable means of accounting for the full complexity of human behavior, which the behaviorists clearly hoped to do.[50] From Lashley onward, research on the brain mechanisms of learning and memory tended to emphasize the use of instrumental tasks.

By the 1970s, though, classical conditioning procedures had begun to take on a new life. Their rebirth began with the publication of an article by Eric Kandel and Alden Spencer in 1968.[51] Noting the gap that existed between learning (something that involves behavior) and plasticity (something that involves neurons and synapses), they proposed a step-by-step strategy that would allow the discovery of the neural basis of learning at the level of cells and their synaptic connections.[52] This cellular-connection approach was the right stuff. It transformed the field, and is still in place today.

Kandel and Spencer emphasized that the first step was to select an organism that expresses an easily measured and quantifiable behavior that changes with experience. Then, the neural circuit underlying unlearned and learned versions of the behavior should be identified. Next, the cells and synapses in the circuit that change with learning should be pinpointed. Finally, the mechanisms that mediate the neural changes should be determined.

Kandel and Spencer's manifesto challenged researchers to think in terms of circuits rather than large areas of the brain. Lashley had searched for the locus of memory by making lesions in widespread areas, without any sense of where to search—he was really looking for a needle in a haystack. The cellular-connection approach told us how to pinpoint the part of the haystack where the needle might be hiding.

Kandel and Spencer believed that the cellular-connection approach could best be implemented in a simple nervous system containing a relatively small number of well-defined neurons. For this reason, Kandel chose to study behavioral conditioning and plasticity in snails. Since the snail doesn't have a hippocampus, the lack of involvement of the hippocampus in conditioning was not an impediment to the use of these procedures. The emphasis was on conditioning as a tool for studying behaviorally relevant neural changes rather than as a direct model of human memory. The payoff of this approach was fast. By the early 1970s, Kandel and colleagues had implicated specific neurons and synapses in learning in snails. We'll consider some aspects of Kandel's elegant and groundbreaking research in the next chapter, work that led to his being a Nobel Prize recipient in 2000.

The success achieved by Kandel and others in invertebrates using simple conditioning approaches to learning tempted researchers to pursue this approach in mammals and other vertebrates. The fact that the tools for studying neuroanatomy had just undergone a revolution[53] greatly facilitated the effort to trace connections from start to finish (from sensory input to motor output systems) in a complex brain. Now, several decades later, much is known about the neural circuits underlying certain forms of classical conditioning. To the extent that these circuits do not depend on the explicit memory system, they are examples of circuits that engage in implicit learning.

But when this new wave of research on the neural basis of conditioning started in the 1970s, the distinction between explicit and implicit memory did not exist. Researchers pursuing the cellular-connection approach in invertebrates and vertebrates alike were using conditioning as a tool for studying the relation between neural plasticity and behavior. Only later was it realized that these studies were revealing the neural basis of what we now recognize as implicit memory.

Today, we have a detailed understanding of circuitry involved in only a few of the many systems that learn implicitly. The examples discussed below are therefore not necessarily representative of implicit memory in general, but instead are representative of the implicit memory functions that have been characterized, in the tradition of the cellular-connection approach, from sensory input to motor output at the neural level.

BEING DEFENSIVE

One of the first researchers to take up the cellular-connection approach and apply it to the vertebrate brain was David Cohen.[54] While recognizing the importance of studies of simple creatures, he felt that it was also important to pursue this approach in vertebrate species. His animal of choice was the pigeon, and his behavioral paradigm was Pavlovian conditioning.

Pavlov had been studying digestion in dogs when he noticed that they salivated when the person who fed them walked into the lab. He turned this into an experiment, and demonstrated that if a bell was rung just as food was put in the dog's mouth, the dog would salivate when it heard the sound of the bell, even if the food did not follow.[55]

Cohen's research involved the same general procedure I've used in my work, Pavlovian defense or fear conditioning.[56] In Cohen's conditioning procedure, the appearance of a light was followed by a shock. As a result, the onset of the light eventually led to a change in the rate at which the pigeon's heart beat. The change in heart rate, as we've seen, is but one of many responses that occur during defense conditioning. Cohen simply used heart rate as a convenient way of assessing that defense conditioning had taken place. He made impressive progress in his cellular-connection studies, identifying input circuits through which the light was processed and output circuits through which the heartbeat was controlled. However, he did not succeed in connecting the input system with the output system.

It had long been known that the amygdala plays an important role in fear responses and even in learning to fear new stimuli.[57] And one of the regions implicated in heart rate conditioning by Cohen was the bird-brain equivalent of the amygdala. However, Cohen's work fell short of demonstrating that the amygdala is the interface between inputs and outputs, and a likely site of the key neural changes that underlie fear learning.

Studies in the 1980s and 1990s painted a fairly comprehensive picture of the fear-learning circuits in the mammalian brain.[58] Work from my own lab identified the way that inputs get to the amygdala, and studies by Bruce Kapp and his colleagues at the University of Vermont provided evidence for the output pathways exiting from the amygdala. Michael Davis, Michael Fanselow, and Norman Weinberger, and their students and colleagues, also made important contributions to various aspects of the fear-conditioning circuitry. Many of the details of the fear-conditioning pathways were described in *The Emotional Brain,* and some aspects of fear-conditioning circuits have already been discussed in earlier chapters. Here, I'll give only a brief summary of them.

The amygdala contains a dozen or so distinct divisions or areas,[59] but relatively few are important for fear conditioning (fig. 5.6). The lateral nucleus of the amygdala (lateral amygdala) serves as the input zone. It receives information from the various senses,[60] allowing the outside world to be monitored for threatening information. In studies performed together with Alsa Pitkänen, we found that the lateral nucleus has connections with most of the other amygdala regions.[61] Karim Nader, Prin Amorapanth, and I then determined that only the connections with the central nucleus (central amygdala) are es-

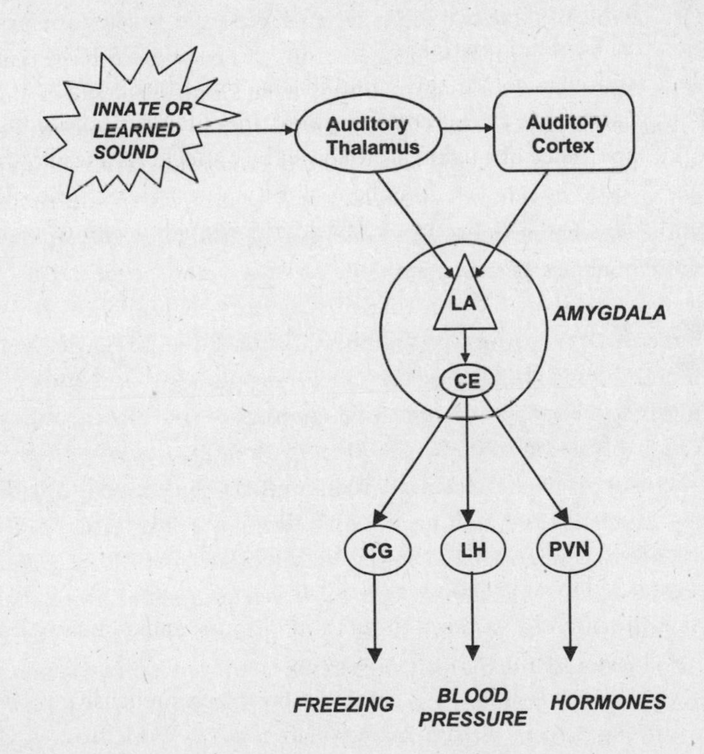

FIGURE 5.6 AUDITORY STIMULI ELICIT DEFENSE RESPONSES THROUGH THE AMYGDALA

Fear-arousing sounds, be they learned or innate, are transmitted through the auditory system to the thalamus and cortex. These regions give rise to connections to the lateral amygdala (LA), which connects with the central amygdala (CE), which connects with brain stem areas that control fear responses (CG, central gray; LH, lateral hypothalamus; PVN, paraventricular hypothalamus). Actual circuits are more complex and involve local regulation by GABA and modulators at each stage (see chapter 3).

sential for fear conditioning.[62] The central nucleus, the output zone, has connections with networks that control fear behavior and associated changes in body physiology. When the lateral nucleus detects some threatening stimulus, the central nucleus initiates the expression of defensive behaviors (like freezing) and other bodily responses associated with fear reactivity (changes in blood pressure and heart rate, stomach contractions, sweat gland activity, etc.).

Studies by Liz Romanski, Claudia Farb, Neot Doron and me show that the lateral amygdala gets inputs about the stimuli from two sources.[63] It receives a crude but fast representation from a subcortical area (the sensory thalamus) and a slower but more complete representation from cortical sensory areas (fig. 5.7). The role of these two input systems to the amygdala was elucidated in studies performed in my lab by Liz Romanski. The path from the thalamus through the cortex to the amygdala, the so-called high road, allows complex information about objects and experiences to initiate fear reactions. But the amygdala also can be activated directly from the thalamus. Since this low road bypasses the neocortex, it only provides the amygdala with a crude representation of the external stimulus. But the arrival of crude information can have important consequences. For example, Fabio Bordi and I found that cells in the amygdala are able to determine the intensity or loudness of a sound through the thalamic pathway.[64] Loudness is a good clue to how close something is, and distance is a good clue to how dangerous that thing is. If you treat loud things as dangerous, even if you don't know the source of the noise, you are probably going to be better off in the long run. So simply by computing intensity from the thalamus, the amygdala can immediately deduce significant details about a stimulus. Intensity is not the only feature gauged by the low road from the thalamus, but it's an important one.

The cortical route includes several more synaptic connections than the thalamic path to the lateral amygdala. Each synaptic link adds time to the transmission process, which is why cells in the lateral amygdala respond to information directly from the thalamus faster than they can respond to information from the cortex.[65] More processing time by the brain means a slower mental and behavioral response from the organism. In situations where rapid responses are required, speed can be more important than accuracy.

Although inputs from the thalamic and cortical paths arrive on different schedules, the inputs reach the same neurons. This means that in addition to jump-starting the system, allowing for rapid initial responses, the thalamic information can also prime lateral amygdala cells to receive the more exacting information from the cortex. As a result, the cells are then more capable of

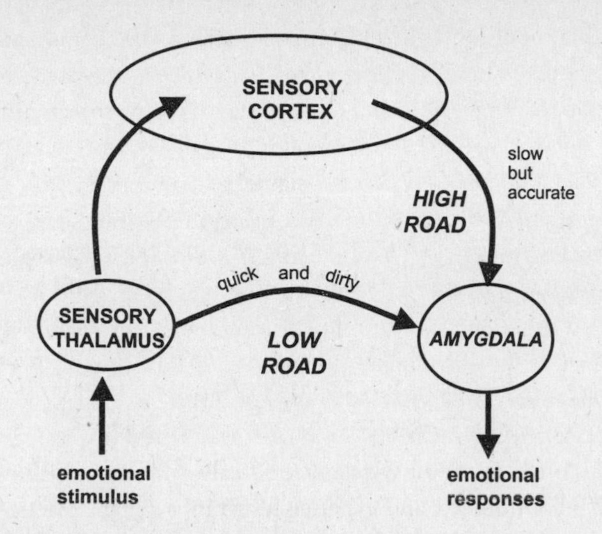

FIGURE 5.7 THE LOW AND HIGH ROADS TO FEAR

Information about external stimuli can reach the amygdala by way of direct pathways from the sensory thalamus to the amygdala (the low road), as well as by way of an indirect route through the cortex (the high road). The low road is shorter and faster but provides the amygdala with a less complete representation of the stimulus. Because the high road passes through the cortex, it can more accurately represent the stimulus, but this takes more time because more connections are involved.

charging ahead if the cortical information confirms the threat, or of putting on the brakes if the cortical information establishes that no danger is present (for example, that the loud crackling sound was from you stepping on a branch rather than from something dangerous, like a bear about to pounce on you, or that the curved shape on the ground is a stick, not a snake).

Just because the cortical route allows a more refined stimulus analysis to trigger the amygdala, it should not be assumed that the high road is a conscious route to the amygdala. This was often misunderstood by readers of *The Emotional Brain.* The amygdala engages in implicit processing, including implicit learning, regardless of which pathway provides it with sensory information. As with any other stimulus, we become consciously aware of an emotional stimulus only when that stimulus is processed by networks involved in something called working memory, which we'll discuss in detail in chapter 7.

Using the cellular-connection strategy, a likely neural circuit required for fear conditioning has thus been identified from sensory inputs through motor outputs. Work is now under way toward achieving the goal of determining where the essential plasticity is in this circuit. While this research is still in progress, experiments have pointed to the lateral nucleus as a key site of plasticity in situations where neutral sounds are paired with shocks. First, studies by Liz Romanski found that cells in the lateral nucleus receive convergent inputs from pathways that process both sounds and shocks (fig. 5.8).[66] Such convergence is believed to be essential for conditioning to occur. Indeed, Greg Quirk, Chris Repa, and Michael Rogan found that the activity of lateral amygdala cells increases when the tone and shock occur at the same time but not when they are separated (fig. 5.8).[67] These changes in the neural responses precede the development of conditioned fear behavior, suggesting that the neural changes might well be responsible for the behavioral learning. Studies by Steve Maren and Denis Pare have now also shown plasticity in the lateral amygdala.[68] Further, conditioning does not occur if the lateral nucleus is damaged.[69] Conditioning is also prevented if synaptic activity and/or plasticity is disrupted by injecting certain drugs in the lateral nucleus, as shown by Jeff Muller, Ann Wilensky, and Glenn Schafe in my lab, as well as by research from other labs.[70] Together, these findings indicate that the lateral amygdala is a key site of plasticity during fear learning.[71] As we will see in the next chapter, progress has also been made in uncovering some of the precise cellular and molecular mechanisms in the lateral amygdala that underlie fear conditioning.

As Pavlov suspected, defense conditioning plays an important role in the everyday life of people and other animals. It occurs quickly (one pairing of the neutral and aversive stimulus is often sufficient) and endures (possibly for a lifetime). These features have no doubt become a part of the brain's circuitry due to the fact that an animal usually does not have the opportunity to learn about predators over the course of many experiences. If an animal is lucky enough to survive one dangerous encounter, its brain should store as much about the experience as possible, and this learning should not decay over time, since a predator will always be a predator. In modern life, we sometimes suffer from the exquisite operation of this system, since it is difficult to get rid of this kind of conditioning once it is no longer applicable to our lives, and we sometimes become conditioned to fear things that are in fact harmless. Evolution's wisdom sometimes comes at a cost.

A BLINK IN TIME

One of the most thorough applications of the cellular-connection approach to the study of memory in the mammalian brain has come from Richard

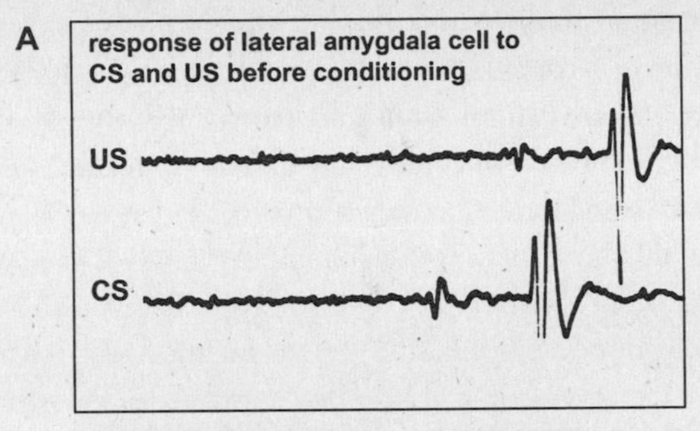

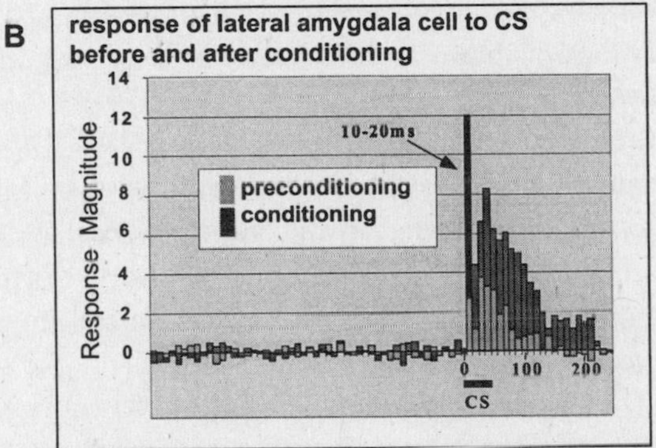

FIGURE 5.8 NEURAL RESPONSES OF LATERAL AMYGDALA CELLS DURING FEAR CONDITIONING

(A) Individual cells in the lateral amygdala respond to both the conditioned stimulus (CS) and unconditioned stimulus (US). (B) Pairing of the US with the CS alters the response of lateral amygdala cells to the CS. The largest increase occurs in the earliest component of the amygdala's response to the CS (10–20 ms), suggesting that the US alters processing of the CS in the low road (figs. 5.6, 5.7).

Thompson and his students. Thompson had worked with Alden Spencer, and his studies of plasticity in the spinal cord[72] had contributed significantly to the genesis of the cellular-connection approach subsequently developed by Kandel and Spencer. In the 1970s, Thompson began to study brain mechanisms underlying the Pavlovian conditioning of eye blink responses in rabbits as a simple mammalian preparation for pursuing the locus of learning in the brain. We encountered this kind of conditioning in our discussion of H. M.

Earlier studies by Isadore Gormezano and others had worked out in precise detail many of the basic principles underlying eye blink conditioning to a sound that was followed by an annoying stimulus (a shock or a puff of air to the eye).[73] Before conditioning, the shock or puff causes the eye to blink but the sound does not. After conditioning, though, the sound elicits a blink, and the blink is precisely timed from the onset of the sound—the eye closes right before the time when the puff or shock would occur. The blink therefore appears to be an adaptive response that protects the eye from irritation.[74]

Building on this behavioral work, Thompson and other researchers pursued the neural basis of eye blink conditioning.[75] It was discovered that removal of the entire neocortex had no effect on the ability of rabbits to be conditioned. In fact, eye blink conditioning could occur following removal of most of the forebrain, including the amygdala, hippocampus, and other areas. This led to the conclusion that the essential plasticity underlying this form of conditioning must be in the brain stem rather than the forebrain. Indeed, studies by John Moore, a former student of Gormezano's, found that damage in the lower brain stem did prevent conditioning, while at the same time leaving the response elicited by the unconditioned stimulus unaffected.[76] The rabbits could therefore blink their eye in response to a direct stimulus normally, but could not learn to blink to the warning sound.

Through a series of studies involving the use of many different techniques, Thompson and others (including Joseph Steinmetz, Mike Maur, and Chris Yeo) have elaborated the brain stem mechanisms of eye blink conditioning.[77] They have been led to the conclusion that the critical site in the brain stem is the cerebellum, a wrinkled mass of tissue that sits on top of the brain stem and that had long been believed to be involved in the control of posture and movement,[78] and that had previously been hypothesized to be a site of learning.[79] The researchers showed that lesions of certain areas of the cerebellum prevented conditioning, and that the activity of neurons in these areas changed during the course of conditioning. Pathway-tracing studies determined that the cerebellum has neural connections that allow it to integrate the sound and the air puff, and to control the eye blink response. In fact, the eye blink can be conditioned by substituting the tone and the puff with direct electrical stimulation of the neural pathways that transmit the tone and the puff into the cerebellum. Although not everyone favors the view that plasticity is an important feature of cerebellar function,[80] studies by Steve Lisberger using a different training procedure have also implicated cerebellar circuits in motor learning.[81]

Why should we care about how rabbits blink? As with many approaches in neuroscience, it is important to judge this work in a broader context. The research is not just showing how the rabbit brain blinks the rabbit eye, but instead really concerns how movements come under the control of extrinsic signals. Control of precise movements at just the right time is a fundamental aspect of many kinds of acquired behavioral skill, from learning to ride a bike, to hitting and catching baseballs, to driving cars, and many other things that we do every day. The cerebellum, as we have seen, is also involved in posture and balance, and both of these functions are modified by experience—astronauts, for example, readily adapt to the changed balance and postural requirements imposed by altered gravity. The way we hold our body at rest, the way we move, and the grace (or lack thereof) with which we do things contribute significantly to the physical image we project. They are part of who we are. Studies of the role of the cerebellum in motor learning are thus revealing clues about an important class of implicit learning capacities.

THE TASTE OF THINGS TO COME

Most people like sweet and dislike bitter-tasting foods. This preference is present very early in life—an infant will spit out bitter foods on the first encounter.[82] Infants born with massive parts of the forebrain missing[83] and rats with lesions of all of the forebrain and midbrain[84] still exhibit taste preferences, suggesting that taste is mediated by the hindbrain. The early appearance of taste preferences suggests that they are built into the hindbrain during prenatal development and are possibly innate.[85] Nevertheless, taste preferences are subject to modification by experience. Many adults enjoy eating broccoli and spinach, even if they were disliked in childhood. And some foods, once enjoyed, can become highly aversive if they make you sick.

Studies of food poisoning in animals, especially rats, have helped us understand a fundamental kind of learning, so-called conditioned taste aversion (CTA), a special form of classical conditioning. Farmers had long known that cattle avoid foods that make them sick, but this avoidance was not studied seriously until the 1960s, when John Garcia and his students found that rats made ill by injection of a mild poison strongly avoided the last food eaten before getting sick.[86] What was particularly striking about this experiment was the fact that the avoidance would develop even if sickness was induced hours after the food was consumed. The finding violated one of the cardinal principles of learning—that associations form between stimuli that occur at the same time, and not between stimuli separated by hours. The fact that CTA violated this law was one of the major challenges to the universal nature of learning promoted by the behaviorists (chap. 4). And because these data were viewed by strict behaviorists as impossible, Garcia had a great deal of trouble publishing his studies.[87] But he persisted and eventually succeeded.

CTA has become a major area of research in behavioral neuroscience. Although the circuits involved in it are not as well characterized as those that govern fear and eye blink conditioning, a good deal of progress has been made.[88] Most of the work has used saccharin, a novel sweet taste, as the conditioned stimulus (this is like the tone in a fear-conditioning study). Rats prefer drinking saccharin-flavored water over plain water, but if they get sick after drinking saccharin, they will subsequently avoid it. Often the sickness is induced by an injection of lithium chloride, a nausea-producing substance that serves as the unconditioned stimulus (like the shock in fear conditioning).

Like other forms of classical conditioning, CTA requires that the neutral taste and the nausea-producing stimulus come together synaptically at individual neurons. The taste pathway from the tongue goes to a region of the hindbrain called the nucleus of the solitary tract, which is involved in taste preferences. The same general region of the hindbrain receives fibers from the gut, telling the brain about nausea and other gastrointestinal conditions. However, because different regions of the hindbrain receive the nausea and taste signals, conditioning is not likely mediated by synaptic integration in this area.[89] The two parts of the nucleus of the solitary tract do project onto a common area in the midbrain called the parabrachial nucleus. Damage to this region, where taste- and nausea-stimulus processing overlap, prevents CTA from occurring. Furthermore, in normal animals, cellular activity elicited in this region by taste stimuli increases following exposure to the nausea-inducing stimulus, indicating that cells in the region are involved in the conditioning of taste by nausea. The parabrachial nucleus thus is likely to be an important area of plasticity underlying CTA.

But this is not the whole story. Removal of the forebrain, which leaves only the brain stem to run the show, disrupts the acquisition of CTA, in spite of the fact that the same procedure leaves taste preferences undisturbed.[90] The parabrachial nucleus thus does not work alone in the mediation of CTA and instead depends on the forebrain areas. Anatomical tracing studies give clues as to which areas of the forebrain are important here. The parabrachial nucleus sends fibers to the taste area of the thalamus, which in turn sends fibers to the taste area of the cortex. Further, the parabrachial nucleus, taste area of the thalamus, and taste cortex all send fibers to the central nucleus of the amygdala. Using experimental lesions, it has been shown that damage to two of these regions, the cortical taste area and the central amygdala, interferes with CTA. The taste cortex is believed to be involved in detecting and discriminating novel tastes, which is an important aspect of learning about the consequences of new tastes. The role of the central amygdala is controversial. Some believe it is not required for learning, whereas others believe it is essential.[91] Differences in the way CTA is established or tested may account for the different views.[92] While more work is needed to resolve these issues, CTA has proven to be a useful model system for studying memory circuits, and has also yielded some important findings about the molecular basis of memory, as we will see in the next chapter.

WHENCE SUCCESS?

Early research on learning mechanisms was inspired by the behaviorist tradition in psychology. By the 1970s, though, researchers came to realize that less is more when it came to the choice of behavioral tasks used to study learning, and began taking their inspiration more from neurobiology, especially from the cellular-connectionist perspective. The emphasis therefore shifted away from complex instrumental learning procedures to simpler conditioning tasks, since, as Kandel and Spencer noted, it would be easier to discover the circuits and specific synaptic changes involved in simple than in more complex forms of learning.

The advantages of classical over instrumental conditioning as a starting point for a neural analysis of learning are readily illustrated by the history of research on fear learning. Early work in this area emphasized avoidance-conditioning procedures, instrumental tasks in which the subject learns to avoid an aversive event, like foot shock, on the basis of a warning signal. It turns out, though, that avoidance conditioning begins as classical conditioning: the warning signal produces the avoidance response because it was first associated with shock. For example, a rat can be taught that if it runs from one end of a maze to the other when the tone comes on, the shock will be prevented. But the first step is that the rat has to learn, through classical conditioning, that the tone predicts the shock. Only then can it use the tone-shock association to learn the instrumental response of avoiding the shock by running through the maze in the presence of the tone. It makes good sense, in retrospect, that the way to figure out how the brain mediates avoidance conditioning would be to first understand how the tone-shock association is formed, and then to ask how the tone-shock association is used to establish avoidance. But this was not realized in the early days, and progress in understanding fear learning was slow until researchers decided to simply focus on the classical conditioning of fear.

Obviously, memory researchers will not be satisfied with explaining only the simple forms of learning. But armed with knowledge of how those simple forms work, we may well be able to take on more complex ones. For example, now that the classical conditioning of fear is well understood, it might be possible to figure out the neural basis of avoidance. This is an important avenue of research since the hallmark of anxiety disorders is pathological avoidance of anxiety-producing situations, which greatly constricts the lives of anxious people. As we will see in chapter 9, recent studies have made considerable progress in using knowledge obtained about the neural basis of fear conditioning as a stepping-stone to understanding avoidance learning.

CONSCIOUS AND UNCONSCIOUS MEMORY RECONSIDERED

What exactly do we mean when we say that the hippocampus is involved in conscious memory and that other forms of memory, which are independent of the hippocampus, are not directly accessible to consciousness? Does it imply that the hippocampus makes consciousness possible? Probably not, since hippocampal damage does not disrupt normal conscious awareness.

Examples of memory deficits produced by hippocampal damage in humans have indeed mostly involved failures in the ability to consciously retrieve information about past experiences. When tested on learning tasks that do not require conscious retrieval, like priming or conditioning, hippocampal-damaged patients basically perform acceptably. But when required to retrieve information consciously, often within the same task they performed well on implicitly, they do poorly. For example, as we've seen, though hippocampal-damaged patients respond to a tone previously paired with an air puff to their eye, they have no conscious memory of having been conditioned.

The importance of conscious retrieval in defining the hippocampal deficit was highlighted in a study performed by Larry Squire and colleagues.[93] They had normal subjects and persons with amnesia due to hippocampal damage watch a silent movie. The subjects were asked to pay attention to the movie while ignoring sounds and air puffs that would occur throughout. This type of instruction typically biases subjects to form explicit memories about the attended event, and was aimed at keeping the tones and puffs in the background of awareness. The tones and puffs were delivered in two different ways. Some of the subjects received a standard procedure in which the puff occurred at the end of the tone. As expected from prior work, both the normal controls and the amnesic patients conditioned—they learned to blink in response to the sound—in spite of the fact that they were instructed to ignore the tones and puffs. For other subjects, the puff came some time after the tone had ended. This is called trace conditioning and is known from studies of rats and rabbits to be disrupted by hippocampal damage.[94] Indeed, the patients with hippocampal damage did not condition in the trace task. What is particularly significant, though, is what happened to the normal controls. Some of them conditioned and some didn't, and whether they conditioned was closely related to whether they reported, after the experiment, having noticed a relation between the tone and the puff. In other words, if they were conscious of the relation, they underwent trace conditioning; otherwise they did not. This suggested that trace conditioning requires awareness, and that the hippocampus makes this awareness possible.

One problem with this conclusion is that it strains the interpretation of the rat and rabbit data, which reveal the same effect: deficits in trace conditioning following hippocampal damage. If the human hippocampus is involved in trace conditioning because of its role in awareness, then the rat and rabbit hippocampus must also be involved in conscious awareness, since they, too, are required for trace conditioning. If not, the hippocampus is not really involved in awareness.

A recent study by Marvin Chun and Liz Phelps suggests the latter is the case.[95] They tested normal subjects and amnesics on a task that required the identification of a single letter *T* in a sea of *L*'s. The *L*'s thus formed the background or context in which the target *T* appeared. Unbeknownst to the subjects, the context stimuli came in two varieties. Sometimes the *T* was amongst an arrangement of *L*'s that had come before, and sometimes the *T* was in a novel arrangement of *L*'s. Both normal controls and amnesics got better as the task wore on. However, the normals but not the amnesics benefited from the repetition of the background pattern: they were better at finding the *T* in the repeated pattern than in novel ones. Nevertheless, the subjects were not aware, when asked later, of the fact that the background repeated. The hippocampus, in other words, was required to process and learn about the unattended background stimuli.

Chun and Phelps were motivated to perform their study by the fact that studies of rats had long suggested that the hippocampus was involved in contextual processing. For example, as we'll see in chapter 8, when you condition a rat by pairing a tone with a shock, the rat becomes afraid not only of the tone but also of the box in which the conditioning takes place.[96] The hippocampus, it turns out, is necessary for this kind of conditioning[97] because it is able to put together many things at once, to relate or configure them into a context. This is not unlike the way O'Keefe or Eichenbaum or Rudy views the rodent hippocampus: it has a certain arrangement of neurons that allows the creation of a memory involving a complex set of stimuli. There is no need to call upon consciousness in the interpretation of the rodent work.

The Chun and Phelps finding strongly suggests that the human and rodent hippocampus work the same way, forming memories about the relations between stimuli. Regardless of the nature of the relations (whether they are unified configurations/conjunctions of stimuli, relations among different memories, or memories about spatial arrangements), the key function of the hippocampus is probably processing relations. If so, then the hippocampus might not be fundamentally different from other systems that store information. That is, it probably goes about its business implicitly. What is different about the hippocampal and other memory systems is instead the fact that the hippocampus is synaptically connected in such a way that its activity is available to the brain systems that mediate conscious awareness (which we'll consider later in chapter 7), whereas implicit systems are not connected in this fashion.

THE GOOD, THE BAD, AND THE UGLY

Memory is amazing. A simple thought can take you to a past time in your life. And often without thinking at all we "remember" to do what we need to do each day. But with the good come the bad, and even the ugly, aspects of memory, its failures. No one knows more about this than Dan Schacter, who put the terms explicit and implicit memory on the map. His recent book, *The Seven Sins of Memory,* lays out in graphic detail the ways memory fails us. Schacter's sins are transience, absent-mindedness, blocking, misattribution, suggestibility, bias, and persistence.[98]

Transience is simply the inability to hold onto information. Absent-mindedness is our annoying capacity to fail to pay attention to what we are doing, as when you put your keys down while doing something else, and then can't find them because when you put them down you were mentally involved in some other activity. Blocking is the failure to pull out that fact or name that is on the tip of your tongue. Then there's misattribution, as typified by the belief that a memory formed in one situation when it actually occurred in another. This is particularly important in the context of eyewitness testimony.

Suggestibility is also relevant to eyewitness testimony and to the topic of false memories implanted in the therapeutic process. Bias creeps into memory in several ways, one of which is consistency bias, which leads us to revise our memory of a situation to make it fit what we feel or think now. Finally, there's persistence, which on the surface sounds good. Emotional arousal makes any memory stronger, which can be good, but when the memory is of a traumatic experience, persistence can be debilitating.

Schacter emphasizes that the sins are less design flaws than features. That is, he claims that the sins are by-products of virtues—persistence, as we've seen, can be good. Regardless of the adaptive value, if any, of the sins, at least with knowledge of their existence we can try to protect ourselves from their negative consequences.

REMEMBERING WHO WE ARE

This chapter opened with a quote from the Spanish surrealist filmmaker Luis Buñuel, who said, "Life without memory is no life at all. . . . Our memory is our coherence, our reason, our feeling, even our action. Without it, we are nothing." Memory does indeed make us who we are. But it is important to remember that memory is more than just what we can consciously recall. It therefore follows that a loss of explicit memory, due to damage to the hippocampus, while devastating in many ways, would not eliminate personality. In Alzheimer's disease, for example, the hippocampus and related areas are the first to be destroyed.[99] Even in the face of severe memory problems, its sufferers initially remain much the same person they always were—they walk and talk the same, and have the same basic habits and traits. As the disease spreads widely to other brain areas, those that function implicitly, personality begins to break down.[100]

In order to be yourself, you have to remember who you are. Keep in mind though that the memories involved are distributed across many brain systems, and are not always or even mostly available to you consciously.

CHAPTER SIX

SMALL CHANGE

NOTHING ENDURES BUT CHANGE.

—Heraclitus

A LIVING THING IS DISTINGUISHED FROM A DEAD THING BY THE MULTIPLICITY OF THE CHANGES AT ANY MOMENT TAKING PLACE IN IT.

—Herbert Spencer

Life is change, and the brain is a device for recording changes—for forming memories through learning. Learning and memory, as we've seen, fill in the details of who we are as we become a unique person. But what is the nature of the neural changes that constitute learning and memory? Most neuroscientists today believe that alterations in synaptic connectivity underlie learning, and that memory is the stabilization and maintenance of these changes over time. How, then, does experience actually change synapses, and what makes the changes last?

HEBB'S MAGIC

In an 1894 lecture to the Royal Society of London, Santiago Ramón y Cajal, whom we met in connection with the neuron doctrine, proposed that "the ability of neurons to grow in an adult and their power to create new connections can explain learning."[1] While this statement is often cited as the origin of the synaptic theory of memory, it was anticipated by a number of other ideas. For example, in the mid-1700s, the philosopher David Hartley suggested that mental associations (memories about the relation between stimuli) are the result of vibrations between nerves.[2] More than a century later,

William James, the father of American psychology, wrote in his famous 1890 textbook: "When two elementary brain processes have been active together or in immediate succession, one of them, on reoccurring, tends to propagate its excitement into the other."[3] And in his medical research days, Sigmund Freud argued, "Memory is represented by the facilitations existing between . . . neurons."[4] Still, these suggestions, including Cajal's, were only partially formed when compared to the view proposed by the Canadian psychologist Donald Hebb, in his 1949 book, *The Organization of Behavior*.[5]

Hebb made many seminal contributions to scientific psychology, including pioneering studies on perception, instinctual and emotional behavior, and intelligence. However, he is best known for his synaptic theory of memory, his fire and wire theory, which we encountered in chapter 4 in the context of brain development. And although this theory is revered by brain scientists today, Hebb apparently did not believe it to be particularly important, and certainly didn't consider it to be his best idea or his most significant contribution.[6]

Hebb's notion, as you'll recall, is that "when an axon of cell A is near enough to excite cell B or repeatedly and consistently takes part in firing it, some growth process or metabolic changes take place in one or both cells such that A's efficiency, as one of the cells firing B, is increased." Let's expand this idea a little so we can see how it might apply to memory, and especially to a memory of the fact that two stimuli once occurred together.

In order for two stimuli to be bound together in the mind, to become associated, the neural representations of the two events have to meet up in the brain. This means that there has to be some neuron (or a set of neurons) that receives information about both stimuli. Then, and only then, can the stimuli be linked together and an association be formed between them.

We'll consider a simple example to see how Hebb's theory makes associations possible. Imagine a neuron A that is postsynaptic to two other neurons (fig. 6.1). One of these (S) is strongly connected and the other (W) is weakly connected to A. As a result, when S fires, A also often fires, whereas A is less likely to fire in response to activity in W. Further, imagine that each of the two neurons connected to A is involved in processing distinct stimuli. Because the connection from S to A is stronger than the connection from W to A, A is more likely to fire in response to external stimuli that activate S than those that activate W. If on some occasion stimuli processed by both of the cells occur simultaneously, the weak input to A will likely be occurring at the same time that A is being fired by the strong input. As a result, according to

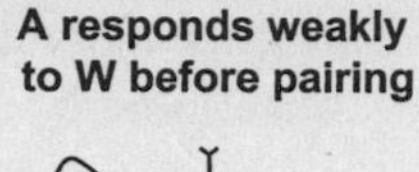

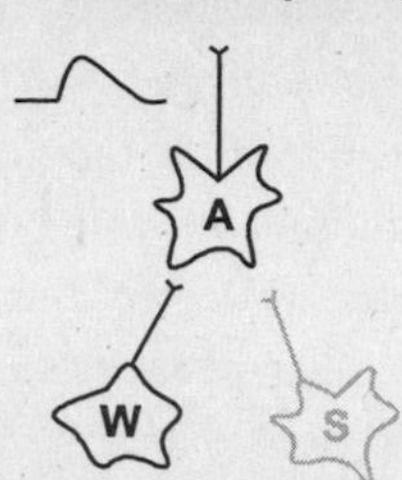

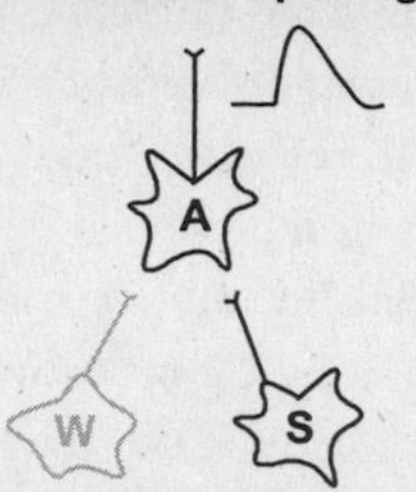

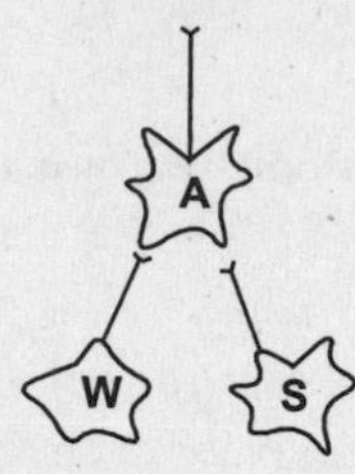

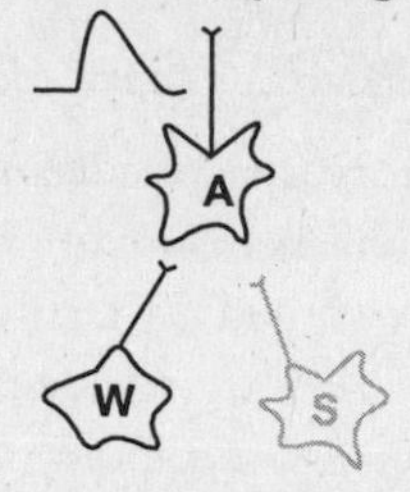

FIGURE 6.1 HEBBIAN PLASTICITY

When weak and strong inputs to a cell are active at the same time, the weak pathway is strengthened by way of its association with the strong pathway. This is called Hebbian plasticity, after the Canadian psychologist Donald Hebb. In the illustration, the connection between cell W and cell A is weak and only elicits a small response from A (*top left*). However, A also receives inputs from S, and these elicit strong responses from A (*top right*). If W stimulates A at the same time as S (this is called pairing, and is shown on the bottom left), the response of A to W will, according to the Hebb learning rule, be strengthened (*compare bottom right to top left*). Hebbian plasticity is believed by many to underlie associative memory (memory about how two stimuli or events are related). Gray shading indicates pathway is inactive.

Hebb's rule, the weak connection will be strengthened. A is now a place in the brain where the strong and weak pathways are related, such that activation of the weak one now has the same effect that only activation of the strong one had before. This sort of thing happens all the time in daily life. If you are walking on the sidewalk in front of your neighbor's house (weak stimulus) and his dog bites you (strong stimulus), you will associate the sidewalk where you were bitten with the dog and be less inclined to walk that way.

Today, neuroscientists use the term *Hebbian learning* to describe changes in the connection strength between two neurons caused by the fact that the postsynaptic cell was active when presynaptic inputs arrived.[7] Hebb realized that his idea was not completely novel, noting that others had proposed synaptic theories of memory, and some historians have suggested that Hebb be given less credit in light of past views.[8] The fact remains, however, that Hebb developed a theory of synaptic strengthening to account for learning, whereas most of the others simply proposed that such changes might occur.[9] And, as we'll see later in the chapter, Hebbian learning actually explains how certain kinds of synaptic changes take place in the brain, and may in fact be a major way that memories are made.

SEARCHING FOR SYNAPSES

Around mid-century, synaptic change was a topic that was in the air. Hebb's book had been published, as had one by the great Polish neuroscientist Jerzy Konorski.[10] Konorski used the term *plasticity* to describe the ability of neurons to be altered by experience and had proposed a theory of synaptic plasticity not too different from Hebb's. And by the early 1950s, a number of studies had successfully shown that repeated delivery of a brief electrical stimulus to a nerve pathway could alter synaptic transmission in that pathway—could, in other words, produce synaptic plasticity.[11]

For example, Sir John Eccles, recipient of the Nobel Prize for his work on synaptic transmission, and one of the legends of modern neuroscience, found that repetitive stimulation of nerves going to the spinal cord led to an increase in the size of the electrical response elicited in postsynaptic neurons in the spinal cord.[12] Electrical stimulation of pathways was used as a simplified approach for activating synapses in a specific area of the nervous system. Though artificial, this method was viewed as a reasonable approximation of direct experience because our experiences in the world can affect the brain only by way of electrical conduction in nerve pathways (chap. 3). As a result, changes like those found by Eccles were interpreted as a step toward understanding the neural basis of learning. However, because the changes were mostly short-lived, they were not judged to be sufficient to account for the persistence of memory. Nevertheless, Eccles remained a true believer in the theory that memory involves synapses.[13]

In 1966, Richard Thompson and Alden Spencer found indirect evidence that synaptic changes might account for a relatively simple example of learning. They studied habituation of the limb-withdrawal reflex in cats.[14] Habituation is a form of learning in which repeated presentation of a stimulus leads to a weakening of a response—you jump the first time you hear a loud noise, but if it is repeated over and over, you jump less.[15] The limb-withdrawal reflex had been known since Sir Charles Sherrington's time to be mediated by synaptic pathways into and out of the spinal cord (chap. 3). Later studies indicated that the actual circuit involved the relay of sensory information from the skin to a group of interneurons in the spinal cord to the motor (or output) neurons that control muscle movement. Thompson and Spencer ruled out changes in the ability of the input and output nerves to transmit signals, and concluded that the plastic changes must have crucially involved the interneurons. This important series of studies helped boost the notion that learning involves changes in synaptic transmission, but did not actually pinpoint the exact synaptic connections that were modified.[16]

Two years later, Spencer and Eric Kandel[17] took an important conceptual step, one that helped close the gap left open by Thompson and Spencer—the gap between behavioral learning and synapses. In their cellular-connection paper (chap. 5), they proposed that changes in synapses induced by learning could be identified if a simple behavior was studied in an animal with a simple nervous system. Cats, rats, or other mammals, whose behavior was more relevant to humans, simply had too many neurons and too many synapses to be studied effectively. Lower vertebrates or, better yet, invertebrates were in their view more suitable subjects.

Kandel and other researchers followed this approach and went on to identify the synaptic basis of several forms of learning in the nervous system of invertebrate species.[18] Especially notable was their ability to pursue synaptic plasticity all the way down to the level of specific molecules required to make memory last. Though truly groundbreaking from the point of view of a biological analysis of a behaviorally relevant form of neural plasticity, it was long unclear how, if at all, this work on lowly creatures like snails might apply to mammals.

In the meantime, though, a means of studying synaptic plasticity in the mammalian brain emerged. As this approach matured over the years, its findings began to converge with those from the invertebrate work, leading to the conclusion that, deep down, synaptic plasticity may be accomplished in similar ways in vastly different animals and in vastly different kinds of learning situations. After we take a look at the story of how it finally became possible to investigate synaptic plasticity in the mammalian brain, we'll return to the invertebrate studies to put the research in perspective.

PRACTICAL MAGIC

In the mid-1960s, Terje Lømo, working on his Ph.D. in Per Andersen's lab in Oslo, made a chance observation. He noticed that a brief burst of electrical stimuli delivered to fibers headed for the hippocampus in a rabbit led to a dramatic and long-lasting increase in transmission (a bigger electrical response to a test stimulus after, as compared to before, the burst) at synapses in the hippocampus,[19] a region believed to be involved in human memory (chap. 5). Although this turns out to have been one of the most significant experimental findings in the history of memory research, when Lømo reported the result in a lecture at a scientific conference in 1966, there was, as he recalls, "no reaction."[20]

Lømo never got around to publishing his results and moved on to subjects of more interest to him. But a few years later, Tim Bliss, a young researcher from England, came to Oslo to work with Andersen. Bliss had been a student at McGill University, where he had attended Hebb's seminars, and he was quite interested in the physiology of memory. In Andersen's lab, Bliss and Lømo pursued the effects of electrical stimulation on hippocampal synaptic transmission that Lømo had noticed earlier, and in 1973, they published a paper describing the phenomenon of long-lasting potentiation, which is now called long-term potentiation (LTP).[21]

In Bliss and Lømo's experiment, a stimulating electrode was put in the fiber pathway going into the hippocampus, and a recording electrode in the hippocampus itself. They then delivered a single electrical stimulus to the pathway, and recorded the electrical response of the postsynaptic neurons. This served as the baseline, the standard against which the rest of the experiment was gauged. Next they gave the potentiating stimulus—a brief burst of many rapidly repeated pulses. Then they started testing again with a single pulse, and continued testing periodically for several hours.

The key finding was that, following the potentiating pulses, the synaptic response got bigger, relative to the baseline response, and remained bigger for hours. According to Bliss, "We knew by the end of the first night that we really had something pretty important."[22]

Although Bliss and Lømo had produced the long-lasting change in the postsynaptic response by electrically stimulating neural pathways, rather than by having the animals actually learn something, they realized that they had identified a mechanism that might be able to translate neural activity generated by environmental stimuli into changes in synaptic efficiency, a mechanism that might be used to record and store information about life's experiences. And the fact that they made this discovery in the hippocampus bolstered their concluding speculation that changes in the efficiency of synaptic transmission might account for memory.

It's interesting to note that the approach used by Bliss and Lømo to produce long-lasting changes in synaptic transmission was very similar to the approach that had failed in the hands of Eccles and others in the spinal cord.[23] Why was it successful with the hippocampus? One possibility is that the hippocampus is especially plastic. While this is true to a certain extent, LTP has since been induced in many areas of the nervous system, including the spinal cord.[24] More likely, LTP was observed in the hippocampus because this part of the brain is organized into neat layers—the inputs come into one layer, and the outputs exit from a different one. This natural segregation made it possible to easily isolate and measure the postsynaptic responses elicited by electrical stimulation of input fibers. Once the phenomenon of LTP was worked out in the hippocampus, researchers knew what to look for and how to look for it in other neural circuits.

In the end, it took a lot of luck to achieve a viable model of plasticity in the mammalian brain. If the hippocampus didn't have such a simple organization, Lømo might not have been able to accidentally discover LTP. And if Bliss hadn't attended Hebb's seminars, or if he hadn't gone to Norway, Lømo's early finding might have remained in a drawer. But somehow the pieces all fell in place, and LTP became a way to practice Hebb's magic.

A SLICE OF LIFE

Bliss and Lømo did their studies in living animals. This was technically challenging, which is why the studies lasted throughout the night. But soon it was discovered that LTP could be studied more easily in thin slices of the hippocampus removed from the brain and submerged in a saltwater bath, with electrodes placed in the relevant regions. Hordes of scientists flocked to studies of hippocampal LTP. In the five-year period beginning in 1975, there were only twelve publications on LTP, whereas in the next five years, after the hippocampal slice technique became available, there were about ninety.[25] But this was just the beginning. Between 1990 and 1994, more than one thousand LTP papers were published, and the number almost doubled in the next five years. The hippocampal slice became one of the preferred ways of studying regular synaptic transmission, as well as the premier way to study synaptic plasticity.

As the properties of LTP unfolded over the years, the possibility that this artificial phenomenon might have something to do with memory only increased.[26] For example, in addition to its rapid induction and persistence over time, LTP was found to involve associative interactions between postsynaptic neurons and the specific presynaptic inputs that were involved in forming the association. Rapid acquisition, persistence, specificity, and associativity are all features one would expect of a memory mechanism (fig. 6.2). Let's consider the meaning of specificity and associativity further.

Bliss and Lømo showed that LTP was *specific* to the stimulated pathway. They started an experiment by stimulating two different pathways that elicit activity from the same population of postsynaptic neurons. They then gave only one of the pathways the potentiating stimulation, and subsequently tested postsynaptic response in both the pathways. Although stimulation of either pathway elicited a response from the postsynaptic cell, only the potentiated pathway changed. Giving potentiating inputs to one pathway, therefore, does not automatically change all of the synapses on the postsynaptic neuron; it only changes those synapses that were stimulated. LTP is thus specific to the synapses involved in the potentiating experience and does not represent a change throughout the entire postsynaptic neuron. This means that a given cell can participate in the storage of information about many different experiences so long as different synapses on the cell are involved in receiving them.

An important prelude to the demonstration of associativity was the finding that LTP involved cooperativity, which was shown by Bruce McNaughton and colleagues.[27] Like Bliss and Lømo, they used two pathways. When each was stimulated in succession with weak stimuli, no LTP resulted. But if the two pathways were stimulated simultaneously with the weak stimuli, the stimuli could combine (cooperate) to produce LTP in both pathways. This suggested that LTP involves some kind of interaction at the cellular level between synaptic inputs, and might therefore be capable of forming associations between different inputs.

The ability to form associations between stimuli is perhaps the benchmark test for a synaptic mechanism of learning. That LTP might be a way to form associations was strongly suggested by the results of a study performed by Chip Levy and Oswald Steward in 1979.[28] They applied weak stimulation to one pathway at the same time that strong stimulation was delivered to another. In contrast to the McNaughton cooperativity experiment, in the Levy and Steward associative experiment, the strong stimulus alone was sufficient to induce LTP, and the weak stimulus added little. However, if the weak in-

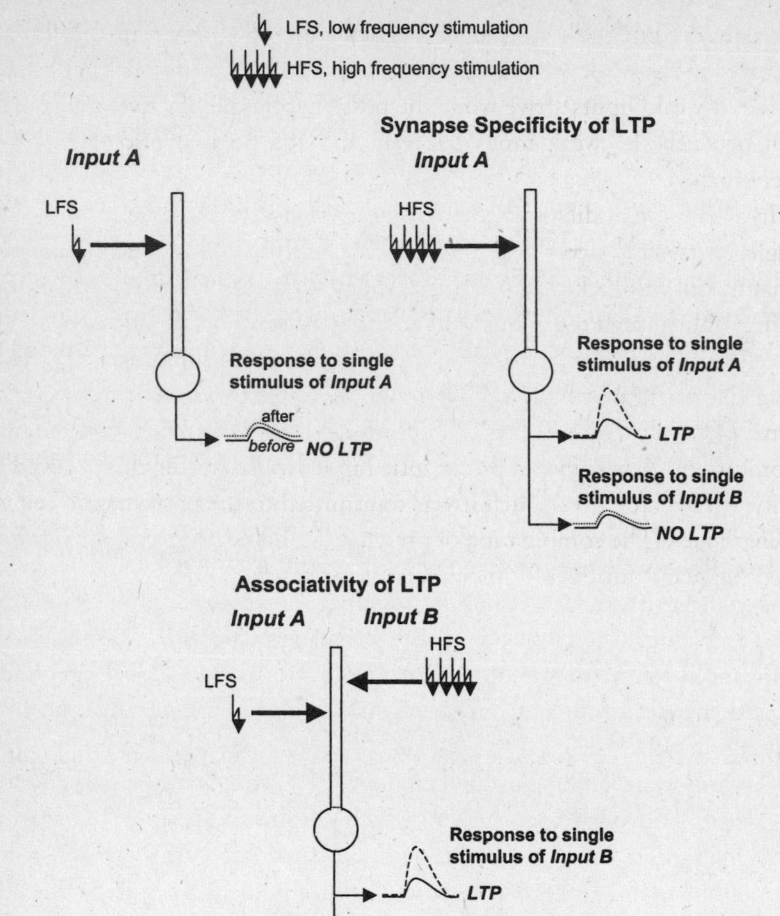

FIGURE 6.2 LONG-TERM POTENTIATION (LTP)

Long-term potentiation (LTP) is a model for studying the synaptic basis of Hebbian plasticity, and thus associative memory. In the illustration, a training stimulus (a series of electrical stimuli) delivered at a low rate (low-frequency stimulus, LFS) or a high rate (high-frequency stimulus, HFS) is applied to one (*top left* and *top right*) or both (*bottom*) of two input pathways (inputs A and B) to a cell. The effects of the training stimulus on the cell's response is assessed by giving a single test stimulus to one or both pathways. The delivery of the test stimuli to the inputs is not shown but the effect of the test stimulus is illustrated at the bottom right as a comparison of the response before and after >

put arrived while the strong input was activating the postsynaptic cells, LTP occurred in the weak pathway as well as the strong one. Just as Hebb had predicted, if weak inputs arrive while the postsynaptic cell is active, the connection between the weak input pathway and the postsynaptic cells will be strengthened.

In 1986, several different research groups reported that the response of a single postsynaptic cell to a weak input could be potentiated if the postsynaptic cell was tricked into acting as though it had received a strong input.[29] Using sophisticated techniques, the researchers were able to artificially reduce the electrical negativity of the cell just before a weak input arrived. (Recall from chapter 3 that when a cell is strongly activated by synaptic inputs, its internal electrical state becomes less negative, which is how an action potential is produced.) When a weak presynaptic input arrived during this induced activity, the synapse over which it was transmitted to the postsynaptic cell was strengthened. The combination of presynaptic and postsynaptic activity is indeed the magic formula for increasing synaptic strength.

ISN'T THAT A COINCIDENCE?

How does the brain actually achieve Hebbian plasticity? How, in other words, is the co-activity of presynaptic and postsynaptic cells registered and stored?

---

the training stimulus. *Top left:* LFS of pathway A does not alter the response to the test stimulus delivered to pathway A (the response to the test stimulus is not different after and before the training stimulus). Thus, no LTP occurs. *Top right:* Delivery of the HFS to input A produces LTP (that is, the response of pathway A to the test stimulus is bigger after than before the training stimulus). However, HFS of input A has no effect on the response to test stimuli given to input B before and after HFS of A, showing that the effects of LTP are specific to the particular input synapses that are trained with the HFS. This is the synapses-specificity property of LTP. *Bottom:* If the LFS is given to one input (A) at the same time that the HFS is given to another input (B), the response of both inputs to their test stimuli is increased (LTP occurs in both). LTP in input A is an instance of Hebbian (associative) plasticity because the response of pathway A to the test stimulus was modified not by its training stimulus (LFS of A does not induce LTP, as shown in top left), but because it was active at the same time that another input to the cell (input B) received a plasticity-inducing HFS. This is the associative property of LTP. Synapse specificity and associativity are two properties that any model of associative memory should possess. Based on figure 55.22 in Beggs et al. 1979.

Two discoveries in the mid-1980s began to clarify this process.[30] The first was by Graham Collingridge, who showed that blockade of a particular type of glutamate receptor (chap. 3) prevented the induction of LTP without interfering with synaptic transmission. Thus, when this receptor was blocked, the synapse still worked just fine (release of transmitter from the presynaptic neuron produced a normal postsynaptic response), but it could not be potentiated by experience. The second discovery was by Gary Lynch and Roger Nicoll, who separately demonstrated that LTP would not occur if calcium was prevented from rising in the postsynaptic cell during action potentials. The two findings actually complement each other, since it is the special glutamate receptor that allows calcium to rise in the postsynaptic cell when an action potential occurs.

You'll recall from chapter 3 that glutamate is the main excitatory transmitter in the brain. When it is released from presynaptic terminals and binds to postsynaptic receptors, the likelihood of the postsynaptic cell's firing is increased. Actually, there are several different kinds of glutamate receptors, and each plays a different role. One (the AMPA receptor) is involved in regular synaptic transmission, while another (the NMDA receptor) is involved in synaptic plasticity (fig. 6.3). There are other glutamate receptors, but these two are the ones most relevant to this discussion.

Presynaptically released glutamate finds its way to both AMPA and NMDA receptors. Binding of glutamate to AMPA receptors is one of the major ways that a postsynaptic cell can be induced to fire an action potential, and is the means by which cells normally get fired up. In contrast, when presynaptically released glutamate reaches NMDA receptor on the postsynaptic cell, it has no effect initially because part of the receptor is blocked.[31] However, once glutamate has activated the postsynaptic cell (caused it to fire an action potential) by binding to AMPA receptors, the block on the NMDA receptors is removed, and glutamate can open the receptor channel and allow calcium to enter the cell. LTP is the result.

For NMDA receptors to pass calcium, both presynaptic and postsynaptic cells must be active. This is the basic requirement for Hebbian plasticity. But how does this set of events form an association between two inputs to a cell? In the next section, we'll explore the chemistry of *how* calcium entry through NMDA receptors produces LTP. For now, though, what we want to consider is *why* calcium entry through NMDA receptors is a means of forming associations between a weak and a strong input.

Activity in the weak input pathway results in the release of glutamate and

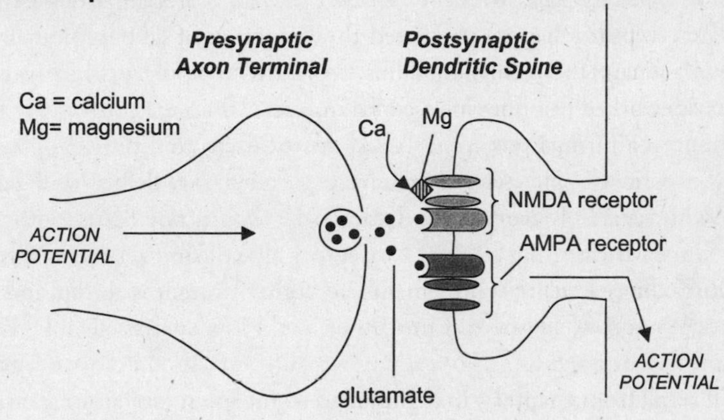

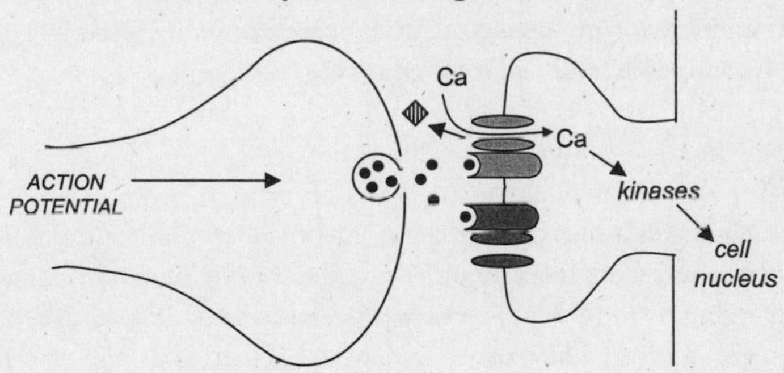

FIGURE 6.3 GLUTAMATE TRANSMISSION

Binding of glutamate to AMPA receptors triggers excitatory responses that contribute to the generation of an action potential from the postsynaptic cell. Although glutamate also binds to NMDA receptors, it has no effect because the receptor channel is blocked by magnesium. However, when the postsynaptic cell fires an action potential, the magnesium block is removed and calcium is able to flow through the NMDA channel. The rise in calcium inside the postsynaptic cell then leads to the activation of kinases that travel to the cell nucleus where additional molecular processes occur, including gene activation, which in turn leads to the synthesis of new proteins that contribute to the strengthening of the synapses (see fig. 6.5).

the binding of glutamate to postsynaptic receptors. Because the connection is weak, though, the input is not capable on its own of making the postsynaptic cell fire an action potential. However, when synaptic activity in the strong pathway activates the postsynaptic cell, the block on NMDA receptors will be removed, even at the weak synapses. Therefore, if the weak pathway is releasing glutamate during this time, NMDA receptors at both the strong and the weak synapses will be able to bind the glutamate. Calcium will flow in through the NMDA receptors, and the weak synapses will be strengthened.[32]

In sum, the reason that NMDA receptors allow LTP to occur is that they are coincidence detectors: they are able to register that presynaptic and postsynaptic neurons were active at the same time. More specifically, NMDA receptors allow the cell to record exactly which presynaptic inputs were active when the postsynaptic cell was firing. This input specificity is key to associativity, and is exactly what Hebb described decades before NMDA receptors were discovered. It is with this in mind that Holger Husi and Seth Grant recently referred to the NMDA receptor and associated molecules as a Hebbosome, a complex set of interacting proteins functioning as a unit in the induction and maintenance of synaptic plasticity.[33]

MAKING CHANGE LAST

The binding of glutamate to its receptors is a brief event, lasting, at most, seconds. But memories can last a lifetime. In order for synaptic changes encoded by NMDA activity to persist, chemical processes that outlast the synaptic action itself are required. These processes have been studied using two different procedures, one that produces a form of Hebbian LTP that lasts only an hour or so, and another that generates a more persistent form of Hebbian LTP.[34] These are called early and late LTP.[35] (There are also forms of LTP in which NMDA receptors play no role, but we will not have much to say about these.)[36]

Early and late LTP are thought of as analogs of short- and long-term memory.[37] From a biological point of view, short- and long-term memory are also distinguished by their chemical requirements, in addition to their longevity. It has been known for several decades that if animals are given drugs that prevent the brain from making new proteins, they are able to learn normally but are unable to form long-lasting memories. (If they are tested within an hour or so of learning some task, they perform well, but show no signs of having learned the task when tested the next day.)[38] This has turned out to be true of most if not all kinds of memory tasks that have been studied, and seems to apply to most if not all species, as well. It is also true of early and late LTP: blockade of protein synthesis has no effect on early LTP but prevents late LTP.[39] These parallels between early LTP and short-term memory, on the one hand, and late LTP and long-term memory, on the other, are consistent with the view that LTP and memory might be mediated by the same molecular mechanisms.[40]

In order to further understand the chemistry of early and late LTP, and to pursue the question of whether LTP and memory have similar molecular underpinnings, we need to examine the concept of second messengers. Neurotransmitters like glutamate are considered first messengers. They are responsible for signaling between neurons. Second messengers pick up where first messengers leave off. Their job is to initiate chemical reactions *within* the cell on the basis of information provided from *outside* the cell during neurotransmission by first messengers.

Calcium is one of the major second messengers. As we've seen, when glutamate binds to its NMDA receptors, calcium flows into a cell (after the block has been removed). Once this occurs, calcium takes over and directs the chemical reactions that strengthen synaptic connections, both in the short run and the long run.

Key to the whole process are enzymes called protein kinases that activate specific proteins. Their job is to phosphorylate certain proteins. Technically, this means that they add a phosphate group to the proteins. But all you really need to know is that phosphorylation turns proteins on, transforming them from an inactive to an active state.

In early LTP, kinases act on preexisting proteins that are present in the cell, waiting to be called upon. As a result of calcium influx during early LTP, several kinases are activated.[41] One important task performed by these kinases appears to be the phosphorylation of AMPA receptor proteins.[42] As a result, after the induction of LTP, the same amount of glutamate released from the presynaptic cell by an action potential will have more AMPA receptors to bind to, and a bigger postsynaptic response will occur, allowing each presynaptic action potential to make a greater contribution toward the firing of the postsynaptic cell. (Recall from chapter 3 that it takes many closely timed action potentials at different synapses on a postsynaptic cell to generate an action potential in that cell, so every little bit helps.)

Although most scientists in this area of research agree that LTP is induced or initiated by calcium-triggered chemical reactions in the postsynaptic cell, some believe that these postsynaptic changes fully account for LTP, while others argue that the presynaptic cell changes as well (fig. 6.4).[43] For example, a

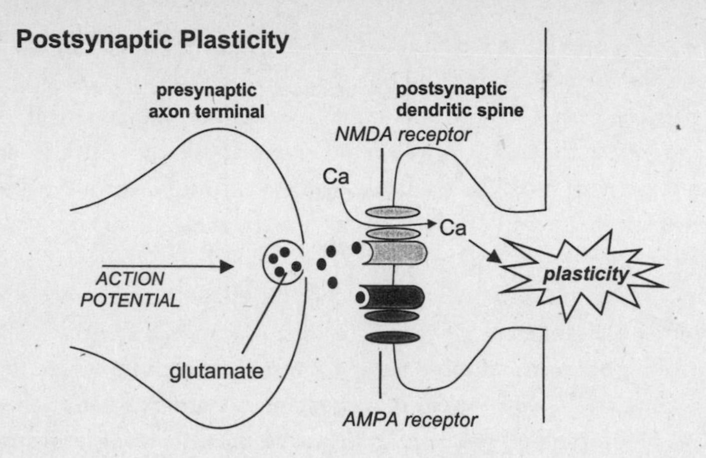

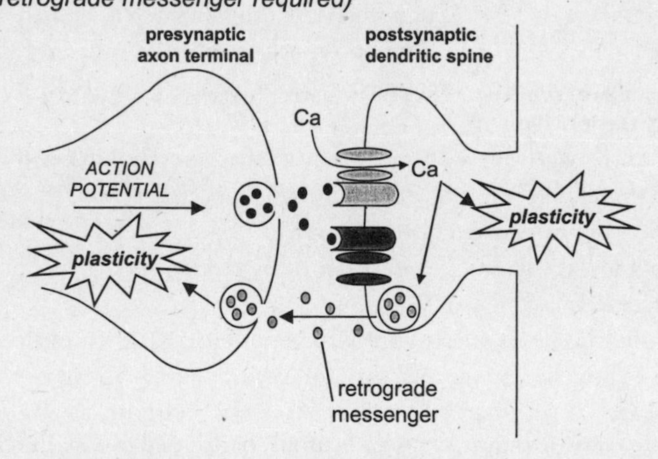

FIGURE 6.4 PRESYNAPTIC AND POSTSYNAPTIC PLASTICITY

LTP is viewed classically as being induced or established postsynaptically following calcium entry through NMDA receptors (*top*). However, in some instances, the maintenance of LTP over time requires that the presynaptic cell be modified as well. In order for postsynaptically induced changes to impact on the presynaptic cell, some message has to be shipped back across the synapse (*bottom*). Chemicals called retrograde messengers are believed to serve this function, though the involvement of retrograde messengers in LTP is still debated.

popular notion is that after LTP has been induced, the presynaptic cell more easily releases glutamate when an action potential occurs in its terminal, leading to a bigger postsynaptic response. Evidence for this theory has been obtained in sophisticated experiments, involving the precise measurement of

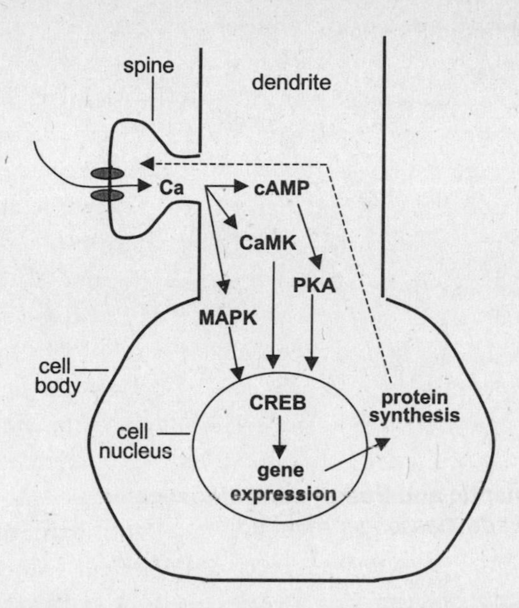

FIGURE 6.5 MOLECULAR CASCADES INITIATED BY CALCIUM DURING MEMORY FORMATION

Calcium influx into the postsynaptic cell leads to the activation of several kinases: cyclic AMP (cAMP)–dependent protein kinase A (or PKA), calcium/calmodulin protein kinase (CaMK), and mitogen-activated protein kinase (MAPK, or MAP kinase). Each of these then activates the gene transcription factor CREB (cAMP response element binding protein), which in turn initiates gene expression. Proteins are then synthesized that are shipped throughout the cell. However, because synapses are tagged during plasticity, only those synapses that have the tag are able to use the new protein (fig. 6.6).

glutamate release from presynaptic neurons, performed by Charles Stevens of the Salk Institute and Richard Tsien of Stanford, among others.[44] However, because LTP is triggered only in the postsynaptic cell, in order for the presynaptic hypothesis to be valid, something has to be communicated from the postsynaptic to the presynaptic neuron after LTP is initiated. One possible way this could happen is by means of retrograde messengers, substances released by the postsynaptic cell after LTP is induced and taken up by the presynaptic terminal, which cause changes in the ease of glutamate release there. Robert Hawkins in Kandel's group and others have found evidence that retrograde messengers do modify the presynaptic terminal in ways suggested by the presynaptic hypothesis,[45] but as other leading LTP researchers, such as Rob Malenka and Roger Nicoll, point out, this notion remains controversial.[46]

While early LTP involves the activation of existing proteins by kinases, the creation of long-lasting or late LTP, like long-term memory, involves the formation of new proteins. One of the key steps in this process is the activation of kinases (fig. 6.5), especially protein kinase A (PKA), MAP kinase (MAPK), and calcium/calmodulin protein kinase (CaMK), in a special way that allows them to move inside the cell nucleus.[47] Once there, they activate a protein called CREB. The job of this gene transcription factor is to activate specific genes that make new proteins that travel back to the synapses that started the whole process and stabilize the connections (see "Tagging Along," below).

Our understanding of the molecular basis of LTP has come from two kinds of studies. The traditional approach has involved the use of drugs that block certain molecular steps, like the flow of calcium through NMDA receptors, or phosphorylation of kinases, or synthesis of proteins. In studies of this type, the drugs are either placed in the bath in which the slice of brain is maintained or are injected directly into the postsynaptic cell. If LTP is disrupted, the chemical step interfered with is implicated in the underlying plasticity. Recently, though, a new approach has emerged. Rather than adding drugs to a brain slice, mice are created that either lack or have an excess of specific molecules, like CREB or certain kinases, as a result of genetic alterations. LTP is then induced in brain slices taken from these mice. As in the drug studies, if LTP is affected, then the missing or exaggerated molecule is implicated.

Most work to date using genetically altered mice has involved permanent deletion of a gene in a line of animals, leading to the absence of the particular molecule in the entire body throughout life.[48] The fact that the molecule is missing throughout life, however, raises some questions about how to interpret the data, since an effect on LTP could be due to the absence of the molecule itself, or could be a secondary effect of the animal's having gone through life without the molecule. The value of this approach is being refined by the ability to selectively alter molecules in specific brain regions, like the hippocampus, leaving other areas intact,[49] and by the ability to turn certain genes on or off at a certain time.[50] As a result, the mouse can grow up with all its molecules intact, and then right before the LTP experiment a chemical switch can be thrown to either decrease or increase the molecule being studied. While the latter approach is obviously more powerful, it is also more difficult to implement. But researchers are working diligently to improve these techniques, hoping to be able to turn genes on or off in specific brain areas on command. Genetically altered mice have already played key roles in brain research, implicating NMDA receptors, protein kinases, and CREB in LTP,[51] and are likely to become even more valuable as the techniques for creating genetically altered organisms improve in the coming years.

I've mentioned only briefly some of the many molecules implicated in Hebbian LTP.[52] Our knowledge of the role of these substances is expanding rapidly. And while there remain many unresolved mysteries about the molecular basis of synaptic plasticity, at a minimum it seems fairly clear that NMDA receptors, calcium, certain kinases, CREB-activated genes, and protein synthesis are involved. The massive amount of data pouring forth on the molecular basis of LTP is a bit overwhelming, so much so that Josh Sanes and Jeff Lichtman, two neuroscientists who work on brain development, have called for a moratorium on molecular studies of LTP.[53] Others, like David Sweatt and Mary Kennedy, see it as a sign of the vitality of the field that so much information is being generated about how LTP works.[54] I share their optimism.

TAGGING ALONG

The synaptic activity that triggers LTP occurs out on the dendrites, while the genes that make the proteins that enable LTP to persist are located far away in the cell nucleus. We've already seen how the gap from the synapse to the nucleus is spanned: calcium comes in through the NMDA receptor and triggers molecular changes that ultimately affect genes in the nucleus. More mysterious has been the way this circle is closed. How, in other words, do proteins made in the cell body know which of the many synapses on the neuron originally were responsible for the initiation of the protein synthesis? It's these synapses, and these alone, that have to be modified in order for the plasticity to be specific to the active synapses.

Recent studies have helped solve some of the mystery. It appears from work by two research groups (Eric Kandel and Kelsey Martin in New York, and Richard Morris and Uwe Frey in Scotland and Germany) that, following some significant experience, the active synapses are given a molecular tag.[55] When the new proteins are subsequently made in the cell body, they are then shipped out to essentially all possible synapses in the cell, but only synapses that were tagged during the initial stimulation are able to use the new proteins to stabilize the connection with the presynaptic terminal (fig. 6.6).

Once they find their way back to the tagged synapses, the proteins stabilize the connection between the presynaptic and postsynaptic neuron. There are two parts to the explanation of how this occurs. The first has to do with the

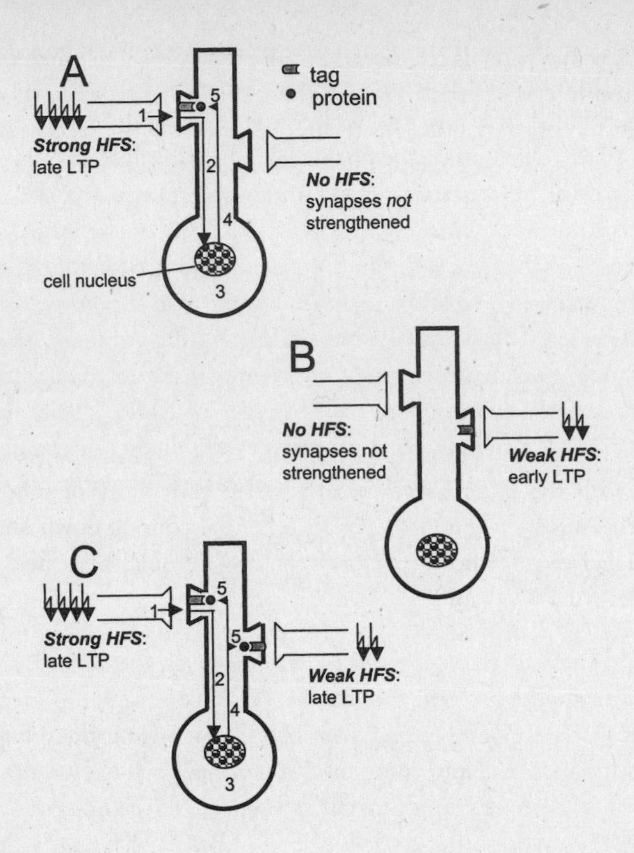

FIGURE 6.6 SYNAPTIC TAGGING

Protein sequestering by a synaptic tag. (A) Strong high-frequency stimulation (HFS) produces long-lasting (late) LTP. Several steps are involved. Strong HFS (1) leads to the creation of a molecular tag at the synapse and triggers molecular processes (2) that activate genes in the cell nucleus (3). Proteins are made and shipped throughout the cell, but are only useful at the sites where the tag was made during HFS. (B) Weak HFS leads to a short-lasting form of LTP (early LTP) that does not require gene expression and protein synthesis. However, it does create the tag. (C) Early LTP can be converted into late LTP by having the weak HFS to one pathway overlap with the strong HFS to the other. This occurs because the weak HFS creates the tag (see part B) that then allows the synapse to make use of the proteins made by strong HFS of the other pathway. Based on Frey and Morris 1997.

stabilization of existing synapses. These are kept in the new, improved, facilitated state after LTP in several different ways. For example, by making more AMPA receptors available on a prolonged basis, a given amount of glutamate released by the presynaptic terminal continues to have a bigger effect. Further, as we saw, terminals connected to the postsynaptic cell may come to release transmitter more efficiently as a result of a retrograde messenger passed back from the postsynaptic cell to the presynaptic terminal. By prolonging temporary changes, the effects of LTP are made to endure.

The other part of the explanation is related to the actual growth of new synapses. Several researchers have shown that LTP does indeed create new connections.[56] These appear to be fostered in part by the release of growth-inducing tonics, neurotrophins, from the postsynaptic cell. We discussed a similar role for neurotrophins in the context of development in chapter 4. Neurotrophins are released from the postsynaptic cell when it is active. They are then taken up by the presynaptic terminals that were also just active, and encourage the terminal to sprout new branches and form new synaptic connections with the postsynaptic neurons. The availability of more synaptic connections means that a given action potential coming down an axon will have a bigger effect, since it has more places (terminal branches) to release transmitter from and more postsynaptic sites to bind to.

Recent studies demonstrate that proteins can be synthesized in dendrites.[57] If dendritic protein synthesis occurs during memory formation it would simplify, to some extent, the problem of achieving synapse specificity since proteins made locally would not have to find their way back from the cell body to the active dendrites. This is an exciting new area of research.

FROM LTP TO MEMORY

In the end, what do LTP, NMDA receptors, kinases, and CREB have to do with memory, as opposed to brain cells living in a dish and being stimulated electrically? Let's start with NMDA receptors.

The first bit of evidence supporting a role for NMDA receptors in memory came from a study performed by Richard Morris, working in collaboration with Gary Lynch in Irvine, California.[58] Morris used a drug that specifically blocks NMDA receptors, leaving AMPA receptors and other aspects of brain function unaffected. He put this drug into the brain of rats in a way that bathed the hippocampus, and then tested the rats in his famous water maze—one of the better tasks for implicating the rat hippocampus in spatial memory (chap. 5). Control rats were able to learn where the submerged platform was located. Rats treated with the drug, though, had great difficulty learning this task in spite of the fact that they were able to swim to the platform if it was above water, showing that the drug did not affect normal vision or swimming ability. Subsequently, when the brain was removed from these rats and slices of the hippocampus kept alive in a dish, LTP could be readily induced in the controls, but LTP induction was impaired in the drug-treated brains. Since hippocampal LTP and hippocampal memory were both disrupted by the same treatment, the conclusion was that hippocampal LTP might have something to do with hippocampal memory.

Many subsequent studies have followed up on these important observations.[59] There have been successes and failures,[60] as well as alternate interpretations,[61] in the quest to relate NMDA receptors in the hippocampus to spatial memory. Particularly difficult to ascertain has been the issue of whether NMDA blockade produces a specific effect on learning itself, or instead changes other less specific capacities, such as the ability to perceive a given stimulus or the ability or desire to perform the appropriate response. However, the most recent work by Morris and colleagues seems to show fairly convincingly that blockade of NMDA receptors in the hippocampus does interfere with spatial learning.[62]

That hippocampal NMDA receptors are involved in spatial learning is also suggested by the results of studies using genetically altered mice. Susumu Tonegawa and colleagues at MIT created mice lacking important components of NMDA receptors.[63] Particularly significant was the fact that they were able to restrict the loss to the hippocampus—in fact, to a specific part of the hippocampus. The mice performed poorly on spatial learning tasks, and LTP induction was impaired as well. More recently, Joe Tsien, a former colleague of Tonegawa's, took the opposite approach, enhancing the function of NMDA receptors in a specific region of the hippocampus. LTP induction was facilitated, and so, too, were several memory functions dependent on the hippocampus.[64] These studies provide fairly direct evidence for the three-way relation between hippocampal NMDA receptors, hippocampal LTP, and hippocampal-dependent memory.

Similar studies comparing LTP and spatial memory have also been performed for some of the other molecules that have been implicated in LTP.[65] The literature is at this point fairly extensive but incomplete, so I'll mention only a few findings. Alcino Silva and colleagues examined the effects of knocking out CREB in mice,[66] and discovered disruption in both LTP and hippocampal-dependent memory. Further, Ted Abel, Mark Mayford, Eric Kandel, and colleagues have found that genetic alteration of protein kinase A and a calcium/calmodulin-dependent kinase, a major enzyme that phosphorylates CREB, disrupts late LTP and long-term memory without affecting early LTP or short-term memory.[67]

Considerable research thus supports the view that hippocampal LTP and hippocampal-dependent memory operate by similar molecular mechanisms. The parallels, in fact, are quite impressive, but not perfect. For example, in some studies, hippocampal-dependent learning is unaffected by treatments that block LTP.[68] This type of result, though, has been fairly rare, and certainly is not compelling enough to undermine the whole field. Further, we now know that there are a variety of different forms of LTP, some involving NMDA receptors and some not, and some Hebbian (requiring presynaptic and postsynaptic activity) and some not (possibly requiring changes only presynaptically),[69] and there's even long-term depression, where synapses are weakened by uncorrelated activity between presynaptic and postsynaptic cells.[70] While we have a long way to go before we fully understand how the synapses make memories, at a minimum, LTP has been an excellent tool for pursuing the topic.

A MISSING LINK

Underlying much of the work on LTP is the assumption that LTP is not just *a* way to study how experience changes synapses, but is, in fact, *the* way that synapses are changed when we learn. The various findings indicating that the same molecules are involved in LTP and memory are consistent with the view that LTP occurs during learning, but this evidence is, as critics have noted, circumstantial.[71] In an effort to silence these critics, LTP researchers have attempted to demonstrate that something like LTP does take place in the hippocampus during learning.[72] However, these studies have, for various reasons, come up short.[73]

For a long time, I myself was ambivalent about LTP. I thought it was an interesting phenomenon but was not convinced that it was the answer to how memories are created. I became more intrigued when Marie-Christine Clugnet and I found that we could induce LTP in a neural pathway carrying auditory information to the amygdala, a pathway my lab had been studying because of its role in fear conditioning. But what finally made me a convert were studies performed in my lab by Michael Rogan while he was working toward his PhD. Now I am a proselytizer.

Rogan decided to take on the challenge of trying to show definitively that LTP occurs during learning. While most of the earlier work in this area had involved the hippocampus, the lack of understanding of the actual circuits involved in hippocampal-dependent memory (spatial memory) was a barrier.

Rogan instead started with a circuit known to be involved in fear conditioning and then asked whether LTP occurred in that circuit. This reversal of the strategy (starting with a learning circuit and looking for plasticity in it, rather than starting with a form of plasticity and asking how it might relate to learning) turned out to be very advantageous.

My lab had been studying the anatomy of fear conditioning in rats and had shown that in order for a sound to acquire aversive properties when associated with a shock, the sound had to be relayed from the auditory thalamus to the lateral nucleus of the amygdala (chap. 5). And, as I noted above, we had also found that we could induce LTP in this same pathway. Rogan put these findings together, and then took two additional steps.

First, given that we used sound stimuli in conditioning studies, and that sounds get to the amygdala through the thalamoamygdala pathway, Rogan asked whether induction of LTP in this pathway would change the way sound stimuli are processed in the amygdala. He induced LTP in the usual manner with rapid electrical stimuli, but made a key change in the way LTP was tested—using a natural stimulus (in this case, a sound) rather than electrical stimulation of nerve fibers. His findings showed that induction of LTP in the pathway that carries the sound to the amygdala enhanced the amygdala's response to that sound. When this article was published, it was accompanied by commentaries from several leaders in the field who noted that this was indeed an important step forward in the long quest to relate LTP and memory.[74]

Rogan's first study showed that artificial alteration of synaptic transmission in the pathway changes the way that pathway processes external stimuli. This was an interesting discovery, but it left unanswered the question of whether, during natural learning, something like LTP occurs in the brain. In the second study, he therefore substituted fear condititioning for LTP induction. A few years earlier, Liz Romanski had shown that single cells in the lateral amygdala receive both sound and shock information, thus demonstrating that they might be coincidence detectors.[75] It thus seemed reasonable that fear conditioning with a shock might change the responses elicited from lateral amygdala cells by sound stimuli. Indeed, Rogan's results showed that fear conditioning and LTP induction produced very similar changes in the electrical responses of amygdala cells to sound stimuli. Fear conditioning, in other words, seems to induce LTP (fig. 6.7). This work, which earned Rogan the prestigious Lindsley Prize in 1999 for the best dissertation in behavioral neuroscience,[76] prompted a commentary by Charles Stevens, one of the top researchers in LTP, who asked whether the "million-dollar question" about memory had finally been answered.[77] He concluded that though it had not, we had taken a key step toward doing so.

Fear conditioning has thus been more successful in closing the gap between LTP and memory than studies of spatial learning or other hippocampal-dependent tasks.[78] On the basis of careful anatomical and behavioral studies, we had developed a clear idea about which synapses might be changing during fear conditioning, and we could easily stimulate them with both natural and electrical stimuli. While the circuitry of the hippocampus is well understood, the relation of learning to specific circuits in the hippocampus has been difficult to determine because of the complexity of the learning tasks. In spatial learning, a rat is free to learn about any of the many stimuli present in the environment, making it difficult to determine, in Hebbian terms, precisely what's being associated with what. In contrast, fear conditioning is a straightforward form of associative learning that is easy to define in terms of Hebbian plasticity, where strong (foot shock) and weak (sound) stimuli interact at synapses on the same cells.[79]

An important question is whether the huge amount of work that has been done on the molecular basis of hippocampal LTP applies equally to amygdala LTP and fear conditioning. Fortunately, much of this research does seem relevant. For example, studies by several researchers, including Sarina Rodrigues in my lab, and Mike Davis and colleagues and Mike Fanselow and colleagues (especially Jeansok Kim and Steve Maren) have shown that blockade of NMDA receptors in the lateral amygdala prevents the acquisition of fear conditioning.[80] Further, Glenn Schafe in my lab has shown that disruption in the lateral amygdala of protein synthesis or of some of the same kinases that prevent the induction of late LTP in the hippocampus interferes with the formation of long-term memory for fear conditioning without affecting short-term memory.[81] Some of the main kinases involved are PKA and MAP kinase. PKA has also been implicated in context conditioning by Schafe and by Rusiko Bourtchouladze in Kandel's lab, but the exact site of action in the brain is not known.[82] Also, when CREB levels are increased in the region of the lateral amygdala, weak training produces strong learning.[83] CREB, PKA, and other molecules have also been implicated in aspects of fear conditioning through studies of genetically altered mice by Eric Kandel, Alcion Silva, Mark Mayford, Tsunmo Tonegawa, Joe Tsien, and others.[84] Importantly, Y. Y. Huang and Kandel have shown that late-phase LTP in the lateral amaygdala also depends in part on NMDA receptors, PKA, MAP kinase, and protein synthesis,[85] further closing the gap between memory formation and LTP in the lateral amygdala.[86] Fear conditioning and lateral amygdala LTP thus seem to require the same mix of molecules that has been implicated in synaptic plasticity and memory in the hippocampus: Hebbian changes in synaptic strength, which are triggered by calcium entry through NMDA receptors, which are stabilized by protein kinases that induce CREB-related genes to make new proteins. NMDA receptors, MAP kinase, CREB, and protein synthesis also have been implicated in conditioned taste-aversion learning, another form of aversive

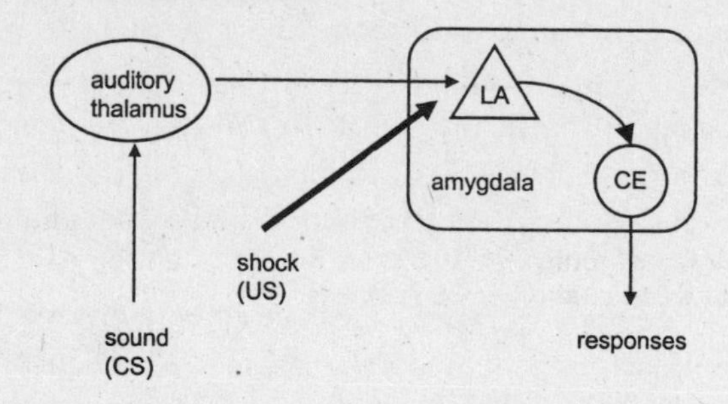

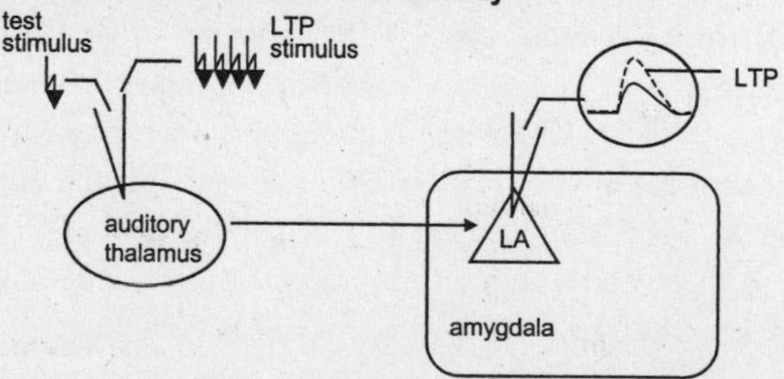

FIGURE 6.7 DOES AMYGDALA LTP ACCOUNT FOR FEAR CONDITIONING?
Fear conditioning is believed to occur as a result of the convergence of a weak conditioned stimulus (CS) input to the lateral amygdala (LA) that overlaps with the arrival of a strong unconditioned stimulus (US) input. As a result, the processing of the CS is modified such that it gains access to circuits that control fear responses. In short, the ability of the weak CS to activate the lateral amygdala (LA) is potentiated by the strong US. This is very similar to what occurs during associative LTP, suggesting that associative LTP may underlie fear conditioning (see also fig. 6.8). One line of evidence sup- >

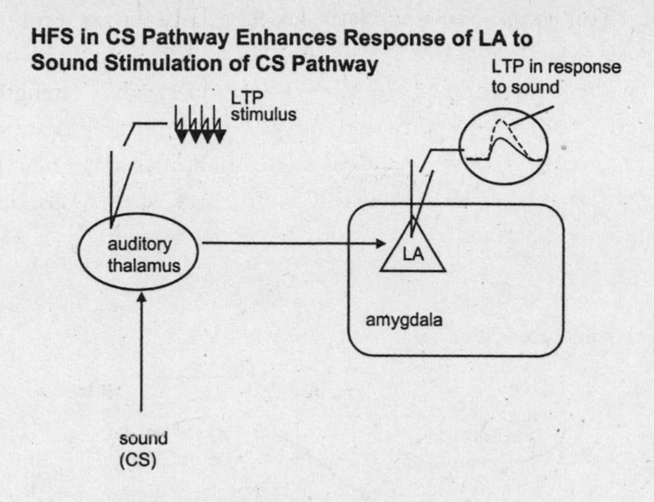

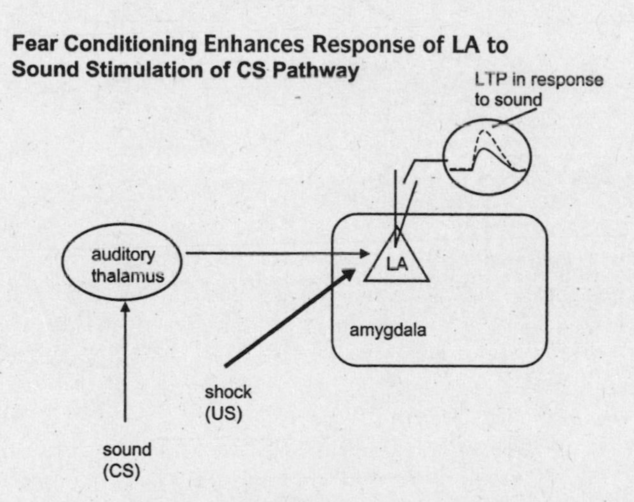

porting this notion comes from studies showing that LTP can be induced by high frequency electrical stimulation (HFS) of the CS pathway in an anesthetized rat. LTP can thus occur in the CS pathway, at least under these artificial conditions involving electrical stimulation of pathways rather than a real CS. However, HSF of the CS pathway in an anesthetized rat also enhances the response of LA cells to an auditory CS, showing that natural stimuli, such as those used as a CS in a fear-conditioning study, can access artificially induced LTP. Most important, though, is the fact that fear conditioning (pairing of an actual CS and US) under natural conditions (the rat is awake and unrestrained) also potentiates the response of amygdala cells to the CS, showing that LTP occurs during fear conditioning. Further, the acquisition of behavioral fear responses parallels the development of LTP, strongly suggesting that LTP underlies fear conditioning.

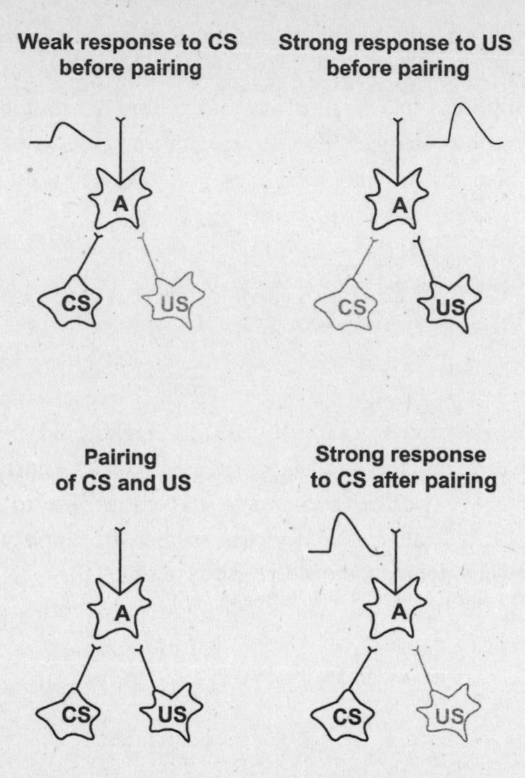

FIGURE 6.8 CLASSICAL CONDITIONING AS HEBBIAN PLASTICITY

Classical conditioning involves the convergence of information about the CS and US onto individual neurons and can be thought of in terms of Hebbian plasticity (see fig. 6.1). The CS only weakly activates A (*top left*) while the US strongly activates A (*top right*). If the CS and US occur together (*bottom left*), the strength of the CS connection increases such that the CS elicits a stronger response after conditioning than before it (*bottom right*). Gray shading indicates a pathway is inactive.

classical conditioning, by Yadin Dudai and colleagues at the Weizmann Institute in Israel (especially Raffi Lamprecht, Kobi Rosenblum, and Diego Berman).[87]

What is the significance of the fact that fear conditioning and taste-aversion learning by the amygdala and relational (spatial) learning by the hippocampus all involve similar molecular changes? Obviously, the circuits are very different at several levels. For example, the fear-conditioning circuits of the amygdala are one synapse removed from sensory systems, while the hippocampus is several synaptic steps removed from them. In addition, the consequences of activating these circuits are quite distinct. Activation of fear-conditioning circuits in the amygdala leads to the expression of hard-wired bodily responses by way of connections with the brain stem, whereas activation of hippocampal circuits leads to representations that allow for a multitude of responses through widespread connections with cortical areas. Different forms of memory are thus distinguished not by the molecules that make them but by the circuits in which those molecules act.

There is still quite a lot of work to be done on the molecular basis of fear conditioning to pin down some of the particulars. For example, recent studies in brain slices of the amygdala by Marc Weisskopf and Liz Bauer in my lab have suggested that at least some of the plasticity may be mediated by calcium entry through special calcium channels.[88] Following up on this work, Bauer and Glenn Schafe showed that fear conditioning could be prevented by blocking these calcium channels, suggesting that calcium entry through both NMDA receptors and calcium channels may contribute to fear learning.[89] The theory is being evaluated in ongoing studies. In hippocampal LTP, this combined involvement of NMDA receptors and calcium channels has been seen.[90] Other research indicates that presynaptic plasticity may occur at some synapses in the amygdala during LTP induction and during fear conditioning.[91] If so, classical fear conditioning in the rat might turn out to be similar to classical conditioning in invertebrates, in that both presynaptic and postsynaptic changes take place. We'll revisit this issue below when we consider invertebrate learning. Also unresolved is the locus of memory storage—plasticity in the lateral amygdala is clearly involved, but whether this is the only site of change that is maintained over time remains to be determined.[92]

Studies of fear conditioning have also revived interest in a strange but possibly very significant phenomenon in memory research—reconsolidation.[93] The recent discovery, made by Karim Nader and Glenn Schafe in my lab, is that protein synthesis in the amygdala seems necessary for a recently activated memory to be kept as a memory. That is, if you take a memory out of storage you have to make new proteins (you have to restore, or reconsolidate it) in order for the memory to remain a memory. One way of thinking about this is that the brain that does the remembering is not the brain that formed the initial memory. In order for the old memory to make sense in the current brain, the memory has to be updated. This work has stimulated a lot of interest from both scientists and lay persons. One man called and asked whether it might be possible for him to eliminate the memory of his ex-wife by blocking protein synthesis in his brain while thinking of her. The practical side of this is that it might be possible some day to have trauma victims recall their trauma in the presence of some drug or other brain alteration that reduces the stranglehold of the memory on the person's psyche. After we proposed this, though, a therapist made a very good point. What would it mean to a Holocaust survivor, for example, to lose such memories after having lived for many years and having developed an identity based in part on them? This is a very important concern, and touches on the deep ethical issues that scientific discoveries can raise.

Fear-conditioning research has capitalized on breakthroughs in modern neuroscience to move us forward in our understanding of how synaptic changes underlie memory. Consistent with the cellular-connection strategy outlined by Kandel and Spencer in 1968, a simple behavior was chosen and the circuit was mapped. The plasticity of cells in the circuit was studied. The cellular changes have been related to specific synapses, and the synaptic changes have been related to specific molecular events. It was inconceivable in 1968 that this could be done in a mammalian brain, and it is a testimony to how fast the field of neuroscience has advanced that so much has been achieved in recent years.

SNAIL TALES

The molecules of memory are conserved not just across different kinds of memory within a species but also across vastly different kinds of organisms. In fact, many of the basic facts about the molecular basis of plasticity were first revealed in studies of invertebrate organisms, and only later were they found to be applicable to plasticity in the mammalian brain through studies of LTP and behavioral learning. It is time to put the mammalian brain aside and take a look at some of the research on invertebrates that led to our current molecular understanding of memory.

The neural basis of learning and memory has been investigated in many different invertebrates (bees, grasshoppers, crayfish, slugs, flies, and various mollusks),[94] but studies of the mollusk *Aplysia californica* have been particularly thorough and informative. Much of the work on this marine snail has been conducted by Eric Kandel and his students and colleagues at Columbia University and elsewhere.[95] This pioneering research was a major factor in Kandel's receipt of the Nobel Prize in 2000.

BEHAVIORAL STUDIES

Aplysia breathe through gills that are covered by a piece of skin called the mantle. If the mantle is lightly touched, the gill retracts. This defensive reflex protects the gill from injury, and has been the subject of extensive investigation as a behavioral model of learning and memory. The plasticity of other reflexes in *Aplysia* has also been studied,[96] but the gill-withdrawal reflex will be used to illustrate the basic findings.

The central nervous system of an *Aplysia* has only about twenty thousand neurons.[97] It has been estimated that fewer than one thousand neurons (and possibly as few as four hundred) are active during the gill-withdrawal reflex, and many of these are not believed to be essential. Essential ones include sensory neurons that process the touch of the mantle skin, and motor or output neurons that control the gill withdrawal. The sensory neurons make synaptic connections with the motor neurons. Also important are sets of excitatory and inhibitory interneurons that make synapses with the sensory or motor neurons and regulate the reflex. When you compare this with the billions of neurons in the mammalian brain, it is clear why it might be easier to investigate the neural basis of behavioral functions in the *Aplysia.*

The gill-withdrawal reflex exhibits several forms of learning, including habituation: the gill retracts less if the the mantle is touched repeatedly. Habituation is a form of nonassociative learning: a single stimulus is involved, and it is not associated with anything else. Habituation can be reversed rapidly by giving a strong stimulus, such as an electric shock, to some other part of the mollusk's body, like the tail. Touch of the mantle after a shock results in a big response. This is an independent form of nonassociative learning called sensitization and is not simply a recovery from habituation, since the response to the touch of a second area of skin, one that was not habituated, is also amplified by the shock. Sensitization is considered nonassociative because the touch-test stimulus and the shock were never related, and the learning is not specific to the test stimulus—many stimuli given after the shock produce bigger effects. The effects of sensitization can be short-lived or long-lasting, depending on the strength of training. A weak shock given only once will produce short-lasting changes in the reflex that disappear within a matter of hours, whereas repeated presentation of the shock leads to changes that can last for days. Like habituation, sensitization is also regarded as a form of nonassociative learning because a single stimulus (the shock) causes the behavioral change.

Associative learning—namely, classical defense or fear conditioning—also occurs in the gill-withdrawal reflex. This is essentially the same kind of conditioning task that we've already encountered several times in mammals. See, for example, the discussion of implicit learning in chapter 5. If a shock is delivered to the tail while the mantle is touched, subsequently the touch alone

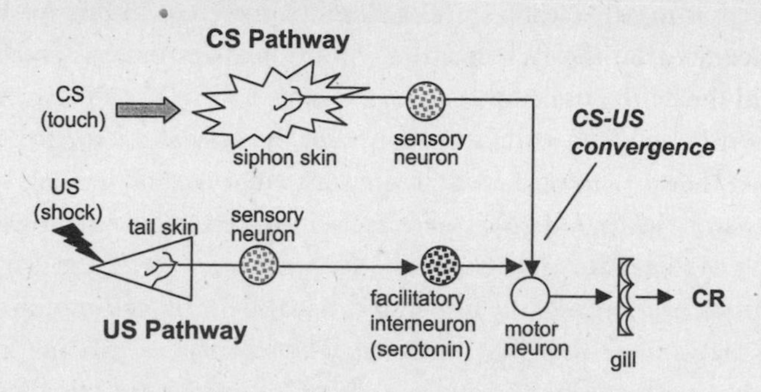

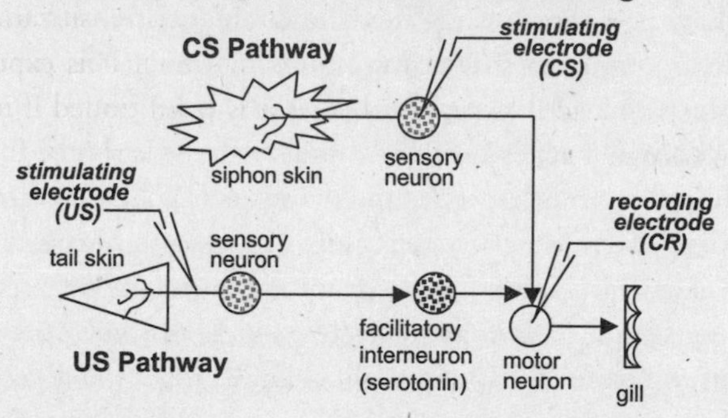

FIGURE 6.9 CLASSICAL CONDITIONING IN THE *APLYSIA*

Top: In the *Aplysia,* a light touch to the siphon skin (conditioned stimulus, CS) activates sensory neurons that then send inputs to motor neurons while an electric shock (unconditioned stimulus, US) to the tail skin activates sensory neurons that send inputs to facilitatory interneurons. The terminals of the facilitatory interneurons end on the terminals of sensory neurons transmitting the CS to the motor neurons. The US pathway thus forms a presynaptic synapse (a synapse on a terminal). The release of serotonin from the facilitatory interneuron thus facilitates transmission of the CS to the motor neuron. For more details on cellular mechanisms, see figures 6.10 and 6.11. *Bottom:* The process of classical conditioning can be simulated by replacing external CS and US events with electrical stimuli of the sensory neurons (or the pathways into the sensory neurons) that process the CS and the US. Similarly, rather than measuring overt behavior (conditioned response, CR), the activity of the motor neuron is recorded. Through the use of this and other cellular analogs of learning, researchers have learned much about the biology of memory.

will lead to a stronger withdrawal of the gill than it did before conditioning. That this is associative learning (and not just sensitization due to the shock) is demonstrated by the fact that the conditioned response is smaller if the touch and shock do not occur at the same time. The relation between the two stimuli is thus key. The associative nature of the learning is further indicated by the fact that if two touch stimuli are used (that is, if different parts of the mantle are touched), and only one is paired with the shock, the paired touch elicits a bigger response than the unpaired one.

Conditioning and sensitization are similar in that a strong stimulus changes the response to a weak one. But they differ in specificity: in associative conditioning, the amplified response only occurs in reaction to a stimulus that was paired with the shock, whereas in sensitization the response to stimuli that have no relation to the shock is also bigger. Sensitization basically makes the snail jumpy, so that any stimulus to which it is exposed after a strong stimulus will lead it to react, whereas it is conditioned if it reacts to a stimulus that occurred at the same time as a strong stimulus but not to other novel stimuli that occur afterward. Studies of classical conditioning often use unpaired training (sensitization) as a control to ensure that the conditioned response is due to the associative relation between the strong and weak stimulus, as opposed to the nonassociative effects of the strong stimulus alone.

The big advantage of studies of animals like the *Aplysia* is, of course, that the small number of neurons involved in the control of a behavioral response like the gill-withdrawal reflex makes it relatively easy to pursue questions about exactly which neurons and synapses are changed by experience, and what sorts of molecular events might underlie the synaptic alterations that constitute the memory. But even in a simple nervous system like that of the *Aplysia,*[98] it has proven useful to narrow the focus down even further. Because the actual behavioral response is controlled by a number of different neurons, some studies sidestepped behavior completely and instead examined the electrical activity of a single motor neuron as an indicator of whether learning occurred. In this experimental setup, the electrical activity of motor neurons in response to natural stimuli (touches and shocks) or to electrical stimulation of the sensory pathways that transmit natural stimuli can be assessed. The sensory or motor neurons also can be directly treated with chemicals, like drugs that mimic or block the effects of neurotransmission or other aspects of cell function. A further reductive step has involved complete elimination of the *Aplysia,* except for a single sensory and a single motor neuron that grow synapses between each other in a dish. The response of the motor neuron to

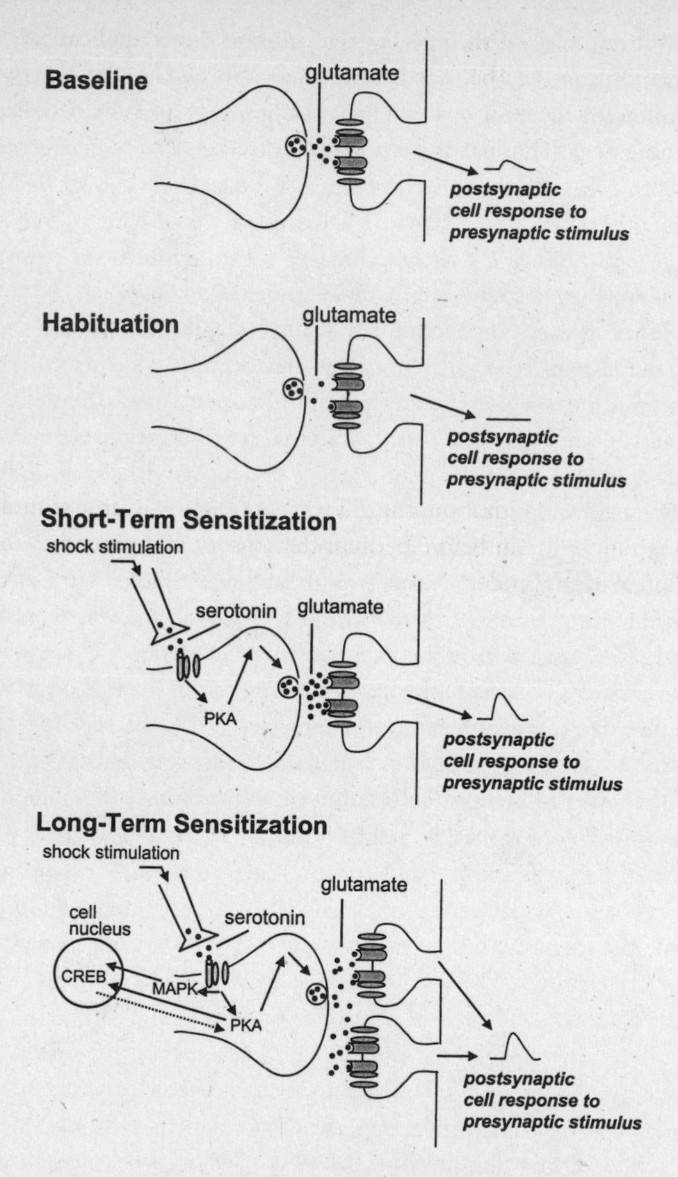

FIGURE 6.10 CELLULAR MECHANISMS OF NONASSOCIATIVE LEARNING (HABITUATION AND SENSITIZATION) IN THE *APLYSIA*

The first time a stimulus is delivered to the presynaptic neuron (baseline), it leads to the release of a certain amount of transmitter (glutamate) from the terminal and produces a postsynaptic response (*shown on right*). When the stimulus is repeated, the amount of glutamate released decreases and the postsynaptic response decreases (habituation). However, the same stimulus leads to a greater amount of glutamate >

electrical stimulation of the sensory neuron or to direct application of drugs to the neuron can then be tested. With these approaches, *Aplysia* researchers have learned much about how synaptic changes such as those occurring during learning are established and maintained.[99]

MEMORY MECHANISMS

All forms of gill-reflex learning involve changes in synapses between sensory neurons that receive inputs from the mantle skin and motor neurons that control the gill response. In habituation, for example, the response of the postsynaptic neuron to a presynaptic input weakens, and the gill response gets smaller, because the presynaptic terminal comes to release less glutamate. It simply gets depleted.

In contrast, in sensitization, the gill reacts more to the same stimulus after the tail is shocked than before because the sensory neuron comes to release more glutamate. To understand why more glutamate comes out after the tail is shocked requires that we consider the way the shock pathway interacts with the synapses that connect the sensory and motor neurons.

The shock pathway forms synapses on the terminals of the sensory neurons. These are called axoaxonic synapses, since the axon terminals of the shock pathway make synapses with other axon terminals, in this case, terminals of the sensory pathway. This contrasts with the situation we've consid-

---

release, and a larger postsynaptic response, when it occurs after an electric shock (sensitization). Because the enhanced response to a single shock only lasts for several hours, it is called short-term sensitization. Short-term sensitization of the presynaptic response results from the release of serotonin from the shock pathway. The serotonin binds to serotonin receptors located on the presynaptic axon terminal, and leads to the activation of protein kinase A (PKA), which results (through steps not shown) in a greater amount of glutamate release and thus a bigger postsynaptic response. Short-term sensitization is thus said to occur by way of presynaptic facilitation. If repeated electric shocks are given, long-term sensitization, lasting days, occurs. While presynaptic facilitation is again involved, so are other processes. PKA and MAP kinase are both activated by serotonin. These enter the cell nucleus and activate the gene transcription factor CREB. CREB then initiates the synthesis of proteins that then amplify the ability of PKA to enhance glutamate release for a prolonged time (several days). Additionally, growth processes are initiated postsynaptically such that new spines are created. Glutamate now has more places to affect the postsynaptic site, further enhancing the ability of a presynaptic stimulus to affect the postsynaptic cell.

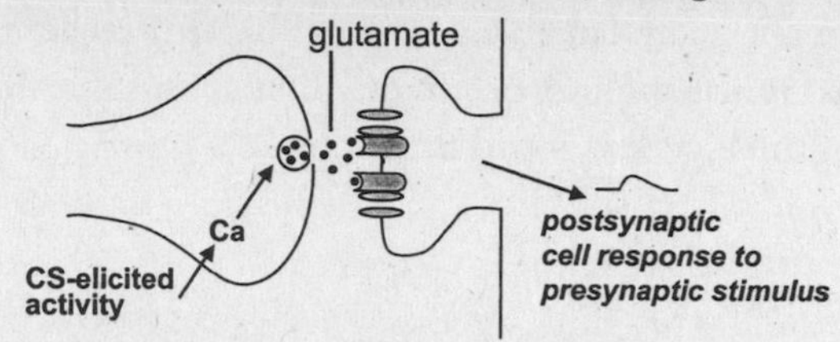

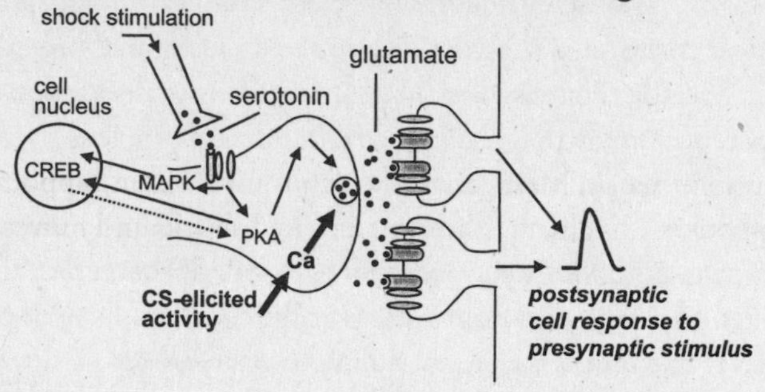

FIGURE 6.11 CELLULAR MECHANISMS OF ASSOCIATIVE LEARNING (CLASSICAL CONDITIONING) IN THE *APLYSIA*

Before conditioning (*top*), CS-elicited activity leads to a small postsynaptic response (*shown on right*) that quickly habituates if nothing further happens. However, if the CS is followed by a shock (*bottom*), the postsynaptic response is larger (*shown on right*) than before pairing (compare with response above). Comparison of the bottom diagram with the diagram at the bottom of figure 6.10 shows the difference between sensitization (i.e., presynaptic facilitation) and conditioning (activity-dependent presynaptic facilitation). The main difference is due to the presence of the CS during conditioning—the CS enhances the molecular effects of the US, leading to a bigger postsynaptic response. For further details about the mechanisms, see figure 6.10 and main text.

ered most so far, in which axon terminals contact dendrites. The tail shock pathway thus ends on the presynaptic terminal of the sensory neuron and causes more transmitter to be released. Because the increase in the efficiency of transmission between the sensory and motor neuron is due entirely to alterations in the sensory neuron terminal, it is referred to as presynaptic facilitation. Because the state of activity of the postsynaptic neuron is irrelevant, sensitization by definition is a form of non-Hebbian plasticity (recall that Hebbian plasticity requires presynaptic and postsynaptic activity).

How does the tail shock cause presynaptic facilitation? The tail shock pathway releases the modulator serotonin at the sensory terminal. When serotonin binds to its receptors on the terminal, it activates second messengers there that ultimately lead to the activation of a protein kinase, namely PKA, which is also involved in LTP.[100] PKA in turn leads to other changes that cause action potentials to last a little longer than usual. As a result, more glutamate is released from the terminal after the shock than before.[101] This causes a bigger response in the motor neuron, and therefore behaviors controlled by motor neurons, like gill withdrawal, are more strongly expressed.

The sensitizing effects of shock are short-lived unless the shock is given repeatedly. When a repeated shock is administered, additional processes are activated, and the effects of sensitization can last for days. In particular, PKA is activated in a special way that allows it to enter the cell nucleus.[102] In addition, a second protein kinase, MAP kinase (which is involved in hippocampal and amygdala plasticity), is also activated by repeated shocks and moves inside the nucleus. PKA and MAP kinase then phosphorylate the gene transcription factor, CREB, which we also encountered earlier in our survey of LTP.

CREB-activated genes make new proteins that end up facilitating transmission between the sensory and motor neuron in various ways. One effect of these proteins is that PKA becomes persistently activated, which means that the short-term sensitizing effects of PKA are lengthened, continuing the effects on transmitter release described above. Presynaptic facilitation simply continues. But there is at least one other effect that is also important: some of the proteins stimulate the sensory terminal to grow new axonal branches that make new synapses with the motor neuron.[103] With more synaptic connections between two cells, a given action potential in the presynaptic axon has more ways to influence the postsynaptic cell, leading to a stronger response. The two effects (persistent activation of PKA and formation of new connections) combine to produce long-lasting increases in transmission between the sensory and motor neurons, and stronger reflex responses.

Classical conditioning, though obviously associative in nature, was long believed to be mediated by a non-Hebbian synaptic mechanism, namely, an amplification of presynaptic facilitation. That is, when synaptic activity occurred in the presynaptic terminal while the transmitter release was being facilitated (sensitized) by the shock, the degree of facilitation increased and came to be coupled to the stimulus that elicited the activity. This was considered associative, since it involved changes restricted to the stimulus that was activating the terminal, but it was non-Hebbian because the postsynaptic cell was not believed to be involved. However, recent work has shown that the initial studies on which this view was based were incomplete. It is now known that the postsynaptic cell does participate in associative classical conditioning. Researchers like David Glanzman and Jack Byrne argue that classical conditioning in the *Aplysia* does indeed involve Hebbian synaptic changes, and probably also involves NMDA receptors, in addition to an activity-dependent enhancement of presynaptic facilitation.[104] One way of understanding this is that nonassociative (non-Hebbian) presynaptic facilitation functions as the presynaptic component of associative (Hebbian) plasticity.

Considerable work has also been conducted on the molecular basis of classical conditioning in the *Aplysia.* While some of the early steps are different from those involved in sensitization, the long-term changes involve phosphorylation of CREB by protein kinases and synthesis of new proteins by CREB-activated genes. The establishment of long-term plasticity seems to follow a common pattern even when it is triggered by different kinds of short-term changes.

The most obvious differences between associative conditioning in the *Aplysia* and plasticity in the mammalian brain is thus the importance of presynaptic plasticity in the *Aplysia.* The recent finding that NMDA-mediated plasticity in the postsynaptic cell is also significant in *Aplysia* plasticity helps close the gap. And the possibility that fear conditioning,[105] amygdala LTP,[106] and several forms of hippocampal plasticity[107] might, like plasticity in *Aplysia,* involve presynaptic as well as postsynaptic changes further strengthens the notion that similar mechanisms are used to make memories in diverse species, suggesting that memory mechanisms are conserved across many levels of evolution.

FLY GENES

Research on the fruit fly, *Drosophila melanogaster,* has also been instrumental in the accumulation of our knowledge of memory and its molecules. But in contrast to the highly focused cellular-connection approach used to study the molecular basis of plasticity in *Aplysia,* the fly work has focused on behavioral studies of mutant strains of flies that lack specific molecules, and has for the most part not related the molecular effects on learning and memory to specific cells and synapses.[108]

The studies of genetically altered mice we discussed earlier are descendants of fly mutation studies started several decades earlier by Seymour Benzer and his colleagues. Although Benzer started investigating fly behavior in the

1960s, the use of flies in research on learning began in the following decade, when Chip Quinn and Yadin Dudai in Benzer's lab showed that normal flies could remember a smell associated with electric shock.[109] When given the choice between going into a chamber that contained an odor associated with the shock and a chamber that held a different odor, the flies often avoided the former chamber.

Most work on mutant flies has involved this kind of learning task, which is simply a variation of classical defense or fear conditioning. These flies are called learning mutants, since they are altered genetically in a way that causes them to fail to learn or remember. Learning mutants are created by treating normal flies with a chemical that produces more or less random mutations in their DNA. The flies are then tested for deficits in learning and memory. Those that are deficient are bred following precise mating schemes that create lines of genetically identical flies that have learning and memory impairments.

Genetic mutations can disrupt learning for a variety of different reasons that have nothing to do directly with learning and memory, and it is necessary to rule these out when drawing conclusions about whether a mutant has a specific deficit in these areas. For example, a fly that cannot smell or that is defective in perceiving a shock, or one that is chronically ill, will not learn. These are concerns that apply to the studies of genetically altered mice described earlier, as well. In addition, as in the mouse studies, it is important to keep in mind that a mutant animal goes through its entire life with the genetic alteration, sometimes making it difficult to distinguish effects on learning and memory from other more general consequences of having grown up without certain proteins and their products. Behavioral changes in a mutant could, in other words, be due either to the mutation itself or to the consequences of surviving with the mutation.

The first fly learning mutant to be identified was called dunce. Then came amnesiac, cabbage, rutabaga, turnip. All of these variants performed poorly on the fear-conditioning task. In each of these mutants, learning actually did take place to some extent, but then decayed rapidly after training. Memory that remains after the initial decay, however, is fairly stable. This suggests that the mutations affect the early more than the late phase of memory.

Most of the initial work on learning mutants focused on relatively short-lasting memories. However, recent studies by Tim Tully, Jerry Yin, and Chip Quinn showed that long-lasting memory that persists for days can be created by giving normal *Drosophila* multiple training trials that are spread out over time (this is called spaced training).[110] Disruption of protein synthesis prevents this long-term memory from forming but has no effect on a shorter-lasting kind of memory induced by a few training trials given all at once (called massed training). Like a college student, the fly does better if it takes in information at a regular pace over time than if it crams.

Yin and Tully made use of this difference in massed vs. spaced training to examine whether cAMP-mediated gene induction and protein synthesis underlies long-lasting memory. They employed genetic techniques (see above) to increase the level of the gene transcription factor CREB available in flies at the time just before training. This step overcomes many concerns about whether the deficit is due to a failure to form memory or to some irrelevant consequence of growing up without the gene. The researchers found that in the animals with extra CREB, a single training trial could accomplish what required multiple spaced trials in normal flies—create a memory that lasts for days. This suggests that training-induced CREB activation (and the consequent induction of CREB-related genes) is an essential step in the production of proteins essential for long-term memory. The molecules leading to induction of CREB-related genes and their proteins are thus the stars of memory across species and training procedures.

Flies are ideal in many ways for performing genetic studies of learning and memory. They exhibit a robust form of learning, and because of their brief life span, many generations can be studied in a short time. But they are not especially well-suited specimens for pursuing the individual cells and synapses involved in learning and memory. Although recent work has made some progress in pinpointing the locus of plasticity,[111] circuit analysis in the fly will probably remain difficult due to the small size of the nervous system in this tiny creature.

I HAVE TO ADMIT IT'S GETTING BETTER

A key question you may now be asking yourself is whether all this hard-core neuroscience has, in fact, any practical application. In other words, might it be possible to use this kind of work to help improve normal memory and, even more important, to rescue or prevent age-related memory loss?

One of the pioneers in LTP research, Gary Lynch of the University of California, has been looking into ways of altering the molecular basis of synaptic plasticity in the hope of finding a means of improving memory. Several years ago, he and his colleagues discovered a class of molecules they called ampakines.[112] These drugs increase the efficiency of glutamate transmission at AMPA receptors, thereby allowing weaker stimuli to activate NMDA receptors. When hippocampal slices were treated with these drugs, LTP induction was facilitated. And when rats were given the drug, hippocampal-dependent learning speeded up. Michael Rogan and I teamed up with Ursula Staubli to test the effect of these substances in fear conditioning in rats.[113] We found that in the presence of ampakine treatment, the rats learned faster. It thus appears that enhancing glutamate transmission in both the amygdala and hippocampus facilitates learning. Lynch is currently working on ways to apply these potentially powerful agents to human memory in the hope of developing memory-enhancing drugs.

Recently, Joe Tsien of Princeton genetically engineered mice so that their NMDA receptors worked more efficiently.[114] These mice showed faster learning of both a spatial task and fear conditioning. Tsien nicknamed the strain "Doogie" (after the precocious doctor of prime-time fame). This finding provides powerful support for the role of NMDA receptors in learning, and also highlights the possibility that the genetic manipulations might someday make possible the rescue of the loss of memory function in aging humans.

The essence of who you are is stored as synaptic interactions in and between the various systems of your brain. As we learn more about the synaptic mechanisms of memory, we learn more about the neural basis of the self.

CHAPTER SEVEN

THE MENTAL TRILOGY

BETWEEN KNOWLEDGE AND DESIRE STANDS . . . FEELING.

—Immanuel Kant

Throughout much of history, the mind has been viewed as a trilogy, a tripartite amalgam that includes cognition, emotion, and motivation.[1] For some, the trilogy was a description of different aspects of a single mental faculty, whereas for others, it represented three distinct, separate capacities. During most of the twentieth century, both versions of the mental trilogy were out of favor.[2] When the behaviorists reigned, psychology ignored the mind altogether, making the mental trilogy moot.[3] Later, the cognitive revolution brought the mind back to psychology, but thinking and related cognitive processes were (and for the most part still are) emphasized at the expense of emotion and motivation.[4] Clearly, however, it is important to understand not just *how* we attend to, remember, or reason, but also *why* we attend to, remember, or reason about some things rather than others. Thinking cannot be fully comprehended if emotions and motivations are ignored.

In previous chapters, we've seen how neuronal circuits are assembled during development, and how these circuits are modified when we learn and remember. Now we will begin to use this basic information about circuits and their plastic properties to explore broader aspects of mental function, that is, to begin to assemble a neurobiological view of the self. In doing this, we will consider each component of the mental trilogy, as well as interactions among them. Thinking is the subject of this chapter, and emotion and motivation the following two.

MENTAL JUGGLING

An idea, an image, a sensation, a feeling: each is an example of what psychologists call mental content—stuff that is in the mind. Mental content was the subject matter of experimental psychology when it first emerged as a discipline in the late nineteenth century.[5] But John Watson and fellow behaviorists replaced this focus on subjective states with a mind-less psychology of objectively measurable events (stimuli and responses).[6] When the cognitive revolution later made the mind fair game again, it did not do so by reviving subjective psychology. The thinking process itself, rather than the conscious content that results from thinking, became, and largely remains, the subject matter of cognitive science.

"In order for a mind to think, it has to juggle fragments of its mental states." This simple statement by Marvin Minsky, one of the architects of the branch of cognitive science known as artificial intelligence, gets right to the heart of the matter.[7] Imagine, as Minsky suggests, rearranging the furniture in a room familiar to you. You shift your attention back and forth between locations. Different ideas and images come into focus, and some interrupt others. You compare and contrast alternate arrangements. You may concentrate your entire mind on a small detail one moment, and on the whole room the next. How does the mind do this juggling, and how does it keep track of the imaginary changes? The answer is that the mind uses something called working memory.

How many times have you looked up a phone number and then forgotten it after being momentarily distracted? The reason for this is that you put the number into working memory, a mental workshop that accommodates one task at a time.[8] As soon as a new task engages working memory, the content of the old task is bumped out. For that reason, unless you keep rehearsing the phone number and manage to ignore other things that compete for your attention, it will not remain in your mind.

Working memory is one of the brain's most sophisticated capacities and is involved in all aspects of thinking and problem-solving. It allows you to read a menu and keep its various options in mind while also considering the specials announced by the waiter and then return to the thought you were having before the waiter appeared. It underlies your ability to hold a conversation, play board games like chess, or direct yourself to an unfamiliar destination on the basis of having just looked at a map. In addition to being used in such routine daily activities, working memory also contributes to special human endeavors, like composing music or solving complex mathematical problems, or any other situation in which information has to be held in mind in order to complete a task.

Our understanding of working memory owes much to the pioneering work of Alan Baddeley in the early 1970s.[9] On the basis of his studies of short-term memory, he came to view the mind in terms of two kinds of cognitive systems: a set of specialized systems dedicated to specific mental tasks, and a general-purpose system utilized in all active thinking processes.

Specialized systems come in two flavors. Verbal systems, like systems involved in speech comprehension, are mainly present in the human brain, whereas nonverbal systems are present in all brains. Nonverbal specialized systems are epitomized by sensory systems. Each is involved in processing unique kinds of stimuli (sights, sounds, smells, and so on). As part of their operation, the verbal and nonverbal specialized systems are able to retain what they've just processed for brief amounts of time (seconds). This capacity aids in perception, allowing the system to compare what it is seeing or hearing now to what it saw or heard a moment ago. For example, when listening to a lecture, you have to hold the subject of each sentence in your mind until the verb appears, and sometimes you have to refer back to your memory of earlier sentences to figure out the referent of a pronoun.

The general-purpose system consists of a workspace and a set of mental operations called executive functions that are carried out on information held in the workspace. Although only a limited amount of information can be retained at any one time, the workspace can hold on to and interrelate information of different types from different specialized systems (the way something looks, sounds, and smells can be associated with its location in external space and with its name). This ability to integrate information across systems allows for abstract representation of objects and events. It is especially well-developed in humans, and is likely to contribute to the uniqueness of human cognition.

The information in your working memory is what you are currently thinking about or paying attention to. And because working memory is temporary, its contents have to be constantly updated. But working memory is not a pure product of the here and now. It also depends on what we know and what kinds of experiences we've had in the past. In other words, it depends on long-term memory.

The importance of long-term memory to thought cannot be overemphasized. One of the earliest examples of its significance is still one of the best. In the 1930s, Sir Frederic Bartlett had people listen to folktales from foreign countries and later asked them to recount the stories.[10] Not surprisingly, he found that these unfamiliar stories were not remembered very accurately. What was surprising was that the errors of recall were not random but were quite systematic. The subjects often rewrote the stories in their own minds, especially parts that were particularly foreign to them, revising the plot to the point where it resembled a more familiar Western narrative.[11] To explain his findings, Bartlett proposed that "Remembering is . . . an imaginative reconstruction, or construction, built out of the relation of our attitude towards a whole active mass of past experiences." He concluded that when we face a problem, we draw upon mental schemata, organized bundles of stored knowledge. For example, if you are asked a question about how baseball is played, you would draw upon a baseball schema, your collective knowledge of baseball obtained from specific direct experiences you've had with the game as spectator or player, as well as things you've heard or read about baseball. Bartlett's findings do not just concern the personal, idiosyncratic, and fallible nature of memory, but also emphasize how long-term memories, when retrieved into the temporary workspace of working memory, can guide our thoughts and actions, as well.

It has been known for centuries that we can only keep a few things active in our minds (in working memory) at once.[12] George Miller, one of the pioneers in cognitive psychology,[13] figured out, through psychological experiments, that the magic number is about seven pieces of information. Some people can hang on to eight or nine, whereas others manage only five, but, on average, temporary storage can hold about seven items. (It's probably no coincidence that telephone numbers within an area code were designed to have seven digits.) But, as Miller noted, we can effectively expand that capacity by chunking or grouping information—it's about as easy to remember seven letters as seven words or ideas. No doubt one of the reasons human cognition is so powerful is because we have language in our brains, which exponentially increases the ability to categorize information, to chunk. A whole culture, for instance, can be implied by a name.

The concept of working memory subsumes what used to be called short-term memory. But as the term *workspace* implies, working memory is more than just an area for temporary storage. It underlies mental work. As Minsky noted, thinking involves juggling of mental items—comparing, contrasting, judging, predicting. It is the job of the executive functions of working memory to do the juggling.

In the spirit of viewing the mind in terms of computer-like operations, some cognitive scientists like Tim Shallice and Phillip Johnson-Laird have referred to executive functions as supervisory[14] or operating[15] system functions. A computer operating system is responsible for controlling the flow of information processing, moving information from permanent memory (ROM) to a central processing unit with active memory (RAM), scheduling tasks to be performed using the active memory, and so on. Similarly, executive functions are involved in the constant updating of temporary memory, selecting which specialized systems to work with (pay attention to) at the moment, and then moving relevant information into the workspace from long-term storage by retrieving specific memories or activating schemata pertinent to the immediate situation. Through executive functions, specialized systems are also directed to attend to certain specific stimuli and to ignore others, depending on what working memory is working on. In complex tasks involving multiple kinds of mental activities, executive functions plan the sequence of mental steps and schedule the participation of the different activities, switching the focus of attention between activities as needed.[16] Executive functions are crucially involved in decision-making, allowing you to choose between different courses of action given what is happening in the present, what you know about such situations, and what you can expect to happen if you do different things in this particular situation. Executive functions, in short, make practical thinking and reasoning possible.

The executive represents a powerful mental capacity, but is not all-powerful. Like the workspace, it has its limits. It basically can do one or at most a few things at a time. This is why you forget a phone number if you are distracted while dialing. With practice and training, we can learn to divide our attention between two mental tasks simultaneously, but only with difficulty.[17] In this sense, the executive is more like an old-fashioned DOS operating system that can only run one program at a time than like a multitasking Windows operating system that can concurrently run word processing, spreadsheet, e-mail, calendar, and other programs.

But there's also a sort of chunking that takes place in executive functions. As we've seen, the executive is involved in scheduling the sequence of steps in a complex task. Here, the executive is doing more than one thing at a time, but the things are all related to the overall goal. If the executive has to work on multiple unrelated goals at the same time, however, the system begins to fall apart, especially if the goals conflict with one another. An easy way to stress people is to make them do too much at once. Planning, decision-making, and other aspects of mental life suffer when the executive is overloaded.

HEADQUARTERS

Aleksandr Romanovich Luria, the great Russian psychologist, developed broad-ranging and influential theories of brain function on the basis of his studies of World War II soldiers who had sustained gunshot wounds to the head.[18] One of his major conclusions was that damage to the frontal lobe interferes with the ability to plan and execute goal-directed behavior. Other studies of neurological patients with frontal lobe damage, as well as of psychiatric patients who underwent frontal lobotomy, supported the conclusion that the frontal lobes are involved in executive functions (planning, problem-solving, and behavioral control), as well as in short-term or temporary memory.[19] More recently, it has been shown that neural activity, as measured with devices such as PET and MRI scans, increases in the frontal cortex when humans perform tasks that require temporary storage and executive functions.[20] Working memory thus has come to be thought of as a function of neural circuits in the frontal lobes.

The frontal lobes (one sits on each side of the brain) are huge, and account for about one-third of the mass of the human brain.[21] All mammals have frontal cortex, but, for most creatures, its main job is movement control. The prefrontal cortex, which is located in front of the movement-control regions,

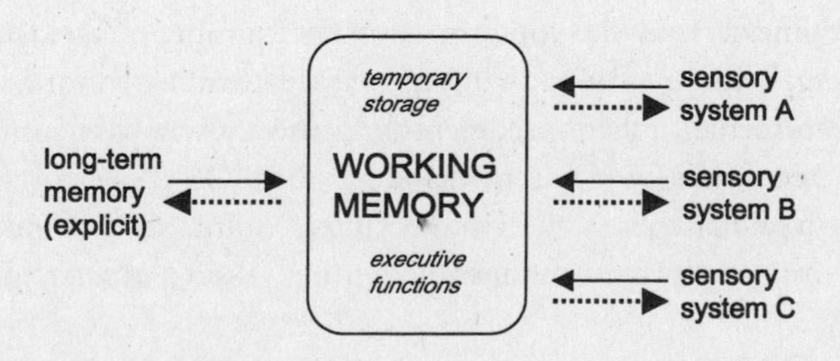

FIGURE 7.1 WORKING MEMORY

Working memory can process information from diverse sources, allowing the information to be compared, contrasted, integrated, and otherwise cognitively manipulated by so-called executive functions. In order to perform these mental operations, working memory has to be able to store the information temporarily.

is especially well developed in primates (some say this region doesn't even exist in nonprimate mammals).[22] This is the region classically implicated in working memory.

As in most areas of neuroscience, animal research is crucial for determining exactly how neural systems participate in psychological functions, and much of our current understanding of the role of the prefrontal cortex in working memory has come from studies of monkeys. The first clues emerged in the 1930s, when it was found that damage to the monkey prefrontal cortex interfered with the ability to remember under which of two objects a reward (for example, a raisin) was hidden during a delay period in which the objects were out of sight.[23] This temporary storage deficit in the so-called delayed response task in animals with prefrontal damage has been extensively studied and has been repeatedly confirmed in various ways using different kinds of stimuli.

The role of the monkey prefrontal cortex in working memory has been elaborated in work by the laboratories of Joaquin Fuster at UCLA and Pat Goldman-Rakic at Yale.[24] Both researchers have recorded the electrical activity of prefrontal neurons while monkeys perform delayed response tasks and other tests requiring temporary storage and have shown that cells in this region become particularly active during the delay periods. It is likely that these cells are directly involved in retaining information during the delay.

The synaptic connections of the prefrontal cortex, also elucidated through studies of monkeys, explain its ability to participate in working memory tasks involving many different kinds of stimuli.[25] The prefrontal cortex is a convergence zone. It receives connections from various specialized systems (like the visual and auditory sensory systems), enabling it to be aware of what's going on in the outside world and to integrate the information it gathers. It also receives connections from the hippocampus and other cortical areas involved in long-term explicit memory, allowing it to retrieve stored information (facts, personal experiences, schemata) relevant to the task at hand. In addition, it sends connections to areas involved in movement control (including movement-control areas in the frontal cortex as well as in subcortical regions), allowing executive decisions to be turned into voluntary actions.

A VISION OF WORKING MEMORY

The visual system is one of the most thoroughly studied systems in the brain. For this reason, research on working memory and other cognitive processes often uses visual stimuli. Studies of visual processing, combined with the vast

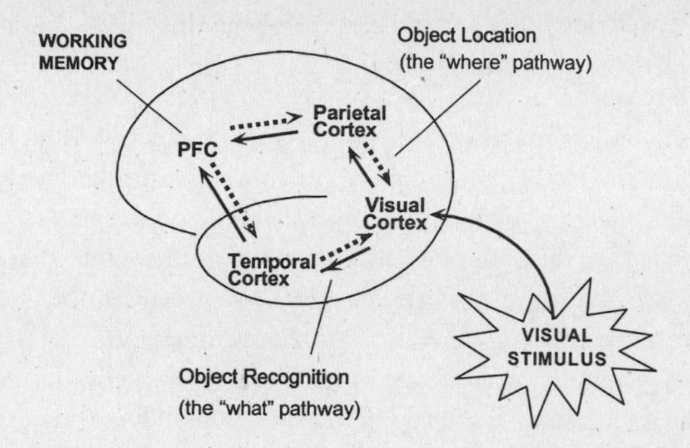

FIGURE 7.2 "WHAT" AND "WHERE" VISUAL INPUTS TO WORKING MEMORY

Working memory is mediated by neural networks in the prefrontal cortex (PFC). Much of what has been learned about working memory has come from studies of visual processing. In the visual system, several subsystems play important roles in stimulus processing. One of these is involved in recognizing what an object is and another in locating its position in space. In working memory, these two kinds of representations come together so that objects not only have identities but also locations when we "see" them.

amount of knowledge about the anatomical connections and function of the visual system, have thus helped construct a fairly detailed understanding of the synaptic connections underlying working memory.

Cortical visual processing begins in the primary visual area located in the occipital lobe (the rearmost part of the cortex). This area receives visual information from the visual thalamus, processes it, and then distributes its outputs to a variety of other cortical regions. The work of Leslie Ungerleider, Semir Zeki, and David Van Essen, among others, has been responsible for elucidating the pathways of visual processing in the cortex.[26] Although these circuits are enormously complex,[27] the pathways responsible for two broad aspects of visual processing are fairly well understood. Thanks to Ungerleider and Mort Mishkin, these have come to be known as the "what" and "where" visual pathways.[28] The "what" pathway is involved in object recognition, and the "where" pathway in figuring out the spatial location of that object relative to other stimuli in the outside world. The "what" pathway involves a processing stream that travels from the primary visual cortex to the temporal cortex, and the "where" pathway goes from the primary cortex to the parietal cortex.[29]

The end stage of the "where" pathway in the parietal cortex is directly connected with the prefrontal cortex.[30] As Goldman-Rakic, Fuster, and others have shown, cells in the prefrontal region are active during delay periods in spatial tasks (which involve "where" processing), indicating that these cells may be involved in the temporary storage of the stimulus information during the time when the stimulus is absent.[31] Moreover, cells in the "where" processing area in the parietal cortex are also active during the delay.[32] Synaptic interactions between the parietal "where" area and prefrontal cortex may therefore allow the animals to keep in mind the spatial location of the reward.

The "what" area in the temporal cortex is also connected with the prefrontal cortex.[33] Earl Miller and Bob Desimone therefore asked whether cells in the "what" visual pathway and the prefrontal cortex might both be active during a delay period in which monkeys had to remember which of two similar stimuli was associated with reward.[34] In this experiment, the reward was paired with a particular stimulus, which is why the "what" stream of processing was involved. They found that cells in both the prefrontal cortex and the temporal cortex "what" processing area were active during the delay.

The maintenance of visual information in working memory thus appears to depend crucially on information transfer over synaptic pathways between specialized areas of the visual cortex and the prefrontal region. The pathways from the specialized visual areas tell the prefrontal cortex "what" is out there and "where" it is located. Moreover, these are two-way streets: the prefrontal cortex, by way of synaptic pathways back to the visual areas, instructs the visual areas to attend to and stay focused on those objects and spatial locations that are being processed in working memory. Recent studies by Goldman-Rakic and Liz Romanski have found that auditory working memory involves similar relations between auditory processing streams and prefrontal areas, suggesting that this scheme involving linkages between specialized sensory processing systems and prefrontal cortex may be generally applicable across many systems.[35]

•

ORDERS FROM THE TOP

The flow of information from lower- to higher-level processing stations is what cognitive psychologists call bottom-up processing, and the flow from higher to lower stations is called top-down processing.[36] Thus, the pathway from the "what" area to the prefrontal cortex transmits bottom-up information (placing object-identifying information in the workspace), and the pathway from prefrontal cortex back to the "what" area transmits top-down information (executive control signals that help keep attention focused on the object that is represented in the workspace). Top-down activities are just another name for executive functions.

Jonathan Cohen and David Servan-Schreiber used the so-called Stroop task to study top-down executive influences on lower-order processing.[37] In this procedure, people are given brief visual presentations of simple words (color names) and are asked to either name the word or state the color of ink the word is printed in. On some trials, the word name and color are in conflict (the word *green* is presented in red letters) and on others they are congruent (the word *green* is presented in green letters). It takes much longer to state the ink color in the conflict than in the congruent situation, but subjects tend to get it right. Unlike normal subjects, patients with damage to the prefrontal cortex fail this task.[38] They are unable to use the instruction given by the experimenter to suppress the more commonly given response of word-naming. Schizophrenics, who are believed to have frontal dysfunction, also do poorly.[39] Functional imaging studies by several groups have shown that the prefrontal cortex is activated during the conflict condition in normal subjects but not in schizophrenics.[40] These findings suggest that executive functions, by way of connections from the prefrontal cortex back to lower-order processors (color and word processors), use the instructions given by the experimenter (and kept active in working memory) to inhibit the more natural word-naming response and facilitate the less common response of stating the ink-color name. Cognitive scientists often find it useful to program computer simulations to help them understand how psychological and sometimes neural processes work. Cohen and Servan-Schreiber did this in an effort to account for the role of the prefrontal cortex in the Stroop task.[41]

Although it generally has been assumed that temporary storage and executive functions are carried out by the same networks in the prefrontal cortex,[42] most studies have focused on the former rather than the latter. However, Mark D'Esposito and colleagues designed an experiment to explore explicitly the role of prefrontal areas in executive functions. They had human subjects perform a verbal identification task (e.g., respond if you hear a vegetable name) and a visual task (respond if a certain spatial arrangement of stimuli occurs).[43] The tasks were performed either separately (which was easy and did not require executive functions) or simultaneously (which was difficult and did require executive functions, since the subjects had to keep track of two different kinds of stimuli at once). The results showed that the prefrontal cortex was activated when the two tasks were performed simultaneously, but not when the tasks were performed individually. Since there was no temporary storage requirement in the task (no delay between stimulus and response), the results specifically implicate the prefrontal region in executive functions. A number of other studies have also now clearly demonstrated that the human prefrontal cortex is indeed activated when executive operations are performed, as well as when information is stored temporarily.[44]

For obviously reasons, it is more difficult to study executive functions in nonhuman primates. You can't just give verbal instructions to a monkey, which has to be trained nonverbally on each step of a given task. But it has long been thought that deficit in delayed response performance by monkeys with frontal lesions was not simply due to a loss of temporary storage, since the deficit was worse when there were distracting cues.[45] It was as if the monkeys couldn't ignore irrelevant, distracting information.

Recently, studies have begun to provide evidence for the operation of executive-like functions in the monkey brain. For example, John Reynolds and Bob Desimone trained the monkeys to focus their attention on the location on a screen where a dot appears. This is, in essence, the training of an executive function, attention. Then they presented a picture in that same area (that is, within the field of attention). Cells in the early stages of the visual cortex responded to the picture. When a second stimulus was presented on the screen with the first, though, the cells did not respond. Executive functions continued to focus cellular activity on the first picture, and the new information about the second picture was ignored.[46] This work, like the studies described above, suggests that attentional signals are involved in directing traffic in the "what" pathway, controlling which stimuli are processed. In related work, Leslie Ungerleider and colleagues[47] had human subjects perform similar tasks while functional brain activity was measured. They reached a very similar conclusion: that top-down activity regulates lower visual areas.

Studies by Earl Miller's lab suggest that the prefrontal cortex in the monkey brain, as in the human brain, is the source of top-down signals.[48] Monkeys were first shown a target stimulus. That stimulus then appeared in an array with two irrelevant stimuli. After a delay, the three-stimulus array appeared again. If the target stimulus was in the same location, the monkey had to make a response on a lever. The results showed that prefrontal cells were particularly responsive to the selection of the target stimulus from the array.

Further, although cells in the visual cortex are also sensitive to target selection,[49] their activity occurred after prefrontal activity, suggesting that the attentional signal originates in the prefrontal region and is transmitted in top-down fashion to the visual cortex.[50]

But is the activity occurring in the prefrontal cortex necessary in cognition, or just along for the ride? In other words, is the frontal cortex activity the result or the cause of top-down cognition? Results obtained by Robert Knight suggest that it is necessary, that it is causal.[51] Knight recorded electrical activity in the visual cortex of humans using electrodes placed on the scalp. Although this technique does not provide detailed information about specific neural circuits, it is sufficient to detect activity in areas like the visual cortex. Normally, if subjects are instructed to expect a certain stimulus, the electrical response to that stimulus in the visual cortex will be enhanced. However, this attentional modulation of lower-level sensory processing did not occur in patients with damage in the prefrontal region. The prefrontal cortex appears to be necessary for top-down cognition to work.

The importance of top-down signals from the prefrontal cortex in selecting neural activity in lower processors was also demonstrated in an elegant study by Yasushi Miyashita's team in Japan.[52] They made lesions in the brains of monkeys in such a way that allowed them to present visual stimuli so that the "where" area in the temporal cortex received information either bottom-up (from the lower areas of the visual system) or top-down (from the prefrontal cortex), but not both. They found that top-down information from the prefrontal area was sufficient to selectively activate cells in the "where" pathway.

A central aspect of executive functions is decision-making. Studies by several research groups have begun to reveal how this works in the brain. Most of the investigation has focused on a simple behavior, eye movement, which is one that offers many advantages to the experimenter. A monkey is first instructed, through behavioral training, to respond to certain stimulus conditions by moving its eyes to a certain location. This response is then used to assess how the monkey makes decisions about what to respond to on the basis of what it sees. For example, Michael Shadlen and Jong-Nam Kim showed monkeys visual stimuli moving in one direction or another.[53] The moving stimuli were embedded in a busy background, making it impossible to respond correctly every time. The monkey, in other words, had to use the imperfect information available to make a decision about the stimulus. Further, a delay was imposed between the stimulus and the response. Earlier work had shown that the motion stimulus is processed in a special area of the visual cortex called MT, which has connections with the prefrontal cortex.[54] In the new experiment, Kim and Shadlen recorded neural responses in the prefrontal cortex and found that neural activity in this region predicted, during the delay, which direction the eyes were about to move in. They concluded that these prefrontal cells are involved in the conversion of stimulus information into response plans. These cells are, in their view, part of the decision-making process.

Some investigators argue that circuits in the so-called "where" area of the parietal cortex, like prefrontal cells, are involved in planning and decision-making about movements.[55] In many ways, these cells apparently function similarly to prefrontal cortex cells. However, one difference seems to be that the parietal circuits are involved in planning the next movement, whereas prefrontal circuits participate in planning several steps ahead.[56] Later, when we consider emotion and motivation, we will have the opportunity to explore other aspects of decision-making.

ONE SYSTEM OR MANY?

Studies by Joaquin Fuster first suggested that the prefrontal cortex might integrate spatial ("where") and object ("what") information, since cells responding to these two kinds of information were intermixed in the prefrontal cortex.[57] This result was consistent with the view that the prefrontal cortex engages in general-purpose temporary storage across many processing domains. More recently, though, Pat Goldman-Rakic has questioned the assumption that the prefrontal cortex is a general-purpose working memory processor.[58]

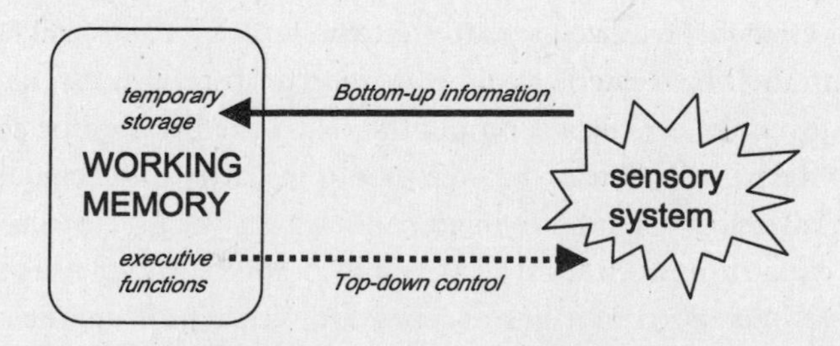

FIGURE 7.3 BOTTOM-UP AND TOP-DOWN PROCESSING

The process of sending information to working memory from sensory systems is an example of bottom-up processing, whereas the control of sensory processing by working memory is an example of top-down processing.

She and Fraser Wilson found evidence that cells involved in object and spatial processing are clustered in different areas of the prefrontal cortex.[59] Recently, Liz Romanski and Goldman-Rakic found additional areas that they believe are involved in auditory working memory.[60] These findings suggest that temporary storage is carried out by domain-specific regions in the prefrontal cortex and question the notion of a universal workspace. Ed Smith and John Jonides reached a similar conclusion, on the basis of a survey of functional imaging studies of working memory in humans.[61] However, Earl Miller has pointed out that there are a number of cells in each frontal area that do not respond exclusively to one kind of stimulus. Further, the different areas are interconnected extensively with one another. It is quite possible that the various areas work together to integrate information across domains, and could, in fact, constitute a single distributed system (a system that is spread over several brain regions) that mediates temporary storage, rather than a collection of independent systems. Additional work is needed to resolve this issue.

Regardless of how temporary storage functions are laid out, it appears that executive functions are not partitioned in the prefrontal cortex on the basis of the stimulus domain.[62] This does not mean, however, that there is a single executive area that takes care of all executive functions. Executive functions, in fact, seem to be spread across multiple regions of the frontal cortex. For example, the classic area of the prefrontal cortex implicated in working memory has been the lateral prefrontal cortex, located on the outside surface of the frontal cortex.[63] However, another area, located on the medial or inside surface of the frontal cortex, has also been consistently activated in functional imaging studies of working memory, and is one of the main regions implicated in executive functions.[64] This region, the anterior cingulate cortex, like the classic working memory area in the lateral prefrontal region, receives inputs from various specialized sensory systems, and is anatomically interconnected with the classic lateral area.[65] Moreover, both regions are part of what cognitive psychologist Michael Posner has called the frontal lobe attentional network,[66] a cognitive system involved in selective attention, mental resource allocation, decision-making processes, voluntary movement control, and/or resolving conflict between competing stimuli.[67] It is tempting to think of the executive aspects of working memory as involving synaptic interactions between neurons in the lateral prefrontal and anterior cingulate regions, and perhaps other areas.[68] Nevertheless, the various aspects of executive functions may not be equally distributed across the various areas. There is, in fact, some evidence for localization of particular executive functions—that is, in some tasks different aspects of executive functions (for example, stimulus or response selection vs. conflict resolution vs. decision-making) have been found to engage different areas of the prefrontal cortex to different degrees.[69] This in no way implies the existence of multiple executives, each with their own planning and decision-making capacities. Instead, it suggests that the various component executive functions are achieved by a set of interconnected circuits that are spread over several brain regions in the frontal cortex, and even other regions, as discussed more in chapter 9.

THE NUTS AND BOLTS OF THOUGHT

How does working memory function at the level of cells and synapses? While this process is still not fully understood, we can at least begin to piece together an explanation.

The prefrontal cortex, like other areas of the neocortex, has six layers.[70] And, as in other areas, the middle layers tend to receive inputs from other regions, while the deep layers tend to send outputs to other regions. So axons from other cortical areas, such as areas involved in "what" and "where" processing, form synapses on cells in the middle layers of prefrontal cortex. These input cells then send axons to cells in the deep layers, which give rise to connections that go back to the middle-layer cells, or to other cortical or subcortical areas, especially areas involved in the control of movement, and thus of behavioral responses. In this manner, the deep and middle layers can influence each other. In addition, though, input cells in the middle layers and output cells in the deep layers each give rise to local connections to other cells in the same layer. This arrangement allows the input cells to influence other input cells, and output cells to influence other output cells.

Transmission of inputs to and outputs from the prefrontal cortex, and between cells within and between layers within the prefrontal cortex, is mediated by the binding of presynaptically released glutamate to postsynaptic receptors. Interestingly, extrinsic inputs to these circuits account for only a small part of the excitatory synaptic connectivity of the prefrontal cortex. The connections within the prefrontal cortex, both within and between layers, are far more numerous than the connections coming in from other areas, such as sensory processing regions. The mutual excitations mediated by the internal connections enable input signals from the outside to be amplified and kept active, and may well contribute to the sustained activity that has been observed during delay periods.[71]

Of course, there is also inhibition in these circuits. Most of the inhibitory connections are on the deep-layer output cells. However, because these cells also have excitatory projections back to upper layers and to other cells in the same layer, local inhibition in the lower layers can regulate much of the excitatory flow through the network.

Much of what we know about the contribution of excitatory and inhibitory circuits in the prefrontal cortex to working-memory centers around the neuromodulator dopamine.[72] The prefrontal cortex receives a rich supply of axons containing dopamine,[73] and depletion of the substance in the prefrontal cortex (by injecting dopamine-destroying drugs) is as effective as complete removal of the prefrontal cortex in disrupting delayed-response performance in monkeys. In dopamine-depleted animals, performance recovers if a drug that activates a certain class of dopamine receptors is infused into the prefrontal cortex, thus tricking the receptors and the postsynaptic cells into acting as if dopamine fibers were still there releasing their transmitters. Infusion of dopamine into the prefrontal cortex of young monkeys enhances working memory capacity, and in older ones, it reverses the age-related decrement in working memory.[74]

Dopamine cell bodies are located in the brain stem, in a region called the ventral tegmental area. The axons of these cells then branch extensively and reach many areas of the forebrain, including the prefrontal cortex, where their terminals release dopamine.[75] In primates, the dopamine terminals are fairly evenly distributed throughout the layers, allowing dopamine to bind to receptors and then modulate excitatory and inhibitory transmission in both the input and output layers. Although there are many subclasses of dopamine receptors, the D1 family (which includes D1 and D5 receptors) has been most clearly implicated in working memory.[76] These receptors are located on the spines and shafts of dendrites of excitatory cells and seem to reduce the transfer of excitation from the dendrites to the cell bodies, allowing only especially strong excitatory inputs to get through to the cell body and elicit excitation.[77] Dopamine release in the prefrontal cortex also seems to facilitate GABA inhibition, possibly by way of presynaptic facilitation of transmitter release, leading to a further reduction of excitation through prefrontal circuits.[78] Some of these effects appear to involve the triggering of protein kinase A in cells containing dopamine receptors. Integrating these findings, Amy Arnsten has proposed that dopamine participates in working memory by biasing cells to mainly respond to strong inputs and thereby focusing attention on active current goals and away from distracting stimuli.[79]

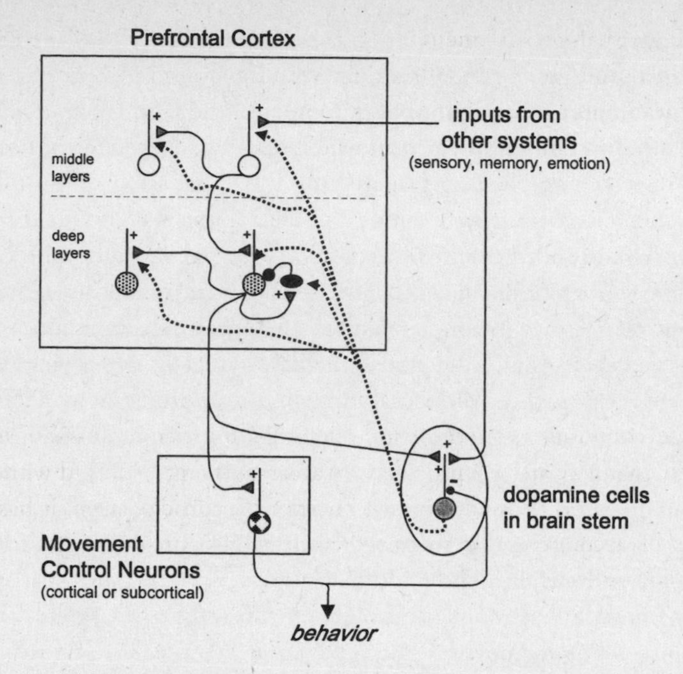

FIGURE 7.4 CELLULAR MECHANISM OF WORKING MEMORY

Inputs to the prefrontal cortex from sensory, memory, or other systems terminate on the dendrites of cells located in the middle layers. These cells have extensive excitatory (+) connections with other cells in the same layer and also connect with cells in the deep layers. Deep-layer cells are connected with other deep-layer cells, as well as with cortical and subcortical motor regions that control behavior. Deep-layer cells also connect with inhibitory interneurons. Dopamine cells in the brain stem modulate all aspects of the circuitry in the prefrontal cortex, enhancing or facilitating the excitation. The extensive excitatory connectivity in this circuitry, and its enhancement by dopamine, might underlie the ability of working memory to hold stimuli in working memory as long as the organism remains engaged in the task. The output of motor systems inhibits dopamine cells, suggesting that once behavior is produced, the facilitation by dopamine terminates and working memory is released to do other things. Based on Durstewitz et al. 1999.

THE WIZARD OF OZ

Most of us have an intuitive sense that thinking (cognition) and consciousness are closely related. This intuition is correct but incomplete. Virginia Woolf was probably referring to this when she pointed out, "We all indulge in the strange, pleasant process called thinking, but when it comes to saying . . . what we think, then how little we are able to convey!"[80]

Imagine that you are mentally engaged in some activity, like reading the newspaper. You ignore the other things going on around you. But if something meaningful occurs in the background, like the sound of your name, you stop reading and turn to the person who called you. Although your conscious mind was ignoring everything apart from the visual signals from the paper, your brain was not. Inputs from other sensory systems continued to be actively processed, otherwise the mention of your name could not have interrupted you. Cognition and consciousness, therefore, are not the same. What, then, is consciousness, and what determines what we are conscious of?

Consciousness can be thought of as the product of underlying cognitive processes. We've actually already seen how the key processes work. The stuff were are conscious of is the stuff that working memory is working on.[81] Though cognitive science provided a way of studying the mind without getting entangled on the controversial question of consciousness, it has, in the process of accounting for working memory, also provided a practical approach for understanding how consciousness works.

Let's translate the example above into cognitive language. During focused attention, mental resources are allocated to the task at hand. The executive, in other words, keeps lower-level processors engaged in activities that support the task being worked on. However, if lower-level processors that are not being worked with at the moment detect some event that is unrelated to the current task but is more important than the current task, resources are allocated to processing the new event. The task management, scheduling, and conflict resolution functions of the executive shift attention to the new event and move information relevant to it into working memory. This forces out the existing information. Behaviors relevant to the new task are initiated: you turn toward the source of the sound. If, after saying your name, the person asked, "What's that smell?" you would probably start sniffing in an effort to answer the question. During these operations, memories are retrieved into working memory, enabling you to recognize that the sound spoken was your name and that it was your friend who spoke it. (How else would you know your name except by virtue of the fact that you remember it; and how else do you know that it was your friend who said your name except that you remember her voice and appearance?) Retrieved episodic and semantic memories, as well as schemata, then aid in executive decision-making.

Although the end result of executive processes (monitoring, resource allocation, task management, conflict resolution, memory retrieval, etc.) was the representation of your name as conscious content in working memory, it's important to recognize that the executive processes that made this possible functioned unconsciously. As neuroscience pioneer Karl Lashley pointed out in the early 1950s, we are never aware of processing, but only of the consequences of processing.[82] Like the Wizard of Oz, executive processes work behind the scenes.

The working-memory model of consciousness suggests that the prefrontal cortex should play some role in consciousness. This is not a novel idea.[83] For example, the Russian psychologist Luria, whom we encountered earlier, pointed out two characteristic deficits of patients with damage to the frontal lobes: they lack spontaneous, purposeful behavior and they lack an understanding of their deficits.[84] And Arthur Benton, an American neurologist who long studied the effects of frontal lobe damage, described a typical patient in the following way: "He has never been seen to sit self-absorbed, to daydream or to indulge in introspection."[85] These patients are not, strictly speaking, "unconscious," but rather seem to have a fractured understanding of what they experience.[86] At the same time, they do not typically suffer total destruction of the prefrontal cortex, and the remnants of their conscious states may be due to the undamaged parts of this large and complex brain region—as we've seen, working-memory functions are not located in a single region of the prefrontal cortex but are distributed across widespread regions.

On the other hand, there may exist primitive levels of consciousness, especially involving the passive awareness of events as opposed to the active use of on-line information to guide decision-making and behavior, that do not depend on the prefrontal cortex.[87] These kinds of mental states may typify consciousness, to the extent that it does exist, in organisms that have less or no prefrontal cortex, as discussed below.

If the prefrontal cortex plays an essential role in human consciousness, it should be involved in explicit memory, memory to which we have conscious access. Indeed, damage to the human prefrontal cortex disrupts the conscious retrieval of long-term memories, especially episodic memories.[88] The prefrontal cortex is activated during episodic memory retrieval,[89] and there is also evidence that the prefrontal cortex is involved in the encoding (formation) of episodic memory.[90] These findings suggest that, in order to have an explicit, conscious memory, in addition to the involvement of the medial temporal lobe (chap. 5), two conditions need to be satisfied: you have to be conscious of the information constituting it at the time of the original experience (that is, the experience had to be represented in working memory at the time it occurred), and during retrieval, you have to transfer the information from cortical storage circuits (chap. 5) into working memory.

Just as researchers studying working memory often use visual stimuli, theorists interested in the brain and consciousness tend to focus on visual awareness as a window on conscious experience. For example, Nobel laureate Francis Crick and his colleague Christof Koch have proposed an explanation of visual awareness that builds upon knowledge of the role of the visual system in working memory.[91] They argue that we have access to information processing occurring in visual-processing areas connected with the working memory areas of the prefrontal cortex but do not have access to information processed by areas not connected to these areas. Indeed, the primary area of the visual cortex is not connected with the prefrontal cortex, and we do not experience the contours and shadings of stimuli that it processes. However, the later processing areas are connected with the prefrontal cortex and we experience the products of these areas—global features of objects, like their shape, color, motion, and location.

Several studies provide support for the Crick/Koch theory.[92] For example, Roger Tootell and colleagues examined whether the brain networks involved in the processing of visual motion would be active when subjects were tricked into seeing motion where none exists.[93] They used the waterfall illusion, which is produced by the presentation of stationary stimuli at slightly different times. The result was indeed that the motion-processing area, which is an advanced visual area that is connected with the prefrontal cortex, was particularly active during the illusion of movement.[94]

Given that different features of a visual stimulus (shape, color, location, motion) are processed in different cortical areas, in order to have a conscious perception of the stimulus as a whole object, rather than as a collection of features, the elements have to be bound together.[95] Although we can separately identify these features, we usually experience them together as the coherent object rather than as component parts. Understanding how the integration occurs is called the binding problem.[96]

Binding is taken care of, at least in part, by the transfer of information from the individual processing regions to areas that integrate the information. As we've seen, this kind of convergence occurs within each sensory system. For example, the earliest stages of the visual cortex process primitive features of objects (contours and brightness, for example) and the later stages process more complex features ("what," "where," and movement, for example). Further, different complex features are processed in different circuits, requiring integration or binding across circuits as well as within them. And visual stimuli do not occur in isolation but in the context of other kinds of stimuli (sounds and smells, for example), requiring binding across sensory modalities. Further, the conscious experience of a sensory stimulus is not just a sensory experience. We usually experience stimuli as meaningful objects rather than as raw sensations. Thus, the way the stimulus looks, sounds, and smells has to be integrated with relevant information stored in memory, including facts and past experiences, as well as stored information about the emotional and motivational significance of the stimulus. Because prefrontal regions receive convergent inputs from sensory, memory, emotional, and motivational circuits, they are believed to be capable of performing the complex kind of information integration (binding) that must occur in the brain during a conscious experience.

But some researchers feel that information convergence in working memory is not sufficient to account for the conscious experience of a stimulus, or for our ability to make decisions and take action relevant to that stimulus.[97] They argue that the additional ingredient needed is neuronal synchrony. This term refers to the simultaneous firing of populations of neurons. Synchrony is proposed to achieve two important goals—enhanced activation of postsynaptic cells and coordination within local areas across widespread regions. It is well-established that a group of postsynaptic cells are more strongly activated when they receive synchronous inputs from presynaptic cells. Thus, if the brain is processing a salient visual stimulus, cells in the visual areas will fire synchronously. As a result, the information processed in these areas will activate prefrontal areas more strongly, and the visual stimulus will make it into working memory more easily. The other proposed role of synchrony is coordination. That is, synchrony is believed by some to help solve those aspects of the binding problem that information convergence cannot. According to this view, neurons that fire together in widespread brain regions are temporarily bound together, and this coherence of firing, when combined in just the right way across the brain, facilitates the representation in working memory of momentarily relevant information from diverse regions. While considerable evidence exists showing that synchrony occurs and enhances postsynaptic firing,[98] the second role of synchrony, coordination across widespread regions, is controversial, especially as an explanation of conscious perception. (We'll encounter a less controversial role of synchrony in chapter 11, namely in the coordination of synaptic plasticity across brain regions.)[99]

The working memory theory of consciousness, even without the assistance of synchronous information coordination across brain regions, gives us a neural handle on consciousness. But how far can we go with this pursuit of consciousness and the brain without butting heads with the mind-body problem (chap. 2)? After all, it's one thing to say that the information in working memory is what we consciously experience, and another to explain how those experiences emerge from working memory. Similarly, it's one thing to say that voluntary behavior results when conscious content in working memory initiates executive functions that then lead to muscle movements, and quite another to explain how conscious content is translated into executive control. To a dualist, one who believes in a separate mind substance that has to somehow be merged with physical matter, these are indeed different phenomena. But if, instead, you, like me, believe that the things we describe in mental terms are, in fact, processes going on in the brain, a working memory point of view is an excellent way of attacking the problem. While it may not at this point fully account for all the subtleties of how consciousness and other aspects of the mind arise from the brain, working memory is clearly a concept we can work with well.

THE REVOLUTION OF THE THINKING BRAIN

As I discussed in chapter 3, the prefrontal cortex is especially well-developed in humans, is present in other primates, rudimentary in nonprimate mammals, and doesn't even exist in other creatures. What are the implications of this for understanding working memory and consciousness?

Let's focus first on the difference between nonhuman primates and other mammals. To do so, we have to consider the organization of the prefrontal cortex in more detail—specifically, the distinction between different prefrontal areas. We've already discussed the lateral prefrontal cortex, which also has divisions,[100] above. The medial prefrontal cortex is located on the inside wall of the hemisphere. Imagine the brain's hemispheres as the two parts of a hot dog bun. The brown part on the outside is like the lateral cortex and the white part on the inside is like the medial cortex. One region of the medial prefrontal cortex is the anterior cingulate, which, as we've seen, is also involved in executive functions. In addition, the ventral prefrontal cortex, especially the orbital cortex, appears to play a role in working memory, particularly for emotional information, as we'll discuss in chapters 8 and 9. While other mammals have medial and ventral prefrontal cortices, primates alone appear to have lateral prefrontal cortex.[101] Thus, one of the major regions involved in working memory in primates clearly does not exist in other animals. The fact that the cognitive capacities of these creatures do not compare with those of primates suggests that the unique features of primate cognition came with the development of the lateral prefrontal region and its integration with existing networks involving the medial and ventral areas. For example, rats can engage in temporary storage, especially about emotionally relevant information, and can focus attention on specific stimuli and so forth, but are far more limited than primates in their capacity to categorize the world, to discriminate among different stimuli and events, to relate or associate things with one another on-line, and to use the results of these cognitive analyses to guide problem-solving and decision-making.

It's important to point out that temporary storage can be carried out in domain-specific systems, like sensory or emotional systems, which accounts for short-term memory in such animals as birds and reptiles. It also accounts for domain-specific short-term memory in mammals. These domain-specific kinds of short-term memory processes may underlie primitive conscious representations in animals that lack elaborate working memory functions. Domain-specific temporary storage may allow, for example, an awareness of significant stimuli, like the sight of a predator, the pain of being injured, the taste of food, or the joy of sex. When the arousal elicited by the temporarily stored information is sufficiently intense, the activity of other systems may be inhibited, allowing the active system to dominate brain function until the arousing situation subsides. It is known, for example, that sexual impulses are inhibited in threatening situations.[102] This kind of arousal could be achieved by way of interactions between, for example, a predator detection system and brain stem arousal circuits (as described in the next chapter). Indeed, Gary Aston-Jones of the University of Pennsylvania and others have shown that brain stem arousal systems underlie vigilance, sustained behavioral engagement in a task.[103] Such connections could make possible focused attention even in creatures lacking a prefrontal cortex and its multimodal integrative capacity and executive functions.

It is, of course, difficult to know if other animals experience anything, since we can't ask them outright, but if they do, it may be something akin to domain-specific sensory consciousness. The key element that distinguishes working memory from sensory consciousness is that the former allows for the simultaneous interrelation of temporarily stored information across domains and the flexible use of such information in decision-making, capacities that prefrontal circuits seem to make possible. By this analysis, something akin to human consciousness would be present in other animals with well-developed working memory systems (nonhuman primates) but not in other creatures.

But even nonhuman primates lack the unique features of human consciousness. As the novelist and naturalist Annie Dillard says, "It is ironic that the one thing that all religions recognize as separating us from our creator—our very self-consciousness—is also the one thing that divides us from our fellow creatures. It was a bitter birthday present from evolution."[104]

Other than size, which in this case is significant, the human prefrontal cortex has another important advantage over the prefrontal cortex of nonhuman primates: it has access to a processing module specialized for language use. Increased size adds power, but the presence of language does more than just soup up the cortex. (I'm referring here to the kind of grammatical natural language that characterizes essentially every human brain, rather than the other communicative capacities that can occur in other kinds of animals, like chimps and even parrots.) Language radically alters the brain's ability to compare, contrast, discriminate, and associate on-line, in real time, and to use such information to guide thinking and problem-solving. The difference between having only nonverbal working memory and having both verbal and nonverbal working memory is enormous for how the cognitive system works.

It is, in my opinion, the structuring of cognition around language that confers on the human brain its unique qualities. Other animals may be consciously aware, in some sense, of events going on in their world. They may have domain-specific consciousness, or in the case of nonhuman primates, domain-independent nonverbal consciousness, but lacking language and its cognitive manifestations, they are unlikely to be able to represent complex, abstract concepts (like "me" or "mine" or "ours"), to relate external events to these abstractions, and to use these representations to guide decision-making and control behavior. Some advanced primates do have the ability to visually recognize themselves in a mirror,[105] which suggests a sense of self-recognition in the absence of natural language. Interestingly, recent studies in dolphins and whales show something similar, suggesting parallel evolution of this capacity—but questions have also been raised about the interpretation of these findings.[106] In any event, self-recognition does not necessarily imply self-awareness.[107] We know from many kinds of experiments that stimulus recognition is not the same as being consciously aware of the stimulus.[108] Stimulus recognition requires only that an immediately present stimulus match some representation of a similar stimulus in memory. There is no need to have conscious awareness to have recognition.

When I use the term *consciousness,* I usually am referring to the special qualities of human consciousness, especially those made possible by language.

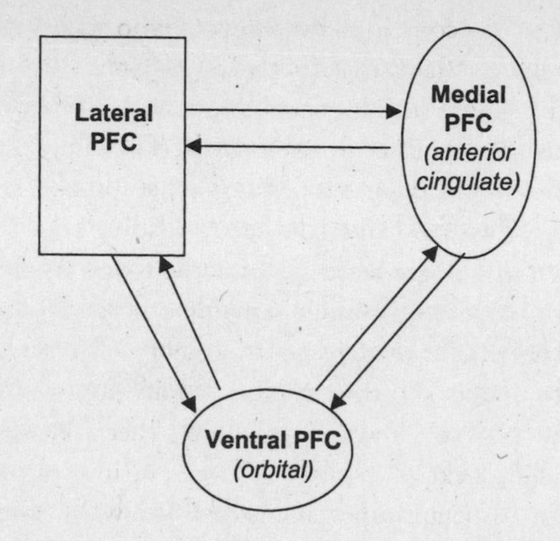

FIGURE 7.5 CONNECTIONS BETWEEN PREFRONTAL CORTEX AREAS INVOLVED IN WORKING MEMORY

Working memory is not the function of one region but of a complex interconnected network in the prefrontal cortex. Some of the areas involved are the lateral prefrontal cortex, which is the classic working memory region, the medial prefrontal cortex (especially the anterior cingulate region), and the ventral prefrontal cortex (especially the orbital region). See the text for descriptions of the contribution of individual areas to various aspects of working memory.

(There may be other ways that the human brain differs as well, but language is a particularly important difference.) This does not imply that we are conscious in English or Chinese or Swahili, nor does it suggest that persons who are deaf and mute should be considered cognitively impaired. My point, rather, is about the way the human brain is wired. That is, the emergence of the cognitive capacities underlying language changed the way the brain works, making it possible for human brains to think and experience events in ways that other brains cannot. The addition of language into the human brain involved a revolution rather than an evolution of function.

P. S.

My idea that language embellishes working memory, and in so doing makes human consciousness unique, bears some relation to Mike Gazzaniga's notion of an interpretive system in the left hemisphere that gives rise to the unique properties of human consciousness.[109] In a nutshell, the essence of the interpreter theory is that our conscious awareness of who we are depends on our linguistic interpretation (labeling, categorizing, explaining) of our experiences as we go through life (this is related to the notion of a narrative or constructed conscious self discussed in chapter 2). The similarity of my ideas about consciousness to Gazzaniga's is perhaps not surprising, since he was my Ph.D. mentor in the 1970s. Mike had been pursuing consciousness in the brain since his own days in graduate school in the 1960s. As part of my Ph.D. research, we did studies of a split-brain patient, known as P. S., that greatly influenced us both.[110] Mike emerged with the theory of the interpreter, and I with a yearning to understand the emotional brain. But we've come full circle. Emotions now play an important role in his theories of consciousness,[111] and, as we see in the next chapter, consciousness, in the form of working memory, has come to be an important part of the way I think about emotions, especially feelings.

CHAPTER EIGHT

THE EMOTIONAL BRAIN REVISITED

THE ADVANTAGE OF THE EMOTIONS IS THAT THEY LEAD US ASTRAY.

—Oscar Wilde

Emotions can, as Wilde says, lead us astray. But their inclusion in the mental trilogy by great thinkers through the ages comes less from their distracting qualities than from their contribution to defining who we are. A purely cognitive view of the mind, one that overlooks the role of emotions, simply won't do. As we begin to assemble the self in terms of synapses, emotions must play a major role.

THE EMOTIONAL BRAIN AND ITS VICISSITUDES

As soon as pioneering brain researchers of the late nineteenth century identified regions of the neocortex involved in sensory perception and movement control (chap. 3), William James asked whether emotions might also be explained in terms of these particular functions, or whether they were rather the business of a separate, yet undiscovered, brain system.[1] Being a pragmatist, he proposed a theory of emotion based solely on functions of sensory and motor mechanisms. Specifically, he argued that emotionally arousing stimuli are perceived by sensory cortex, which activates motor cortex to produce appropriate bodily responses. Emotional feelings then result when our sensory cortex perceives the sensations that accompany bodily responses. Since different emotions involve specific bodily responses, they have distinct sensory signatures and thus feel different. The essence of James's argument is captured by his conclusion that we do not run from a bear because we feel afraid, but instead we feel afraid because we run.

James's theory was quickly refuted by research showing that complete removal of the neocortex failed to disrupt the expression of emotional responses elicited by sensory stimuli, thus suggesting that sensory and motor cortex could not be the key.[2] Subsequent studies then implicated specific brain areas, many involving subcortical or old cortical regions, as opposed to neocortex, a trend that led to the view that the brain does indeed have a special emotion system. This trend ultimately culminated around mid-century in the famous and still popular limbic system theory of emotion, which inspired many subsequent studies (we'll consider this theory in detail below). By the mid-1960s, though, research on the neural basis of emotion had all but come to a halt, or at least a slow crawl—after decades of concentrated attention, neuroscientists forsook the topic.[3]

Like psychologists, brain researchers were strongly influenced by the emergence of cognitive science, and emotions were not part of the cognitive game plan. Emotions seemed more a matter of mental content than of mental processing, and were not pursued by those interested in the thinking process. Although cognitive scientists didn't deny that emotions were important, they believed the subject was not relevant to their field.[4] As interest in cognition rose, research on emotion declined in neuroscience.

In recent years, however, emotion, part two of the mental trilogy, has again become a popular topic for investigation in neuroscience. And enthusiasm by brain researchers has in turn helped revitalize interest in emotions by psychologists. Although I discussed some aspects of this new wave of research in *The Emotional Brain,* much has happened since, especially in terms of how animal studies of emotion relate to the human brain. Before we turn to research on the neural basis of emotion, I therefore want to spend some time discussing the relation of animal and human studies of emotion.

THE CREDIBILITY PROBLEM

Emotion researchers, whether in psychology or brain science, have typically sought to account for what most people think of as the essence of an emotion, the subjective experience that occurs during an emotional state—the feeling of fear when in danger, of anger when mad, or of joy when something good happens. This was clearly the goal of emotional brain theories from James through the limbic system concept.[5] However, most of what we know about the detailed brain mechanisms of emotion comes from studies of emotional behavior rather than from studies of feelings themselves. The explanation for this situation is simple. Feelings can be studied in humans, but, for reasons that will be discussed below, they're more difficult to examine in animals. Since, for both practical and ethical reasons, most brain research is conducted in animals, we end up with a gap between what emotion theories are about (feelings) and what brain researchers actually measure (behavior). This gap, in turn, creates a credibility problem for brain research on emotions.

The credibility issue could be overcome if there were some way to investigate feelings in animals. In humans, the main method by which feelings are studied is that of self-evaluation, often in the form of verbal self-report.[6] However, even self-evaluation is problematic as a scientific tool for studying emotions.[7] As Donald Hebb, father of the Hebb synapse (chap. 6), pointed out, outside observers are often more accurate in characterizing emotional feelings than the experiencing subject.[8] Jealousy, according to Hebb, is often readily detected by impartial observers but denied by the jealous person, who may instead characterize his state as indignation or annoyance.[9] Hebb noted that when observers agree and the subject disagrees about an emotional state, the conclusion of the observers is often a more reliable predictor of future behavior.[10]

The traditional way of using self-report to study feelings is to ask people to reflect back on some previous emotional experience. But assessments from memory are particularly problematic. Daniel Kahneman and colleagues have shown, in a variety of ways, that what we remember about an emotional experience is an imperfect reflection of what was actually experienced.[11] For one thing, people tend to remember how they felt at the end of an emotional episode rather than how they felt about the whole episode. In one study, patients undergoing a painful medical procedure, colonoscopy, were asked to rate their pain every sixty seconds. For some subjects, the procedure was prolonged for one minute, but under more comfortable conditions (the scope was stationary during this time). Later, after it was completed, they were asked to evaluate the entire experience. For subjects who had the stationary scope for the extra minute, and thus experienced less pain at the end, the remembered experience was less aversive, even though the procedure actually lasted longer. On the basis of this and many other studies, Kahneman concluded that the remembered emotional significance (what he calls the remembered utility) of an experience doesn't necessarily reflect the overall experience (its total utility).

Distortions in memory also extend beyond the intensity of experience and include the content of what is remembered. For example, studies by Elizabeth Loftus and others have shown that memories of emotional experiences are often significantly different from what actually happened during them. She reports many instances in which vivid memories of crime scenes turn out to be inaccurate, if unintentionally so; sometimes, again unintentionally, they are completely fabricated.[12] As Bartlett demonstrated long ago, memories are constructions assembled at the time of retrieval, and the information stored during the initial experience is only one of the items used in the construction; other contributors include information already stored in the brain, as well as things the person hears or sees and then stores after the experience.[13]

Because remembered experience is a distortion of actual experience, and because it's difficult to elicit real emotional experience in laboratory studies, some contemporary researchers emphasize the importance of obtaining immediate or on-line evaluations during real emotional episodes (measures of instant utility), especially in natural settings.[14] Though such studies circumvent the errors of remembered utility and the artificiality of laboratory research, they are extremely hard to implement, as well as being time-consuming, and in the end are still subject to the biases and measurement problems inherent in any method that relies on one's introspective evaluation of his or her own mental states.[15] The fallibility and subjectivity of introspection are, after all, what triggered the behaviorist revolution in psychology in the early twentieth century. And recall that the cognitive revolution only succeeded in bringing the mind back to psychology because it figured out how to study the mind without relying on introspection—by focusing on mental processes rather than mental content.

But if you want to assess the introspective content of subjective experience, there just aren't many alternatives to self-evaluation and verbal report. A subjective experience is by definition one that is known directly only by the experiencing person, and verbal descriptions of subjective states are the most direct way of assessing them. While the use of checklists or scales, which allow subjects to indicate their feelings by selecting from choices, sidesteps the need for verbal behavior as the final response, in the end they, too, depend on introspective evaluation of mental states, and inescapably involve the use of words to classify and categorize mental content. For example, the easiest way to identify which emotion you are feeling is to label it verbally as fear, anger, love, or disgust. Some even insist that you don't really know what you are feeling until you have labeled it.[16]

FEELINGS IN ANIMALS

Some have argued that an animal's behavior can be used as a kind of nonverbal self-report to assess its feelings. The neuroscientist Jaak Panksepp takes this approach in *Affective Neuroscience,* as does, more controversially, the psychoanalyst Jeffrey Masson in *When Elephants Weep.*[17] Their logic is as follows: Since animals and humans behave similarly when emotionally aroused (for example, similar fear responses are expressed by rats and people in the presence of danger), they must experience the same subjective states as well. If so, it would be possible to use behavioral responses in animals as indicators of feelings.

The flaw in this approach, which takes us right back to the credibility problem, is revealed by the results of a study of human heroin addicts.[18] The subjects were allowed to press a button to administer either saline or a high or low dose of morphine through an intravenous tube. The subjects did not know what was in the tube at any given point. Periodically, they were asked to rate how they felt. When a high dose of morphine was in the tube, they pressed vigorously and also reported feeling high. When saline was in the tube, the subjects pressed little and reported feeling nothing. But when the dose of morphine was weak, the subjects vigorously pressed the button in spite of the fact that they reported feeling nothing. Clearly, one would be misled by using behavior as a measure of what was felt in this case, since the subjects behaved but didn't feel. Emotional responses are not always external mirrors of internal feelings, but are rather controlled by more fundamental processes.

The problem is only compounded when we examine the relation of behavior to feelings in different species. Just because two creatures act the same does not mean they have the same experiences when they perform those actions. A beetle that finds itself under the approaching footstep of a human does what a human would do—it tries to escape before the foot lands. A robot can be programmed to do what a human would do if an object is sent flying toward his or her head—raise an arm to deflect the object. Do the beetle and robot feel fear, or do they simply express a defensive response? The fact that we have feelings when we act emotional does not mean that every act that looks emotional is accompanied by feelings. The renowned ethologist

Niko Tinbergen reached a similar conclusion: "Although . . . the ethologist does not want to deny the possible existence of subjective phenomena in animals, he claims that it is futile to present them as causes, since they cannot be observed by scientific methods. . . . Hunger, like anger, fear, and so forth, is a phenomenon that can be known only by introspection. When applied to another subject, especially one belonging to another species, it is merely a guess about the possible nature of the animal's subjective states."[19]

Given that we are left with nonverbal behavior as our main tool for assessing emotional states in animals, and that animal studies are the best way to study the brain, how can we escape the credibility problem and gain a richer psychological understanding of the brain's emotional circuits? How can we, in other words, study emotions using nonverbal emotional behavior as a measure without using it as a measure of feelings? Ironically, an important clue about how to do this comes from cognitive science. As we saw in the last chapter, cognitive science was successful because it figured out how to study the mind without getting bogged down in questions about subjective experience. The trick was to treat the mind as an information-processing device rather than as a place where experiences occur. Although early cognitive scientists considered emotions to be more a matter of mental content than of information processing, and thus not subject to cognitive analysis,[20] their processing approach is, in fact, directly applicable to the study of emotion. Just as it is possible to study how the brain processes the color or shape of a stimulus without first figuring out how the conscious experience of color or shape comes about, it is also possible to examine how the brain processes the emotional significance of a stimulus without necessarily first figuring out how that stimulus comes to elicit conscious feelings. Since emotions as processes can be studied in animals and humans alike, and since, as we'll see, emotional processing underlies both emotional behavior and emotional feelings, a processing approach is a way out of the credibility problem.

Even if this method allows us to account for important aspects of emotion in animals without having to resort to subjective states, it should not be taken to mean that subjective states exist only in humans. In the last chapter, I discussed how other animals might have domain-specific forms of consciousness, and in the case of nonhuman primates, domain-independent forms of nonverbal consciousness, but how only humans have verbal working memory, and thus language-based consciousness and the mental frills that language makes possible. The problem is that as soon as we rely on subjective states to explain behavior, we confront our inability to know whether such states really do exist in creatures other than humans. So my approach has been to discuss conscious emotions (feelings) only with respect to humans, and to restrict myself to the notion of emotional processing when I talk about creatures other than humans. In this way, I avoid the construction of a theory that can never be proven, but at the expense of having one that may be incomplete. I'm comfortable with this approach, though others prefer the alternative.[21]

EMOTIONAL PROCESSING

In order to stay alive, remain healthy, and propagate their species, animals have to be able to detect friend and foe, identify safe and nutritious foods, select mates, and respond appropriately to these and other stimuli and situations. Attempting to have sex with rather than defend against a predator would be costly. There must therefore be some mechanism, some circuit, located between input and output systems that translates environmental information into specific responses when certain kinds of stimuli occur. In fact, there must be a number of such circuits in the brain, since different systems are involved in defensive, feeding, and sexual behaviors, to name just a few categories of activities related to survival and well-being. When I use the term *emotional processing,* I have the functions of circuits like these in mind.

Specifically, from this point of view, emotion can be defined as the process by which the brain determines or computes the value of a stimulus.[22] Other aspects of emotion then follow from this computation. First, emotional reactions occur. These overt bodily responses and associated changes in internal body physiology are the advance guard of emotional responsivity. Subsequently (at least in humans), a feeling emerges as we become aware that our brain has determined that something important is present and we are reacting to it. In addition, given that we are in an emotionally arousing situation, we often take action. That is, we do things to cope with or capitalize on the event that is causing us to be emotionally aroused. Emotional actions, in other words, occur when emotions motivate us to do things. I'm going to focus on emotional reactions and feelings here, and save action for the next chapter, which is about motivation.

It's relatively easy to account in neural terms for how emotional reactions follow from emotional processing: information received by sensory systems activates emotional-processing circuits, which evaluate the meaning of the stimulus input and initiate specific emotional responses by triggering output circuits. Defense, food-seeking, and sex circuits receive inputs from the same sensory systems, and thus receive similar information, but a given circuit is only activated when the sensory influx contains stimulus information relevant to its operation (figure 8.1 depicts inputs to the amygdala, the centerpiece of the defense system). These detection and reaction processes take place automatically, independent of conscious awareness of the stimulus and feelings about it. This is why credibility is not a problem from the point of view of emotional processing.

The simplest way to illustrate the independence of emotional processing from consciousness in the control of emotional behavior is to describe an example in which consciousness is not a factor. At some point in your life, you've probably jumped out of the way of something rapidly approaching

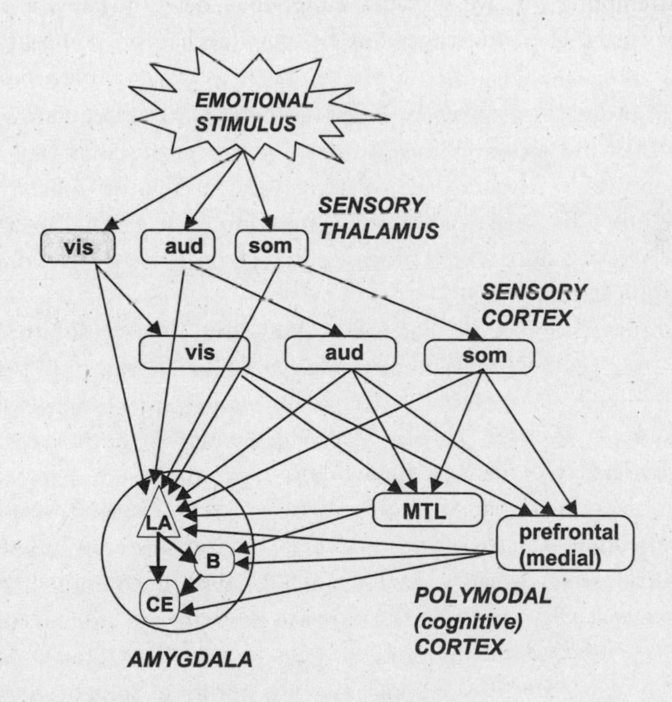

FIGURE 8.1 INFORMATION FLOW TO THE AMYGDALA

The amygdala receives low-level information about objects and events from sensory-processing regions in the thalamus, and more complex information from sensory-processing areas in the cortex. Abbreviations: vis, visual; aud, auditory; som, somatosensory; MTL, medial temporal lobe memory system; LA, lateral amygdala; B, basal amygdala; CE, central amygdala.

and only afterward noticed what it was—a ball thrown at you or a bus flying by, for example—and also only afterward noticed your heart pumping strongly. The feeling of fear came after you jumped and after your heart was already pumping—the feeling itself did not cause the jumping or the pumping. While an anecdote like this doesn't prove anything, there is also quite a lot of scientific evidence for this kind of reaction. The heroin addict study described above is one example. Let's consider another.

It's possible to present stimuli to the brain subliminally (unconsciously). This can be done in a number of different ways, but one commonly used is backwards masking.[23] In this procedure, an emotionally arousing visual stimulus is flashed on a screen very briefly (for a few milliseconds) and is then followed immediately by some neutral stimulus that stays on the screen for several seconds. The second stimulus blanks out the first, preventing it from entering conscious awareness (by preventing it from entering working memory), but it does not prevent the first from eliciting an emotional reaction (the stimulus still changes the beating of the heart or makes the palms sweat). Since the stimulus never reaches awareness (because it is blocked from working memory), the responses must be based on the unconscious processing of the meaning of the stimulus rather than on the conscious experience of it. By short-circuiting the stages necessary for the stimulus to reach consciousness, the masking procedure reveals processes that go on outside of consciousness in the human brain.

In creatures that lack the kinds of working memory processes that enable information to be held in mind consciously (see the discussion of the evolution of thinking in the last chapter), nonconscious processing is the rule rather than the exception. The basic behavioral repertoire of these creatures is regulated nonconsciously because they lack the kind of brain that can have a conscious experience. The human brain has these nonconscious capabilities as well.[24] By using behavioral reactions (and associated changes in body physiology, such as responses of internal organs) that occur in emotional situations to index or measure emotional processing in experimental animals, the brain mechanisms that underlie this processing can be uncovered. Studies of people with brain damage, and functional imaging studies of undamaged people, can be used to verify whether the same systems are involved in emotional processing in the human brain (it's much easier to verify that the same system is involved than to make the initial discovery in humans, which is why the animal studies are essential to lead the way). As long as we don't pursue aspects of emotion that are unique to the human brain, and as long as we choose animals that have emotional processes that are relevant to the human emotion we are interested in studying, we can pursue the brain mechanisms of human emotional processing in animal brains.

It's important to distinguish a processing approach to emotion from two other approaches. On the one hand, since a processing approach relies on overt responses, and sidesteps subjective experience, it may sound fundamentally behavioristic. A few years ago, after giving a lecture in which I argued for this approach, I was, in fact, accused of being a radical behaviorist (one who denies that the mind exists) disguised as a neuroscientist. But that description does not suit me for two reasons. First, I'm interested in what goes on in the black box,[25] and especially in how circuits in the brain represent and evaluate the meaning of emotional stimuli. I don't study behavior to understand behavior so much as to understand how processes in the brain work. This interest in internal processes alone disqualifies me from being considered a radical behaviorist. But, also, I'm very interested in feelings and consciousness, topics that behaviorists shunned. While behaviorists turned away from the analysis of the content of subjective states, I, in contrast, want to understand subjective phenomena in terms of underlying processes rather than as conscious content. Later, in fact, I will consider how feelings emerge from neural processes.

A processing approach may also be confused with the so-called cognitive approach to emotions, which treats emotions as appraisals—that is, as thoughts about a given situation.[26] While some appraisal theorists allow for unconscious appraisals (which is consistent with a processing approach), most emphasize appraisals as conscious thoughts and use verbal self-report to understand the nature of the appraisal process. This approach, obviously, takes us right back to the credibility problem. Conscious appraisals may indeed occur during an emotional state, but there are other, more fundamental processes at work as well. An understanding of these more fundamental processes is the goal of the processing approach.

In sidestepping the credibility problem, a processing approach thus allows us to study unconscious emotional functions similarly in humans and animals, and at the same time offers a method of understanding emotional consciousness (feelings) as well (since feelings themselves result from processes that occur unconsciously). In addition, a processing approach presents another advantage. It allows emotion and cognition to be treated the same (as unconscious processes that can but do not necessarily lead to conscious experiences), and it opens the door for the much needed integration of cognition and emotion (and, as we'll discuss later, motivation)—in short, the reassembly of the mental trilogy.

A QUICK FIX?

One way to begin to reassemble the mental trilogy might be to put all the newly acquired information about the thinking brain that came from the cognitive revolution together with the compelling view of the emotional brain provided long ago by the limbic system concept. Perhaps the notion of the limbic system simply needs to be modernized by treating it as an emotional-processing network rather than as the seat of conscious feelings. However, while the limbic system remains the predominant explanation (both in neuroscience and popular culture) of how the brain makes emotions, it is a flawed and inadequate theory of the emotional brain. I made this point very forcefully in *The Emotional Brain,* but those criticisms bear repeating.

The limbic system concept, brainchild of the pioneering neuroscientist Paul MacLean, was put forth in the context of an evolutionary explanation of mind and behavior.[27] It built upon the view, promoted by comparative anatomists earlier in the twentieth century, that the neocortex is a mammalian specialization—other vertebrates have primordial cortex, but only mammals were believed to have neocortex (chap. 3). And since thinking, reasoning, memory, and problem-solving are especially well developed in mammals, and particularly in humans and other primates that have relatively more neocortical tissue, these cognitive processes were believed to be mediated by the neocortex and not by the old cortex or other brain areas. In contrast, the old cortex and related subcortical regions form the limbic system, which was said to mediate the evolutionarily older aspects of mental life and behavior, our emotions. In this way, cognition came to be thought of as the business of the neocortex, and emotions of the limbic system.

The limbic system theory began to run into trouble almost immediately when it was discovered, in the mid-1950s, that damage to the hippocampus, an old cortical area and the centerpiece of the limbic system, led to severe deficits in a distinctly cognitive function, long-term memory.[28] This finding was incompatible with the original idea that the primitive architecture of the limbic system, and especially of the hippocampus, was poorly suited to participate in cognitive functions.[29] Subsequently, in the late 1960s, it was discovered that the equivalent of mammalian neocortex was present, though in a rudimentary form, in nonmammalian vertebrates (chap. 3). As a result, the old/new cortex distinction broke down, challenging the evolutionary basis of the assignment of emotion to the old cortex (limbic system) and cognition to the neocortex.[30]

The limbic system itself has been a moving target. Within a few years after its inception, the definition expanded from the original notion of the old cortex and related subcortical forebrain nuclei to include some areas of the midbrain,[31] and even some regions of the neocortex.[32] Several attempts have been made to salvage the limbic system by defining it more precisely.[33] Nevertheless, after half a century of debate and discussion, there are still no generally accepted criteria for stipulating which areas of the brain belong to the limbic system. Some scientists have suggested that the limbic system be abandoned.[34]

In spite of these difficulties, the limbic system continues to survive, both as an anatomical concept and as an explanation of emotions, in textbooks, research articles, and scientific lectures. This is in part attributable to the fact that both the anatomical foundation and the emotional function it was supposed to mediate were defined so vaguely as to be irrefutable. For example, in most discussions of how the limbic system mediates emotion, the meaning of the term *emotion* is not defined. Reading between the lines, it seems that the authors are often referring to something akin to the common English-language use of the term, which is to say feelings. However, as we've seen, a conception of emotion in terms of feelings is problematic.

Further, the anatomical criteria for inclusion of brain areas in the limbic system remain undefined, and evidence that any limbic area, in whatever definition, contributes to any aspect of any emotion has tended to be used as validation for the whole concept. For example, because the amygdala was included as one of the limbic areas, studies showing that the amygdala participates in fear were viewed as evidence that the limbic system theory was correct, in spite of the fact that many other limbic areas played little or no obvious role in fear or other emotions. In spite of many hundreds of experiments aimed at elucidating the role of limbic areas in emotion, there is still very little understanding of how our emotions might be the product of the limbic forebrain. Particularly troubling is the fact that one cannot predict, on the basis of the original limbic theory of emotion or any of its descendants, how specific aspects of emotion work in the brain. The explanations are all *post hoc,* concocted after a given experiment to explain the data—scientists typically put more stock in predictions than explanations. This problem is particularly apparent in recent work using functional imaging to study the human brain. Whenever a so-called emotional task is presented, and a limbic area is activated, the activation is explained by reference to the fact that limbic areas mediate emotion. And when a limbic area is activated in a purely cognitive task, it is often assumed that there must have, in fact, been some emotional component to the task. We are, in other words, at a point where the limbic theory has become an off-the-shelf explanation of how the brain works, one grounded in tradition rather than in facts. Deference to the concept is inhibiting creative thought about how mental life is mediated by the brain.

Although the limbic system theory is inadequate as an explanation of the specific brain circuits of emotion, MacLean's original ideas are insightful and quite interesting in the context of a general evolutionary explanation of emotion and the brain. In particular, the notion that emotions involve relatively primitive circuits that are conserved throughout mammalian evolution seems right on target. Further, the argument that cognitive processes might involve other circuits, and might function relatively independent of emotional circuits, at least in some circumstances, also seems correct. These functional ideas are worth preserving, even if we ultimately abandon the limbic system as an anatomical theory of the emotional brain.

LESS IS MORE

The limbic system theory failed in part because it attempted to account for all emotions simultaneously, and in so doing did not adequately account for any one emotion. When I got involved in emotion research in the late 1970s, I decided to take the opposite approach and study one emotion—fear—in detail. Below, I'm going to focus specifically on what we've learned about fear, because it is the emotion about which we know most. But the basic principles that have been uncovered about the fear system are likely to be applicable to other systems as well. Although different brain circuits may be involved in different emotion functions, the relation of specific emotional-processing circuits to sensory, cognitive, motor, and other systems is likely to be similar across emotion categories.

Much of what we've discovered about the neural basis of fear has come from studies of fear conditioning over the past two decades. This procedure, long a standard tool in behavioral psychology,[35] had not been used much to study the brain[36] until I and several other researchers adopted it as way of studying emotional learning circuits. (As mentioned in chapter 5, much of the key work on the neural basis of fear conditioning was done by the labs of Robert and Caroline Blanchard, Bruce Kapp, Michael Davis, and Michael Fanselow, as well as mine;[37] also important were earlier studies by David Cohen in pigeons.)[38] While some limbic areas did turn out to be involved in fear conditioning, the exact locations and the nature of their involvement would never have been predicted by the limbic system theory alone.

In the end, fear conditioning may not tell us everything we would like to know about fear, especially human fear. For example, the neural circuits involved in responding to conditioned fear stimuli may participate in, but are probably not sufficient to account for, more complex aspects of fear-related behavior, especially responses that depend not on specific stimuli but on abstract concepts and thoughts, such as the fear of failing, fear of being afraid, or of falling in love.[39] Nevertheless, fear conditioning has been an excellent way to start understanding some basic facts about fear, especially how fear responses are coupled to specific stimuli that people and other animals encounter in their daily lives.

What makes fear conditioning so useful from a circuit-tracing point of view? For one thing, it's a simple procedure. All it requires to turn a meaningless stimulus, like a tone pip, into a fear-arousing event is a few occurrences of the pip (often only one) at the same time as an aversive event, like a mild shock to the skin. Also, it's versatile—just about any stimulus that predicts shock or many other kinds of dangerous stimuli can serve as a conditioned fear stimulus. In addition, the learning is long-lasting, and maybe even permanent. And it can be administered similarly in humans and rats, making it possible to study the rat brain for the purpose of understanding human fear. Further, the responses are hardwired and automatic. We don't have to learn to freeze or raise blood pressure in the presence of dangerous stimuli, for the brain is programmed by evolution to do these things. We have to learn what to be afraid of, but not how to act afraid.

Implicit in all this, of course, is a strategy for mapping out the fear-processing circuit. All we have to do is trace the pathway forward from the input system (the sensory system that processes the conditioned stimulus, say the tone pip) to the output system (the system that controls freezing or other hardwired responses). The fear-processing circuits, by this logic, should be located at the intersection of the input and output systems.

NUTS IN YOUR BRAIN

Studies of fear conditioning have shown beyond a doubt that the brain region that sits at the intersection of input and output systems of fear, and the key to understanding how danger is processed by the brain, is the amygdala. Actually, there are two of them, one on each side of the brain, but, for convenience, we'll talk about the amygdala in the singular, since the two do pretty much the same thing.[40]

Once an obscure region of the brain, the amygdala is practically a household word these days. The Batman comic series *Shadow of the Bat* featured a monster called "Amygdala" who was named after the "almond-shaped mass of nerves in the brain that controls feelings of rage" ("almond" is indeed the English translation of *amygdala,* the Greek-derived word used to name this structure because of its appearance). Recently, a newspaper column called "Kids' City" discussed the role of the amygdala in childhood fears. A Website exists where you can "click your amygdala," that is, click on certain buttons to expose yourself to stimuli that will supposedly turn on your amygdala. One night, while channel surfing, I came across a Sci-Fi Channel show in which an alien was able to control people's fears by influencing their amygdalae (that's the plural). I have even been contacted by lawyers hoping to mount an "amygdala defense," an argument that the violent crime performed by their client was the fault of the client's amygdala rather than his free will. For better or worse, the amygdala is no longer obscure. But let's put these issues related to the amygdala's popularity aside for now and go a bit deeper into its synaptic organization and function.

As described in chapter 5, the amygdala contains a dozen or so distinct divisions or areas, of which only two are necessary for fear conditioning. Information about the outside world is transmitted to the lateral nucleus from sensory-processing regions in the thalamus and cortex, allowing the amygdala to monitor the outside world for signs of danger. If the lateral nucleus detects danger, it activates the central nucleus, which initiates the expression of behavioral responses and changes in body physiology that characterize states of fear (see figure 5.6 in chapter 5).

But what makes conditioning occur? We've already considered this in chapters 5 and 6, and will just summarize here. When a sound is presented to a naive animal, it reaches the lateral nucleus and mildly activates neurons there. GABA inhibition prevents much from happening in response, and if the sound is repeated without consequence, the cells quickly stop responding. But if the sound is followed by a shock, the weak preexisting response is greatly amplified following the rules of Hebbian plasticity: the shock activates the postsynaptic cell while the sound is causing the presynaptic terminals to release glutamate. During the activation, calcium enters the postsynaptic cell, and then a host of intracellular chemical reactions, involving kinases and transcription factors, activate genes, inducing them to make proteins that then stabilize the relation between the presynaptic and postsynaptic neuron. As a result of conditioning, the sound acquires the ability to elicit strong excitation in the amygdala, making it a more potent stimulus for activating amygdala circuits after conditioning than it was before. A stimulus that would normally not get past the GABA guard in the lateral nucleus thereby comes to travel with ease to the central nucleus, where the floodgates of emotional reactivity are opened.

PUTTING FEAR IN ITS PLACE

Clearly, much has been learned about the role of amygdala circuits in fear conditioning. With this information in hand, we can now begin to ask how processing in the amygdala relates to processing in cortical circuits involved in cognitive processing. We start with a consideration of the connections between the hippocampus and the amygdala, and their contribution to the contextualization of fear—that is, the regulation of fear on the basis of our assessment of the situation we are in.

In an emotionally charged situation, there is often some stimulus that stands out, and others that are less prominent, though nevertheless important. For example, if you are being robbed at gunpoint while visiting a foreign city, the most salient factor is the guy with the pistol pointed at you. But the context in which the robbery occurs is also significant—you may well feel uneasy if you go back to the street corner where it happened, or even back to that city or country. (My son and his friends were on their way to a basketball game at another school when one of the boys turned around and went home so that he would not have to cross an intersection where he tripped and broke two teeth several years earlier.) In the laboratory, when a rat is conditioned to a tone-shock combination in a certain chamber, it will freeze and otherwise act afraid if it simply finds itself back in that chamber at some later point. The tone was the most salient cue, but other cues in the chamber were conditioned as well.

The neural basis of so-called contextual conditioning has been studied extensively in recent years, especially by Russ Phillips in my lab and by Mike Fanselow and colleagues.[41] Like tone conditioning, contextual conditioning is dependent on the amygdala, but unlike tone conditioning, it is also dependent on the hippocampus. Just as the auditory system provides the amygdala with information about a tone, the hippocampus, by virtue of its role in relational/configural/spatial processing (chap. 5), provides the amygdala with information about a context in which the emotional learning is taking place. The context is, in other words, a psychological construction, a kind of memory created on the spot, about the various factors that constitute an emotional situation. And just as tone-shock integration occurs in the amygdala during conditioning, context-shock integration occurs there as well. However, while the lateral nucleus is involved in tone-shock integration, it is not necessary for context-shock integration. The basal nucleus (basal amygdala), which receives the connections from the hippocampus, is critical instead. The basal nucleus communicates with the central nucleus in the control of fear reactions to the context (fig. 8.2). And by way of these hippocampal-amygdala circuits, fear responses can be adjusted depending on the specifics of the situation: a beast in the wild elicits fear, but one in the zoo just fascinates. Beneath it all, though, is a reactive system that is still prepared to respond, as Charles Darwin found out when he tried, without success, to withhold a response to the strike of a poisonous snake behind a protective glass cage in a zoo.[42]

The idea that the hippocampus is involved in context conditioning has been challenged several times.[43] However, these challenges have so far been met, leaving intact the conclusion that connections between the hippocampus and the amygdala account for contextual processing in fear.

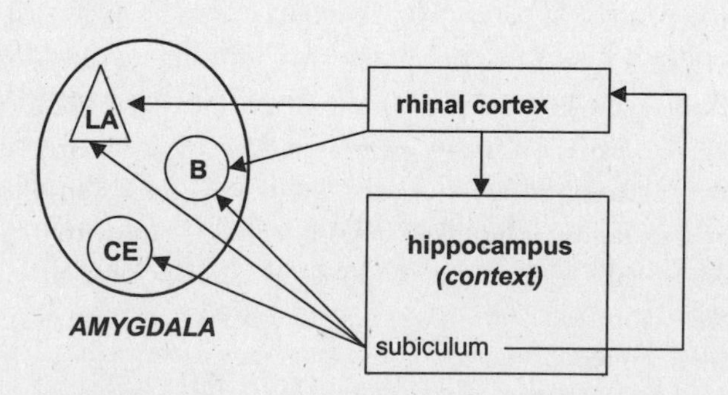

FIGURE 8.2 SOME WAYS HIPPOCAMPAL PROCESSING OF CONTEXT CAN INFLUENCE THE AMYGDALA

Evaluating the context of danger is believed to involve interactions between the hippocampus and amygdala. Information processed by the hippocampus can reach the amygdala through pathways originating in the rhinal areas of the cortex and subiculum. Abbreviations: LA, lateral amygdala; B, basal amygdala; CE, central amygdala.

CHECKS AND BALANCES

The amygdala also interacts with the medial prefrontal cortex. This region includes the anterior cingulate and orbital regions, as well as areas in transition between them (infralimbic/prelimbic cortex). These areas send connections to several amygdala regions, including the central nucleus, as well as to brain stem outputs of the central nucleus, allowing cognitive functions organized in prefrontal regions to regulate the amygdala and its fear reactions.

Several years ago, Maria Morgan of my lab pursued the role of the medial prefrontal cortex in fear regulation. She found that the consequences of damage to this region varied, depending on where the lesion was located. Some lesions led to a marked exaggeration of fear reactions—rats with such damage froze much more than controls each time the conditioned fear stimulus (the sound that had been paired with shock) appeared. In contrast, with other lesions, no such exaggeration of fear intensity occurred. However, when the sound was presented without the shock repeatedly until fear reactions no longer occurred, animals with the lesion required many more exposures to the sound than unlesioned animals to extinguish—to stop acting fearful.[44] The role of the medial prefrontal cortex in fear regulation recently has been confirmed in studies by Greg Quirk, my former colleague now in Puerto Rico,[45] and by Rene Garcia in France.[46]

Collectively, this work suggests that the prefrontal cortex and amygdala are reciprocally related. That is, in order for the amygdala to respond to fear reactions, the prefrontal region has to be shut down. By the same logic, when the prefrontal region is active, the amygdala would be inhibited, making it harder to express fear. Pathological fear, then, may occur when the amygdala is unchecked by the prefrontal cortex, and treatment of pathologic fear may require that the patient learn to increase activity in the prefrontal region so that the amygdala is less free to express fear. Clearly, decision-making ability in emotional situations is impaired in humans with damage to the medial and ventral prefrontal cortex,[47] and abnormalities there also may predispose people to develop fear and anxiety disorders. These abnormalities could be due to genetic or epigenetic organization of prefrontal synapses or to experiences that subtly alter prefrontal synaptic connections. Indeed, the behavior of animals with abnormalities of the medial prefrontal cortex is reminiscent of humans with anxiety disorders: they develop fear reactions that are difficult to regulate. Although objective information about the world may indicate that a situation is not dangerous, because they cannot properly regulate fear circuits, they experience fear and anxiety in these safe situations.

The medial prefrontal cortex may thus serve as an interface between cognitive and emotional systems, allowing cognitive information processing in the prefrontal cortex to regulate emotional processing by the amygdala. In addition, emotional processing by the amygdala may influence decision-making and other cognitive functions of the prefrontal cortex. We'll have more to say about prefrontal-amygdala interactions when we consider the relation between fear circuits and conscious feelings of fear later in this chapter.

OF RATS AND MEN

Bugs, slugs, shellfish, and other invertebrates lack an amygdala. Fear conditioning is taken care of in other ways in these creatures. But for vertebrates (at least reptiles, birds, and all varieties of mammals, including humans), the amygdala is responsible for fear conditioning.

It's much more difficult to study human than rat or other animal brains. However, in the last few years, studies of brain-damaged patients and studies using new brain-imaging techniques have examined the role of the human amygdala in fear and other emotional processes. These investigations still cannot pinpoint the contribution of specific circuits within the amygdala, but have nonetheless been very successful in showing that many of the basic fear-conditioning observations in animals apply equally to the human brain.

The landmark year was 1995, when two studies on the effects of brain damage on human fear conditioning were published. Collaborators of mine (Kevin LaBar, Liz Phelps, Dennis Spencer) working at Yale examined a series of about twenty patients who had undergone unilateral temporal lobectomy, a procedure in which large areas of the temporal lobe (including the amygdala) are removed on one side of the brain in an effort to control severe epilepsy.[48] Regardless of the side of the removal, the patients exhibited impaired fear conditioning. Around the same time, Antonio Damasio and colleagues reported that fear conditioning was disrupted in a patient who had a rare condition that resulted in her sustaining damage restricted to the amygdala on both sides of her brain.[49] In the patients in both our study and theirs, explicit or declarative memory for the conditioning experience was intact, indicating that fear conditioning and declarative memory are separable in the human brain, as in the rat brain (chap. 5). Also, damage to the hippocampus in humans, as it does in rats, disrupts fear conditioning to contextual cues.[50]

These findings on the effects of brain damage on fear in humans have recently been complemented by functional-imaging studies. One study was performed by LaBar, Phelps, and me, and a second was by John Morris, Arne

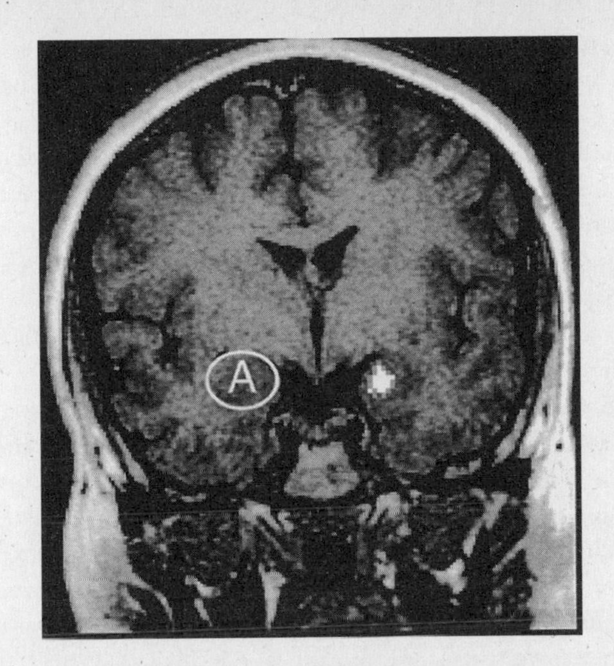

FIGURE 8.3 ACTIVATION OF THE HUMAN AMYGDALA DURING FEAR CONDITIONING

Functional MRI Imaging shows that the human amygdala is activated by the conditioned stimulus during fear conditioning. Increased neural activity is indicated by the white spot on the right side (although activity is only shown on one side, both sides are usually activated, though to different degrees). The approximate location of the amygdala (A) on the left is shown by the white circle. Picture provided by Elizabeth Phelps.

Öhman, and Ray Dolan.[51] Both used functional MRI and both found that the amygdala of humans was activated during fear learning (fig. 8.3). The Morris study had an additional interesting twist—the visual conditioned stimuli were masked, and thus never reached consciousness, showing that the human amygdala can undergo emotional learning to stimuli that are never experienced.

Some researchers, especially those who work on the neocortex, have argued that the direct pathway from the thalamus to the amygdala, the so-called low road, is unlikely to be very significant in the brains of humans and other primates.[52] Their logic is that the cortex is so important in primates that any effect of low-level subcortical processing would be overshadowed. However, a subsequent study by Dolan's group helps put this criticism to rest. Typically, most imaging studies look for areas where neural activity increases or decreases.[53] Suppose activity on the average increases in each of two areas, A and B, during three presentations of a stimulus. Suppose further that in area A the activity was high on the first trial, medium on the second, and low on the third. The average change would be a medium increase. The same average increase could occur in area B if the activity was low on the first, medium on the second, and high on the third. This pattern, though, would not support the conclusion that the two areas are functionally connected, since activity did not change the same way in the two areas on each trial. But if the two areas did change in the same way on each trial, then functional connectivity would be implied. Dolan and colleagues did this kind of analysis on the data they obtained in the masked conditioning study.[54] Specifically, they asked which brain areas changed in a way that would indicate connectivity with the amygdala. They found that across the whole brain amygdala activity during conditioning was most directly related to activity in subcortical visual-processing areas, including an area of the visual thalamus. Particularly significant was the fact that amygdala activity was not at all related to activity in areas of the visual cortex. This finding thus indicates that unconscious emotional learning occurs through the path from visual sensory areas of the thalamus to the amygdala. The low road is indeed used in both the rat and human brain.

Given all the parallels we've seen between the rat and the human, findings from rats should be assumed to apply to humans until they are proven not to, at least in the area of fear reactions. This doesn't mean that the human brain should, like the rat brain, be afraid of cats, but rather that the general wiring plan of the human and rat fear system is the same. As a result, the synaptic circuits that turn on the rat amygdala in the presence of cats will similarly turn on the human amygdala in the presence of stimuli that are dangerous to us.

In addition to studies on the role of the amygdala in conditioning, there have been a number of other imaging studies of humans that are relevant to the topic of fear. For example, it is well known that the expression of emotion on a human face is a potent emotional stimulus. Studies by Dolan's group in London,[55] as well as by Hans Breiter, Paul Whalen, and Scott Rausch,[56] found that exposure of human subjects to fearful or angry faces potently activates the amygdala. Whalen and colleagues found that even masked presentations of such faces cause such activation.[57] Along similar lines, damage to the human amygdala interferes with the ability to judge the emotions expressed in faces and voices.[58] People with amygdala damage, in fact, seem to have trouble in their daily lives deciding whom to trust.[59]

These latter findings in humans are reminiscent of an observation first made in the early 1900s and then popularized in a 1937 report by Kluver and

Bucy: monkeys with temporal lobe lesions that included the amygdala lost their fear of things they were normally quite afraid of, such as humans and snakes.[60] Somewhat later, Edmund Rolls, Taketoshi Ono, and other investigators recorded from neurons in the monkey amygdala in an effort to shed some light on the so-called Kluver-Bucy syndrome.[61] They found cells that responded specifically to faces and other kinds of biologically significant stimuli, like food items and threatening objects.

Recent work in humans by Liz Phelps and Paul Whalen has further implicated the amygdala in social interactions.[62] In separate studies, they found that exposure of white subjects to the faces of unfamiliar African Americans led to amygdala activation, and the degree of activation was directly related to the subjects' score on a test that measures racial biases. Particularly significant is the fact that the bias test was an implicit measure of racial bias. This suggests that implicit (unconscious) tendencies toward racism are reflected in the degree to which the amygdala is activated by stimuli representing the group biased against. This work is taking us into new and provocative areas, but is also raising serious ethical issues for researchers. Given that negative attitudes and biases have their strongest effects on behavior when they are unconscious, and thus cannot be guarded against and compensated for,[63] should researchers inform subjects of these biases? Such studies also force us to confront ethical decisions as a society. How far should we go in using brain imaging to read minds, and how should we use the information we discover? It is testimony to the progress being made that these questions need to be asked.

RELIVING THE EMOTIONAL PAST

Fear conditioning by the amygdala, as I've said many times, is an implicit form of learning, one that does not require conscious participation. During any experience in which we are awake and alert, however, working memory will be aware of what is going on, and if what is going on is significant, the executive will direct the storage of information about the situation in the explicit memory system. We are thus later able to consciously recall (retrieve into working memory) those aspects of the experience that were stored explicitly. While this is true of any kind of explicit memory, explicit memories about emotion are unique.

Explicit memories established during emotional situations are often especially vivid and enduring, and for this reason are called flashbulb memories.[64] The classic example is that most baby-boomers know where they were and what they were doing when they heard the news that JFK had been shot. But we are all aware from our own daily experiences that we remember particularly well those things that are most important to us, those things that arouse our emotions. Emotions, in short, amplify memories.

Studies by Jim McGaugh and colleagues spanning several decades have implicated the amygdala in the emotional amplification of explicit memory.[65] During emotional arousal, outputs of the central amygdala trigger the release of hormones from the adrenal gland that return to the brain. The amygdala, it turns out, is an important target of such feedback. The feedback consists of the direct action of body hormones (like cortisol released from the adrenal cortex) on amygdala neurons, as well as indirect actions whereby hormones (like epinephrine and norepinephrine released from the adrenal medulla) interact with nerves that travel from the body into the brain and ultimately reach and influence neural activity in the amygdala. By way of its connections with the hippocampus and other regions of the explicit memory system, the amygdala then modulates (strengthens) the consolidation of explicit memories being formed during emotional arousal. Later, the memories are more easily retrieved, and the details of the original experience are more vivid. Thus, in addition to *storing* implicit memories about dangerous situations in its own circuits, the amygdala *modulates* the formation of explicit memories in circuits of the hippocampus and related areas. Studies in recent years led by Larry Cahill and Benno Roozendaal in McGaugh's lab have helped to refine the latter conclusion.

According to the mood congruity hypothesis, memories are more easily retrieved when the emotional state at the time of memory formation matches the state at the time of retrieval.[66] For example, we are more likely to remember sad than happy events when depressed. Perhaps amygdala activation during retrieval facilitates remembrance by re-creating, at least in part, the emotional state (the state of the brain resulting from amygdala activation, and all its consequences, as discussed above) that occurred during the original experience—the more similar the pattern of activation is during learning and retrieval, the more efficient retrieval is likely to be. The unreliability of remembered emotion (Kahneman's remembered utility) and the fallibility of eyewitness testimony may be related to the fact that the emotional state at the time of the retrieval will by necessity be somewhat different from the state at the time of the original experience.

As long as the degree of emotional arousal is moderate during memory formation, memory is strengthened. But if the arousal is strong, especially if it is highly stressful, memory is often impaired. Studies by Robert Sapolsky,

Bruce McEwen, Gus Pavlides, David Diamond, Tracy Shors, and Jeansok Kim suggest that stress impairs explicit memory by altering the functioning of the hippocampus. Thus, during highly stressful conditions, the concentration of steroid hormone (cortisol) released from the adrenal cortex rises in the bloodstream.[67] For example, when the stress is induced by threatening stimuli, the amygdala is activated and cortisol is released (fig. 8.4). The hormone then travels to the brain and binds to receptors in the hippocampus, the net effect of which is to disrupt hippocampal activity, weakening the ability of the temporal lobe memory system to form explicit memories. Stressed rats, as a result, do poorly in tasks that require the hippocampus, such as spatial learning.[68] In addition, it is more difficult to induce hippocampal LTP in stressed rats.[69] If the stress continues, hippocampal cells begin to degenerate and ulti-

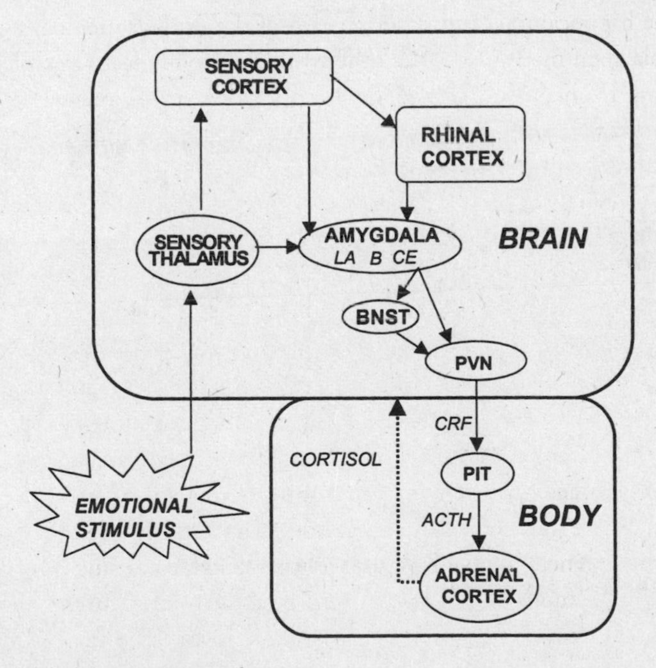

FIGURE 8.4 TURNING STRESS ON

In the presence of stressful stimuli, the central nucleus of the amygdala activates the paraventricular nucleus of the hypothalamus (PVN), either directly or by way of the bed nucleus of the stria terminalis (BNST). Corticotropin-releasing factor (CRF) is released by axons from the PVN into the pituitary gland (PIT), which in turn releases ACTH into the bloodstream, where it travels to the adrenal cortex. The adrenal cortex then releases cortisol, which travels in the bloodstream to various organs and tissue sites in the body, including the brain.

mately die. These changes appear to account in part for the memory disturbances typical of stress-related psychiatric conditions such as posttraumatic stress disorder, or PTSD, and depression.[70] Stress hormones also have an adverse impact on the prefrontal cortex,[71] and may contribute to the fact that people often make bad decisions under stress. In contrast to its effects on the hippocampus and prefrontal cortex, intense stress seems to enhance the amygdala's contribution to fear.[72] That stress hormones are involved in this amplification of fear by stress is suggested by studies in which the level of

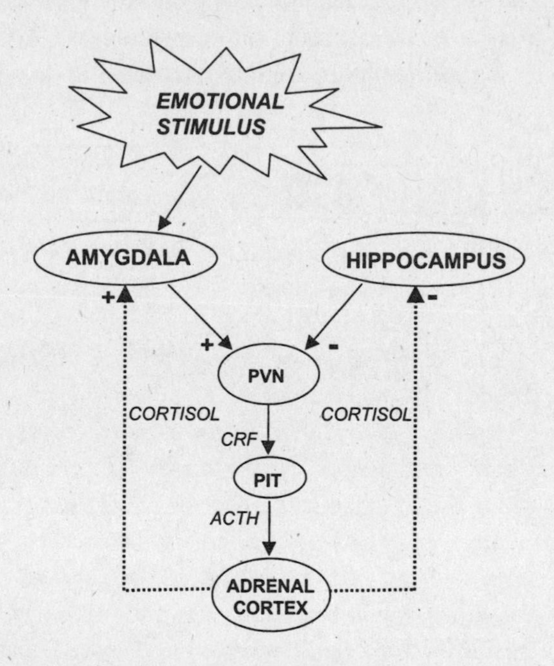

FIGURE 8.5 THE AMYGDALA AND HIPPOCAMPUS PLAY DIFFERENT ROLES IN STRESS

The amygdala initiates stress responses to threatening stimuli by activating the paraventricular nucleus of the hypothalamus (PVN) (fig. 8.4). PVN releases CRF into the pituitary gland (PIT), which in turn releases ACTH into the bloodstream. ACTH leads to the release of cortisol (CORT) from the adrenal cortex. CORT is transported to the brain by way of the bloodstream. CORT impairs hippocampal function (indicated by minus sign) and facilitates amygdala function (indicated by plus sign). Because the hippocampus normally inhibits the PVN (indicated by the minus sign) and the amygdala normally excites it (indicated by plus sign), the effects of CORT can lead to a feed-forward cycle where CORT release leads to more CORT release. That is, the ability of the hippocampus to slow release down is compromised at the same time the ability of the amygdala to stimulate release is facilitated.

stress hormones in the blood is artificially elevated, which has the same effect as stress itself on fear responses.[73]

In short, the exact conditions that lead to a weakened ability to form explicit memories, and to regulate fear by thinking and reasoning, can also amplify fear reactions and enhance our ability to implicitly store information about stressful or traumatic situations (fig. 8.5). There's good and bad news here. The good news is that even when the ability to form explicit memory is impaired, we can store useful information about harmful situations. The bad news is that if we don't know what it is we are learning about, those stimuli might on later occasions trigger fear responses that will be difficult to understand and control, and can lead to pathological rather than adaptive consequences. We'll return to this point when we discuss synaptic sickness in a later chapter.

FEAR ITSELF

Emotional arousal has powerful influences over cognitive processing. Attention, perception, memory, decision-making, and the conscious concomitants of each are all swayed in emotional states. The reason for this is simple: emotional arousal organizes and coordinates brain activity.[74] Here I want to focus on how the emotional coordination of brain activity converts conscious experiences into emotional experiences.

In the last chapter, I argued that our immediate conscious content, the thing we are conscious of at any one moment, is what occupies working memory. If this is correct, then it leads to the conclusion that a feeling (the conscious experience of an emotion) is the representation in working memory of the various elements of an immediate emotional state. In this view, the feeling of being afraid would be a state of consciousness in which working memory integrates the following disparate kinds of information: (1) an immediately present stimulus (say, a snake on the path in front of you); (2) long-term memories about that stimulus (facts you know about snakes and experiences you've had with them); and (3) emotional arousal by the amygdala. The first two are components of any kind of conscious perceptual experience, as the only way to identify an immediately present stimulus is by comparing its physical features (the way it looks or sounds) with memories about the same or similar stimuli. But the third kind of information occurs only during an emotional experience. Amygdala activation, in other words, turns a plain perceptual experience into a fearful one (fig. 8.6). Using Kahneman's term, the amygdala is the source of instant utility in a threatening situation.

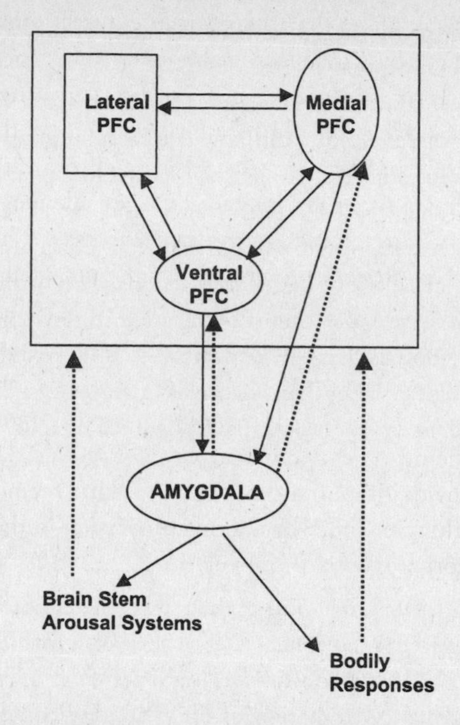

FIGURE 8.6 AMYGDALA CONNECTIONS WITH WORKING MEMORY CIRCUITS

Although the classic working memory area, the lateral prefrontal cortex, does not have direct connections with the amygdala, two other regions implicated in working memory do. These are the medial prefrontal cortex (especially the anterior cingulate region) and the ventral prefrontal cortex (especially the orbital cortex). Further, the three regions are interconnected. The dorsal prefrontal region may therefore have some indirect access to the amygdala through the other regions. In addition, working memory is indirectly influenced by outputs of the amygdala to brain stem arousal systems that release modulatory monoamines in all areas of the prefrontal cortex and by feedback from bodily responses initiated by amygdala activity.

The key question, then, is: How does the amygdala achieve this alteration of consciousness, this transformation of cognition into emotion, or, better yet, this hostile takeover of consciousness by emotion? The answer, I believe, is that emotion comes to monopolize consciousness, at least in the domain of fear, when the amygdala comes to dominate working memory.

The amygdala can influence working memory in a variety of ways, some of which will be described. The first is by altering sensory processing in cortical areas. Working memory finds out about the outside world from sensory-processing areas, so anything that alters how these areas process sensory stimuli will affect the material that is available to working memory. By way of connections with sensory-processing areas in the cortex, amygdala arousal can modify sensory processing. David Amaral has pointed out that while only the latest stages of sensory processing in the cortex send connections to the amygdala, the amygdala sends connections to all stages, allowing the amygdala to influence even very early processing in the neocortex.[75] That sensory cortex areas are influenced by activity occurring in the amygdala is suggested by Norman Weinberger's studies showing that the rate at which cells in the auditory cortex fire to a tone is increased when that tone is paired with a shock in a fear-conditioning situation,[76] together with studies performed by Jorge Armony and Greg Quirk in my lab demonstrating that damage to the amygdala prevents some of the cortical changes from taking place.[77] Because the sensory cortex provides important inputs to working memory, the amygdala can influence working memory by altering processing in the sensory cortex.

The sensory cortex is crucially involved in the activation of the medial temporal lobe memory system (chap. 5). By influencing the sensory cortex, the amygdala can have an impact on the long-term memories that are active and available to working memory. But the amygdala also has strong connections with the rhinal cortex (chap. 5), allowing it to directly influence the medial temporal lobe memory system and thus the memories available to working memory.

The amygdala can also act directly on working memory circuits. Although the amygdala does not have direct connections with the lateral prefrontal cortex, it does have connections with other areas of the prefrontal cortex involved in working memory, including the medial (anterior cingulate) and ventral (orbital) prefrontal cortex.[78] As described in the last chapter, connections within and between these regions constitute distributed circuits that underlie the integrative functions of working memory. Damage to the medial prefrontal cortex in rats, as already discussed, leads to a loss of fear regulation. And studies of monkeys and humans have implicated the orbital region in processing emotional cues (rewards and punishments) and in the temporary storage of information about such cues.[79] The orbital region is connected with the anterior cingulate, and like the anterior cingulate, it also receives information from the amygdala and hippocampus.[80] Humans with orbital cortex damage become oblivious to social and emotional cues, have poor decision-making abilities, and some exhibit sociopathic behavior.[81] In addition to being connected with the amygdala, the anterior cingulate and orbital areas are intimately connected with one another, as well as with the lateral prefrontal cortex, and each of the prefrontal areas receives information from sensory-processing regions and from areas involved in various aspects of implicit and explicit memory processing. The anterior cingulate and orbital areas thus provide a means through which emotional processing by the amygdala might be related in working memory to immediate sensory information and long-term memories processed in other areas of the cortex.

Attention and working memory are closely related (chap. 7), and recent studies by Liz Phelps have shown that amygdala damage interferes with an important aspect of attention.[82] Normally, if we are attending to one stimulus, we ignore others. This is selective attention, and it allows us to focus our thoughts on the task at hand. But if a second stimulus is emotionally significant, it can override the selection process and slip into working memory. Damage to the amygdala, though, prevents this from occurring. The amygdala, in other words, makes it possible for implicitly processed (unattended) emotional stimuli to make their way into working memory and consciousness.

In addition, the amygdala can influence working memory indirectly by way of projections to the various amine cell groups in the brain stem and forebrain that participate in cortical arousal, including cholinergic, dopaminergic, noradrenergic, and serotonergic systems. In the last chapter, we discussed the importance of dopamine and norepinephrine in the regulation of working memory. And as we saw above, norepinephrine plays a critical role in the amplification of explicit memory during emotional states. These arousal pathways are relatively nonspecific since they influence many cortical areas simultaneously. Specificity comes from the fact that the effects of arousal are most significant on circuits that are active. As a result, if the cortex is focused on some threatening stimulus, the circuits involved will be facilitated by the arousal systems. This will help keep attention focused on the threatening situation.

Finally, once the outputs of the amygdala elicit alarm-related behaviors and accompanying changes in body physiology (fight/flight kinds of responses), the brain begins to receive feedback from the bodily responses. Feedback can be in the form of sensory messages from internal organs (visceral sensations) or from the muscles (proprioceptive sensations) or can be in the form of hormones or peptides released by bodily organs that enter the brain from the bloodstream and influence neural activity. Although the exact manner in which bodily feedback influences working memory is not clear, it is likely that working memory has access to this information in one form or another. But the feedback from these responses is relatively slow, on the order of seconds, when compared to the feedback that occurs by way of synaptic transmission within the brain, which transpires within a matter of milliseconds. Body feedback adds at least intensity and duration, but may also help refine our interpretation of the emotion we are experiencing, once the episode has been triggered.[83] As we saw above, bodily feedback in the form of stress hormones can either enhance or impair long-term memory functions of the temporal lobe memory system, which will in turn influence the content of working memory.

In the presence of fear-arousing stimuli, activation of the amygdala will lead working memory to receive a greater number of inputs, and inputs of a greater variety, than in the presence of emotionally neutral stimuli. I propose that these extra inputs add affective charge to working memory representations, and are what make a particular subjective experience a fearful emotional experience.

But what about animals without a well-developed prefrontal cortex? Do they have any kinds of emotional experiences? I argued in the last chapter that it might be possible to have certain kinds of modality-specific conscious states when the activity of one system dominates the brain. This might happen with strong sensory stimulation (loud noise or a painful stimulus), or in response to emotionally charged stimuli (sight of a predator). Modality-specific feelings can be thought of in terms of passive states of awareness, as opposed to the more flexible kind of conscious awareness, complete with on-line decision-making capacities, made possible by working memory.

Although my theory of emotional experience is based on studies of fear, it is meant as a general-purpose theory that is applicable to all kinds of emotional experiences. The particulars will be different, but the overall scheme (whereby working memory integrates sensory information about the immediately present physical stimulus with memories from past experiences with such stimuli and with the current emotional consequences of those stimuli) will apply to all varieties of emotional experience in humans, from fear to anger to joy and dread, and even love.

BEYOND FEAR: THE BRAIN IN LOVE

Whenever I give a lecture on emotions and the brain, the question I am most often asked at the end is: "Is the amygdala involved in emotions besides fear, especially positive emotions?" This is not a question that my work has addressed directly. I've been trying to understand specifically the mechanism of fear rather than how the amygdala works in general. But other researchers have studied the role of the amygdala in processing stimuli that predict desirable things (like tasty foods and sexually receptive partners). Included is the work of Barry Everitt and Trevor Robbins at Cambridge University, David Gaffan and Edmund Rolls at Oxford, Norman White at McGill, Michela Gallagher at Johns Hopkins, and Taketoshi Ono at Toyoma University in Japan. However, the implications of this work in animals for understanding specific human emotions are less clear than the relation of findings on fear-processing in animals to human fear. Nevertheless, this research is extremely important and will be a major part of any thorough understanding of how the brain creates emotions. We'll consider some of this work in the next chapter when we explore how behavior is motivated by emotional processing.

In the rest of this chapter, though, I want to turn to a very interesting set of studies that have begun to approach the neural basis of an emotion that is near and dear to humans—namely, love.

The key to understanding how any mental or behavioral function works in the brain, as I've said often in this book, is to be able to study it in experimental animals. Love might therefore seem to be an unlikely topic for brain researchers, one particularly vulnerable to the credibility problem. The key issue is whether there is some way to study the function in animals that makes sense in terms of human behavior. For fear, we were able to use conditioning because conditioned fear responses are similar in humans and other mammals. In the case of love, though, the situation is more complicated for several reasons, not the least of which is the fact that most animals don't pair up with one another exclusively. Not only did researchers have to find ways to study love behaviorally, they also had to find species that were monogamous.[84]

Only about 3 percent of mammals are monogamous, and even within nonhuman primates, monogamy is fairly rare. One species whose members do remain together as a pair after mating to raise their offspring as a family, even across generations, is the prairie vole, a small rodent living in the midwestern plains in the United States. Given that pair-bonding is so rare, the monogamous prairie vole offers a possible window into the biology of attachment.

Attachment (pair-bond formation) is a key element of love.[85] Perhaps the synaptic mechanisms that underlie attachment in voles are also at work in other species that are fairly monogamous, namely, us.

Vole researchers used a completely different strategy than the one commonly used to study fear. Rather than starting with the circuits and then trying to figure out the chemistry, they started with chemical findings and have attempted to relate them to circuits. Much of the work has been conducted by Tom Insel, Sue Carter, and their colleagues and students.[86] It's also worth noting that Insel credits his mentor, Paul MacLean, for intellectual inspiration. MacLean's ideas thus continue to motivate research on the emotional brain more than half a century after his introduction of the limbic system concept.

In a recent article, Insel pointed out two features of prairie voles that made them attractive for studying pair-bonding.[87] The first is that monogamy even occurs in voles living in laboratory settings. If bonding only occurred in the wild, it would be very difficult to study its neural basis. In the laboratory, bonding can be measured by putting a vole in the middle chamber of a box with three compartments. The vole is free to travel to the compartment on either side. In one of these, it encounters its mate, and in the other, a stranger is present. Voles that have mated choose to be with their partner, whereas unbonded ones have no particular preference. After bonding, if another vole comes into the area, the male engages in mate protection, attacking the intruder.

The second feature of voles that has helped move this work forward is the fact that pair-bonding is present only in prairie voles and not in a closely related creature, montane voles, which are found in the Rockies and live individually rather than in family groups. These animals don't form mate preferences after sexual intercourse, so that when placed in the three-chamber box, they don't spend more time with a vole they mated with than a novel one, and the males don't attack intruders after sex. Differences in the brains of these two kinds of voles might therefore provide important clues about the biology of pair-bonding, family organization, and perhaps love itself.

One of the main discoveries was that receptors for two hormones believed to play an important role in reproductive behavior were located in different circuits in prairie and montane voles.[88] These hormones, called vasopressin and oxytocin, are found only in mammals, and are related to ancestral hormones that play a key role in behaviors like nest-building in nonmammalian species. In mammals, they continue to be significant in reproductive behavior. For example, oxytocin is involved in uterine contractions during labor and in milk release during nursing. In the brain, though, these chemicals function not just as hormones, but also as neurotransmitters and/or modulators, being released from nerve terminals and binding to postsynaptic receptors.

For now, we'll skip over the exact location of the circuits in which the chemicals are present, as the relation of the chemicals, circuits, and behavior is still not perfectly clear. What is clear, though, are the functions of these chemicals in the behavioral differences between the voles. This has been determined by injecting drugs into the brain that either stimulate or inhibit the action of vasopressin or oxytocin. The drugs have not been injected into specific brain areas but into the ventricles, cavities that contain CSF (cerebrospinal fluid), which flows from the ventricles into the spaces surrounding neurons. With such injections, the drug will reach widespread areas of the brain and can then influence neural function in those areas that contain cells with appropriate receptors. When a drug that blocks the action of naturally released oxytocin is placed in the ventricles of a female prairie vole just before mating, she mates but does not bond with the sex partner. The drug disrupts attachment, not sex, suggesting that oxytocin released during mating underlies bond formation in females. In contrast, if a drug that blocks vasopressin is placed in the ventricles of a male prairie vole before mating, the male likewise mates but doesn't bond, and also doesn't engage in mate-guarding aggression. But if the same drug is injected after mating, intruder aggression does occur. This pattern of results in males demonstrates that the drug blocks attachment and not sexual or aggressive responses. Thus, blocking oxytocin in female and vasopressin in male prairie voles causes them to behave like montane voles. Oxytocin only affects bonding in female brains, and vasopressin only affects bonding in male brains; the female sex hormone estrogen is key to oxytocin's action, just as testosterone is essential for the normal function of vasopressin.[89]

While oxytocin and vasopressin are also present in the brains of humans and are released during sexual behavior, they have not yet been proven to underlie attachment. Regardless of whether the vole findings on oxytocin and vasopressin end up being completely applicable to the human brain, this work illustrates important principles that will surely guide research for some time.

An important area for future work is to pinpoint the exact circuits in which vasopressin and oxytocin are acting during sex-related pair-bond formation. Many studies have been performed on the neural basis of sexual behavior, especially in rodents. Implicated are areas within the amygdala (medial and posterior nuclei), the so-called extended amygdala (an extension of the amygdala into the bed nucleus of the stria terminalis), the striatum (especially the nucleus accumbens), and the hypothalamus (including ventromedial, medial preoptic, paraventricular, and supraoptic areas).[90] Given that these regions are intimately connected and that receptors for oxytocin and vasopressin are present on neurons in several of them, it is tempting to think of them as forming synaptic circuits crucial not just for sexual behavior, but also for pair-bonding, but this remains unproven.

Areas of the amygdala are included in both the fear and sex circuits. However, the circuits are otherwise quite distinct. Even within the amygdala, different areas are involved in sex (medial and posterior nuclei) and fear (lateral and central nuclei). This finding emphasizes the importance of mapping the circuit for different kinds of emotional systems rather than assuming that there is a universal circuitry for all emotions. At the same time, different emotion circuits, like the fear and sex circuits, sometimes interact with one another. For example, the medial nucleus sends connections to the central nucleus,[91] where oxytocin receptors are present.[92] This may be related to the ability of both oxytocin and positive social interactions to reduce fear and stress.

Pair-bonding in animals has provided researchers a way of studying something akin to love without having to confront the credibility problem that inevitably arises when any emotion is considered. But in the end we want to know not just about attachment behavior but also about the particular feelings of love. Although we have little research to draw upon at this point, we can use our more detailed understanding of cognitive-emotional interactions in fear to speculate about how the brain does feel love.

Suppose you unexpectedly see a person you care about. Suddenly, you feel the love you have for that person. Let's follow the flow of information from the visual system through the brain to the point of the experience of love as best we can. First of all, the stimulus will flow from the visual system to the prefrontal cortex (putting an image of the loved one in working memory). The stimulus also reaches the explicit memory system of the temporal lobe and activates memories about that person. Working memory then retrieves relevant memories and integrates them with the image of the person. Simultaneously with these processes, the subcortical areas presumed to be involved in attachment will be activated (the exact paths by which the stimulus reaches these areas is not known, however). Activation of attachment circuits then impacts on working memory in several ways. One involves direct connections from the attachment areas to the prefrontal cortex (as with fear, it is the medial prefrontal region that is connected with subcortical attachment areas). Activation of attachment circuits also leads to activation of brain stem arousal networks, which then participate in the focusing of attention on the loved one by working memory. Bodily responses will also be initiated as outputs of attachment circuits, and contrast with the alarm responses initiated by fear and stress circuits. We approach rather than try to escape from or avoid the person, and these behavioral differences are accompanied by different physiological conditions within the body.[93] This pattern of inputs to working memory from within the brain and from the body biases us more toward an open and accepting mode of processing than toward tension and vigilance.[94] The net result in working memory is the feeling of love.[95]

This hypothesis is probably incomplete, but it is probably not completely wrong. It shows how we can build upon research on one emotion to generate hypotheses about others. Given that so much of who we are is defined by our emotions, it is important that we uncover as much as we can about the brain mechanisms of many emotions. This task is just beginning, but the future is bright.

CHAPTER NINE

THE LOST WORLD

EVERY WHY HATH A WHEREFORE.

—Shakespeare, *The Comedy of Errors,* act 2, scene 2

Why do we do the things we do, think the things we think, and make the decisions we make? *Why* questions—questions about our motivations—are fundamental to understanding what makes each of us unique. Following the cognitive revolution, though, motivations, like emotions, were overlooked by many psychologists and brain scientists. The rehabilitation of motivation in both these fields has been slower than that of emotion, but it is also under way. And, as with emotion, the pressure to revive research on and theory about motivation is coming as much if not more from brain science as psychology. So let's rediscover the lost world of motivation, part three of the mental trilogy, from the perspective of how it works in the brain.

ACTING OUT

On a warm summer evening in 1996, a crowd was enjoying a concert in Olympic Park in Atlanta. Suddenly, a bomb exploded. The scene, captured on video, was shown repeatedly on CNN. As soon as the explosion occurred, nearly everyone in the crowd reacted by freezing. After remaining motionless for several seconds, some people starting running away, and soon everyone was in motion.

This short segment of video is a good demonstration of the way an emotional episode unfolds over time. In situations of sudden danger, we initially react by using evolutionarily programmed responses, like freezing, that have long been successful in keeping organisms like us alive. These reactions are elicited, not emitted willfully, and occur automatically, before we have time to think. But we can't freeze forever; sooner or later, we have to take action. In the last chapter, we considered how emotional reactions are automatically triggered by external stimuli. In this chapter, we focus on how emotions, once activated, motivate us to do things, to take actions.

Motivation has many definitions. I use the term here to refer to neural activity that guides us toward goals, outcomes that we desire and for which we will exert effort, or ones that we dread and will exert effort to prevent, escape from, or avoid. Goals direct action, and can be as concrete as a specific stimulus (for example, a particular consumer product) or as abstract as a belief or idea (for example, the belief that hard work will lead to success, or the idea that freedom is worth dying for).

Goal objects are also called incentives. Some are intrinsically motivating (as in the case of food, water, and painful stimulation), while others acquire motivating properties through our experiences with them. In the latter case, they are called secondary incentives. These can arise by association (when a stimulus with low value occurs in connection with one of higher value, as in classical conditioning), by observational learning (seeing the way a stimulus affects other people), by word of mouth (hearing about whether something is good or bad), or by sheer force of imagination.

The view of motivation that I'm going to pursue here is that incentives do their motivating by activating emotion systems. Freezing in reaction to a bomb reflects the activation of an emotional system, whereas running away after several seconds of freezing reflects the motivational consequences of that activation (fig. 9.1). Actions motivated by emotional arousal have a purpose—to deal with the emotion being aroused. Whether all motivated actions are necessarily based on emotional activity is debatable. But that emotions are powerful motivators seems indisputable.

Goal-directed behavior is best thought of in functional terms. While we normally try to escape from or avoid harmful stimuli, in some cases, we have to actively engage a dangerous object in order to achieve protection. With motivated behavior, especially in humans, it often is the relation of the act to the goal, not the act itself, that is key. In fact, for humans, most motivated acts are only arbitrarily related to particular goals. When hungry, we can ob-

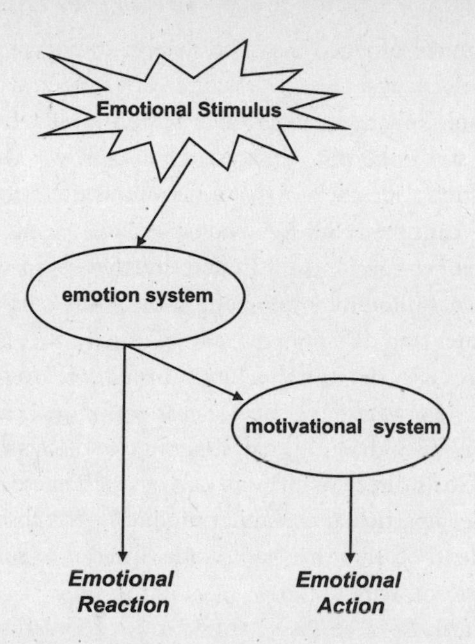

FIGURE 9.1 EMOTIONAL STIMULI ELICIT EMOTIONAL REACTIONS AND MOTIVATE ACTION

Activation of emotional-processing systems by emotional stimuli has two consequences. One is the elicitation of emotional reactions (automatic preprogrammed responses). The second is the activation of motivational systems that in turn guide actions (instrumental responses that are based on either past learning or instantaneous decisions).

tain food in many different ways—walking or driving to a restaurant or a store, calling for a delivery, asking a friend to pick up something. Each type of activity is appropriate under certain circumstances, but none has a predetermined, inevitable relation to food acquisition and consumption. The same general kinds of activities (walking, driving, calling, asking) can be used to achieve many varied kinds of goals. Figuring out what to do under different sets of circumstances in order to achieve your goals is what life is all about. Learning and motivation are thus closely intertwined topics.

PUSH AND PULL

My interest in the topic of motivation was sparked by seeing the freezing crowd take flight in the video of the Olympic bombing. This behavior reminded me of what happened in a classic experiment performed by Neal Miller in the 1940s.[1]

Miller placed rats in an apparatus that consisted of two compartments, one painted white and the other black, separated by an open doorway. The rats started out in the white area, where they received a shock after a delay. At first, they simply froze. But then they figured out that they could escape from the shock by going into the black area.[2] Eventually, they learned that they could avoid the shock altogether by running out of the white compartment as soon as they were placed in it. The rat came to routinely—in fact, habitually—perform the response, even if the shock was turned off. Then, in a later phase of the study, the door between the two areas was closed, rendering the old habit no longer useful. But through trial-and-error behavior, the rats learned that if they turned a small wheel, the door opened. They thus learned a new habit.

Miller interpreted his results in terms of the predominant theory of motivation at the time, Clark Hull's drive theory.[3] In Hull's view, all learning involves the reduction of basic drives (like hunger, thirst, sex, or pain), and current behavior is therefore a product of drive reduction in the past. That is, what we do today in a certain situation is a function of what we did in the past that was successful in reducing drives in similar situations. This is essentially a psychological version of the philosophical position known as hedonism, the idea that we live our lives in such a way as to seek pleasure and avoid pain.[4]

In the late nineteenth century, the pioneer experimental psychologist Edward Thorndike used hedonism as the basis for a psychological methodology.[5] Thorndike showed that hungry cats would learn complex behavioral responses that had no natural relation to feeding but that allowed them to open a door and obtain a bite of food that was visible outside the cage. The cats basically went through a process of trial and error, and those responses that opened the door were then later repeated. Thorndike called this the law of effect: behaviors that are effective in obtaining desirable goals and avoiding undesirable ones are rewarded or reinforced and then repeated, while those that fail to obtain desirable goals or that lead to undesirable ones are punished and are not repeated.

The kind of learning task used by Thorndike came to be called instrumental conditioning—referring to the fact that the behavior is instrumental in achieving the reward or punishment. Because the rewarding or punishing stimulus is associated with the behavioral response, instrumental conditioning is called stimulus-response learning. Instrumental conditioning contrasts with Pavlovian or classical conditioning, where the rewarding or punishing stimulus occurs regardless of what the animal does. The association formed is thus not between the rewarding or punishing stimulus and a response, but between neutral stimuli and the reward or punishment they occur with and predict. Hence, Pavlovian conditioning is called stimulus-stimulus learning.

For example, Pavlov's dog salivated to the sound of the bell because that sound had been previously associated with food. The dog did nothing to obtain the food, which simply appeared when the bell sounded (this is a stimulus-stimulus association). On the other hand, if the dog had learned that walking over to a certain spot and pressing a lever with its paw when the bell sounded would lead to a real piece of food, that behavior would be an instrumental response, rewarded by its success in obtaining food (this would be a stimulus-response association). Together, these two forms of learning became the foundation of behavioral psychology, and behaviorists like John Watson, B. F. Skinner, and Hull hoped to explain all of human behavior in such terms.[6]

Hull's particular contribution to the field was to reinterpret the law of effect in terms of drive reduction, arguing that new instrumental behaviors, new habits, are learned and repeated because they reduce drives. But for Hull, drive only activated or aroused behavior; it did not direct behavior toward specific stimuli that would reduce the drive. Behavioral direction in the presence of drive-elicited behavioral activation was instead based on learned habits, responses that had been successful in reducing drives in previous circumstances.

In this framework, what was the drive that was reduced in the last part of Miller's study, where there was no shock? What moved the rats to learn the new response? Being a drive theorist, Miller assumed some drive must have been reduced, since learning occurred. Because the shock was never presented during the last phase of learning, however, pain prevention could not have been the drive being satisfied. Miller argued that the drive being reduced must have been fear. Fear, in his view, was a learned or acquired drive, as opposed to a biological one (that is, an unlearned, innate or instinctual drive).

The notion of acquired drive added tremendous flexibility to drive theory, since most human behavior is not motivated by pain or deprivation of essential nutrients. Money, for example, has no intrinsic value, and acquires its value only from a mutual acceptance of its worth by those who use it. Similarly, words of praise or scorn are not innately motivating, and achieve their results only by convention.

But drive theory, even with Miller's clever addition, remained problematic for a variety of reasons.[7] One of the main difficulties came from studies showing that rats would learn maze problems to obtain saccharin.[8] Saccharin has incentive value (rats will work to obtain it) due to its sweet taste, but it is non-nutritive and thus cannot satisfy hunger and reduce the supposed drive activated by the biological need for food.[9] As a result of this and other research, drive theory eventually was replaced by the incentive theory. While drives push us from within, incentives pull us from without.[10]

The advantage of an incentive view is that it avoids the postulation of a hypothetical drive state that has to be reduced in order for learning to occur. While incentive theories are therefore less cumbersome, they have their own shortcoming—they often replace the drive problem with the credibility problem (chap. 8). That is, they assume that subjective hedonic experiences, emotional feelings, are what motivate behavior.

The idea I want to develop here is that motivation can be thought of in terms of incentives without assuming that feelings are necessary to translate incentives into actions. I believe that all we need to accept is that in the presence of conditioned (learned) or unconditioned (innate) incentives, emotion systems are activated, placing the brain in a state where an instrumental response becomes a highly probable outcome. In this view, we don't have to postulate the existence of hypothetical concepts like drives or subjective states to explain motivated action. All we need to talk about is real brain systems and their functions.

The brain can be thought of as having a variety of systems that it uses to interact with the environment and keep itself alive. I've used the term *emotion systems* to characterize many of these, but what's relevant here is not the label we give them but the particular function they perform. Included are systems that detect and respond to predators and other dangers, to sexual partners, to suitable food and drink, to safe shelter, and so on. The stimuli (incentives) that activate these systems do so either because of biologically predetermined factors or because of past learning by the individual. When the brain is activated by either an unconditioned or a conditioned incentive, animals, including people, are motivated to perform instrumental responses. These emotionally primed instrumental responses have as their goal, their motive, the alteration of the brain state, the emotional state, that the organism is in.

So long as we resist the temptation to think of emotional states in subjective terms, and instead conceive of them in terms of brain states, we remain on solid ground and avoid the perilous credibility problem discussed in the last chapter.

FEARFUL HABITS IN THE BRAIN

In the years following Miller's classic studies, researchers used aversive instrumental conditioning procedures to study the brain mechanisms of fear.[11] In such studies, rats or other animals learn to do things that avoid the delivery of a shock. Once they learn how to avoid the shock, they perform the response habitually. Given that the hallmark of fear or anxiety disorders is habitual avoidance of situations that might lead to harm or anxiety, this approach seemed to offer the opportunity to understand the neural basis of a clinically relevant kind of learning. However, as mentioned previously, studies of avoidance conditioning failed to lead to a clear understanding of the neural basis of fear. Little effort was made to distinguish the contribution of neural systems to the two kinds of learning that take place in these tasks—initially, the subject undergoes classical fear conditioning, where cues in the apparatus come to be associated with the shock, then an instrumental avoidance habit is learned on the basis of its ability to remove the animal from the situation in which shock is likely.[12] The notion of multiple memory systems (chap. 5) had not yet emerged, and it wasn't fully appreciated that different kinds of learning involve different brain systems. Once researchers started focusing on fear conditioning on its own, stripped out of the context of avoidance conditioning, progress was swift. And as the popularity of fear conditioning increased, the more complex avoidance-conditioning procedures, which were more difficult to relate to brain mechanisms, fell out of favor.

But fear conditioning tells us only about how fear reactions work. If we want to understand how the arousal of an emotion system motivates behavior, we need to turn to instrumental tasks, like avoidance, where the brain learns to take action. With that in mind, after seeing the Olympic bombing video, and being reminded of Miller's rats first freezing and then escaping, I decided several years ago to revisit the question about how fear-motivated instrumental learning—fear habit learning—works in the brain. I didn't, however, want to study just any fear habit-learning task, but one that would enable us to build upon our extensive knowledge of the neural basis of conditioned fear, especially conditioned fear elicited by a tone previously paired with a shock.

Karim Nader, Prin Amorapanth, and I chose to study this[13] using a task called escape from fear.[14] In this procedure, the rats first underwent fear conditioning to a tone-shock combination. As in our other studies, we used the extent to which they froze in the presence of the tone as an index of learning in the fear-conditioning phase. The next day, the animals were placed in a new chamber. After a delay, the tone came on. At first, they froze to the tone. But, eventually, they moved around a bit, and through trial and error, they discovered that movement to the other side of the chamber terminated the tone. Over many trials, the rats learned to escape to the other side immediately to minimize the exposure to the sound. The rats thus learned to take action on the basis of having first learned that tone was associated with shock. The shock, by the way, was never presented in the new chamber. The tone had become a negative secondary incentive, and behaviors that eliminated this stimulus were reinforced and learned as a habit. The design of this experiment was very similar to that of Miller's study, except that we used a tone as the fear-arousing stimulus rather than contextual cues in the apparatus itself. We were now ready to explore the brain mechanisms through which fear system activation motivates action.

Recall from earlier chapters that in order for tone conditioning to occur, the tone has to be transmitted to the lateral nucleus of the amygdala. This region then sends signals to the central nucleus, which controls fear responses. So the first question we asked was whether damage to the lateral or central nucleus would also disrupt the conditioning of the instrumental escape response. The results were clear. Damage to the lateral nucleus prevented the learning of the escape response, but central nucleus lesions had no effect. Thus, the outputs of the central nucleus to the brain stem, required for fear reactions, are not used to learn to do new behaviors, to take action, on the basis of fear system activation. We therefore asked whether damage to one of the other targets of the lateral nucleus—namely, the basal nucleus—would have an effect. This lesion interfered with the ability of the rats to use the information learned during fear conditioning to initiate protective action. It's important to note that basal nucleus damage had no effect on the animal's ability to freeze to the tone in the first part of the study.[15] This pattern of results is what neuroscientists call a double dissociation, where the roles of two brain areas in two behaviors are shown to be distinct: the central nucleus was

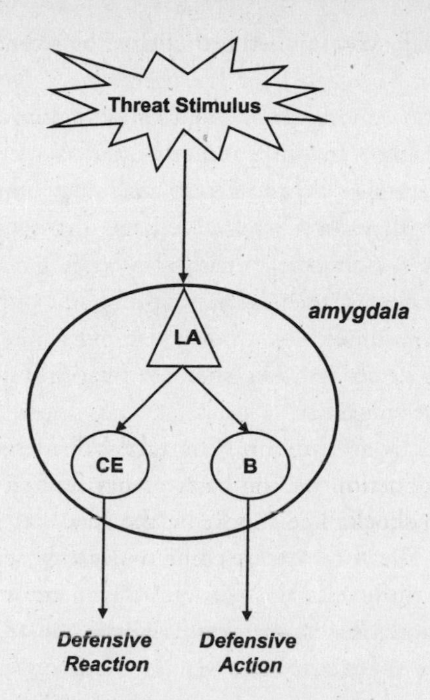

FIGURE 9.2 THE AMYGDALA MEDIATES REACTIONS AND ACTIONS
The lateral nucleus of the amygdala (LA) mediates both automatic reactions (freezing) and instrumental actions (running away after several seconds of freezing) in the presence of a threatening stimulus (like the sound of a bomb). Reactions involve connections from the LA to the central nucleus (CE), while actions involve connections from the LA to the basal nucleus (B) (see text).

found to be involved in fear reaction but not fear action, and the basal nucleus in the fear action but not reaction (fig. 9.2).

So exactly how does a conditioned stimulus, a conditioned incentive, processed by the amygdala motivate and reinforce behavior? To answer this question, we need to consider the nature of reinforcement further.

JUST REWARDS

In the early 1950s, studies by Jim Olds and Peter Milner, two researchers in Donald Hebb's department at McGill, found that rats would return to the location in a chamber where they received a burst of electrical pulses to their brain.[16] The animals, in other words, seemed to be coming back for more.[17]

Olds and Milner knew right away that they were on to something. The stimulation stamped in the behavior, just as rewards do. At the time, behaviorism still dominated psychology, and brain stimulation reward seemed to be a way to figure out the brain mechanisms underlying the topic most important to behaviorism: how responses that lead to rewards are learned.

Olds and Milner devised a way to test whether brain stimulation could reinforce new responses. Electrodes were implanted in the rats' brains and connected to a lever that the subjects themselves could press. Each time they pressed, they got a few pulses. The rats began to press like crazy. This so-called self-stimulation was clearly rewarding—it motivated the acquisition of an arbitrary behavior (pressing a bar).

This fascinating result was due to a mistake.[18] The researchers were trying to stimulate the reticular formation in the brain stem, a region involved in arousal, alertness, and vigilance. This was in the heyday of drive theory, which proposed that drives lead to learning by arousing the organism, and Olds and Milner wanted to investigate whether an increase in arousal, produced by reticular formation stimulation, would enhance learning. The electrode, though, accidentally ended up somewhere in the forebrain rather than in the brain stem. Although the exact area that the electrode was in is not known, in subsequent studies, Olds and Milner and others identified many sites from which self-stimulation could be elicited,[19] the most potent of which were in the hypothalamus. The reason these sites were so effective was not because the hypothalamus is the reward center of the brain, but because a major nerve pathway passes through the hypothalamus. This pathway, called the medial forebrain bundle, was actually the source of the rewarding effect, as I'll explain later.[20]

It wasn't long after the phenomenon of brain stimulation reward was discovered that the idea arose that its effects were due to the stimulation of "pleasure centers" in the brain.[21] This notion was further fueled when a surgeon in New Orleans, Robert Heath, reported that schizophrenic patients found such stimulations pleasurable.[22] Around the same time, Michael Crichton's literary career was jump-started when he popularized pleasure centers in *The Terminal Man*.[23] While many researchers also treated brain stimulation reward in terms of subjectively experienced pleasure, Peter Shizgall, a leading thinker in this field, has argued that the ability of rewards to motivate behavior and to give rise to pleasurable feelings are separate.[24] This is a motivational version of my notion that emotional behaviors are not necessarily caused by emotional feelings, as Shizgall pointed out.

In spite of its immense popularity, brain stimulation reward research eventually ran out of steam.[25] One reason was that the motivational nature of brain stimulation reward was never quite figured out.[26] Did it activate drives, enhance incentives, or both?[27] Was it the same as natural reward, and could it explain learning?[28] These issues remained unresolved, and by the late 1960s, brain stimulation reward, like other topics related to motivation and emotion, was dying out as the influence of cognitive science was increasing. Drives, incentives, and rewards were just not as important to cognitive scientists as they had been to behaviorists.

PLEASE RELEASE ME

Perhaps the most significant outcome of brain stimulation research was the discovery of why medial forebrain bundle stimulation was rewarding.[29] When this pathway is stimulated in the region of the hypothalamus, fibers headed from the forebrain to the brain stem are activated. The main targets of these fibers are neurons that make dopamine,[30] which are located in a region of the brain stem called the ventral tegmental area. These cells, in turn, send their axons throughout the forebrain. As a result, when the dopamine cells are activated by inputs from the medial forebrain bundle, they release dopamine widely in the forebrain.[31]

Dopamine has long been believed to be a critical factor in reward processes.[32] Although there are rewarding conditions that do not depend on dopamine,[33] much of what we know about rewards centers around the role of dopamine. For example, treating rats with drugs that block the effects of dopamine at its receptors in the brain eliminates the rewarding effects of brain stimulation—that is, rats are much less inclined to press to get the brain jolts under such treatment. Further, if hungry rats are given food in one compartment of a two-chambered apparatus, or if satiated rats are given rewarding brain stimulation in one of the compartments, they will later spend more time in that compartment. This is called a place preference. Treatment of rats with drugs that block the action of dopamine prevents the formation of the place preference. A place preference can also be established by giving rats a shot of amphetamine or cocaine, both of which mimic the action of dopamine at its receptors. It's no accident that these widely abused drugs work like rewards in learning situations, and the relation of dopamine to drug addiction was important in sustaining interest in reward and motivation during the years of cognitive domination in neuroscience.

Just as brain stimulation reward was initially thought to be due to activation of pleasure centers, dopamine was believed to be the chemical of pleasure.[34] However, as we've seen, the hedonistic (subjective pleasure) view of brain stimulation reward is incorrect, and the hedonistic interpretation of dopamine's role in reward is incorrect as well.[35] For example, blockade of dopamine interferes with instrumental responses motivated by a sweet reward but does not alter the actual consumption of the tasty stuff when it is obtained—the animals still "like" the reward when they consume it, but they are no longer motivated to work for it. Dopamine is thus more involved in anticipatory behaviors (looking for food or drink or a sexual partner) than in consummatory responses (eating, drinking, having sex). But being hungry or thirsty is unpleasant. Pleasure, to the extent it is experienced (see discussion of the credibility problem in chapter 8), would not come during the anticipatory state but instead during consummation. Since dopamine is involved only in the anticipatory phase, and not in the consummatory phase, its effects (at least in the case of primary need states) cannot be explained in terms of pleasure.

The exact role of dopamine in motivation, reward, and habit learning is still being debated. Although it seems clear that dopamine is not involved in subjective pleasure or in the expression of consummatory responses, there is less agreement about the conditions that depend on dopamine. Some adhere to the classic hypothesis—that it is the basis of reward.[36] Another view is that dopamine release is important for the initiation and maintenance of anticipatory behaviors in the presence of secondary incentive stimuli.[37] Others argue that dopamine release notifies the forebrain that something novel or unexpected has occurred, but not that reward per se has occurred.[38] Still others propose that dopamine is involved in the switching of attention and selection of action.[39] These are not mutually exclusive views, and in fact each appears to correctly characterize certain aspects of what dopamine contributes to motivation.

A MOTIVE CIRCUIT

While activation of dopamine neurons in the ventral tegmental area leads to the release of dopamine in many parts of the forebrain,[40] an area called the nucleus accumbens is particularly germane to reward and motivation.[41] Many of the effects of dopamine-related drugs described above can be achieved by applying the drugs directly to the nucleus accumbens,[42] a region of the striatum located in front of the amygdala near the bottom of the forebrain. For example, animals will press a bar to administer dopamine or related drugs (cocaine or amphetamine) into the nucleus accumbens. Also, dopamine levels rise in the nucleus accumbens in response to natural rewards (food, water, and sexual stimuli), conditioned incentives (stimuli associated with rewards), and brain stimulation reward. Finally, blockade of dopamine receptors in the nucleus accumbens greatly reduces the rewarding effects of medial forebrain bundle stimulation and of natural rewards, and also prevents the development of place preferences.

So how, precisely, does an elevation of dopamine in the nucleus accumbens accomplish all of this? More than twenty years ago, Ann Graybiel[43] and Gordon Mogenson[44] suggested that the nucleus accumbens sits at the crossroads of emotion and movement, and that dopamine release in this region plays a crucial role in motivated or goal-directed behavior. This conclusion was based on four main sets of observation. First, the nucleus accumbens receives massive dopamine inputs from the tegmentum. Second, injection of amphetamine or cocaine into the nucleus accumbens leads to behavioral activation—animals start exploring their environment as if in search of something. Third, the accumbens also receives inputs from areas involved in emotional processing, such as the amygdala. Fourth, the accumbens sends output to areas involved in the control of movement (such as the pallidum, an area that connects with the movement-control regions in the cortex and brain stem). Today, it is widely accepted that the nucleus accumbens and areas with which it is connected constitute key elements of a circuit through which emotional stimuli direct behavior toward goals (fig. 9.3).[45] Let's consider the function of this motive circuit in the broader context of emotional information processing by the brain.[46]

In the presence of an emotionally arousing stimulus, the brain is placed in a state, sometimes called a motive state,[47] that leads to coordinated information processing within and across regions, and results in the invigoration and guidance of behavior toward positive goals and away from aversive ones.[48] Most of what we know about how incentives are learned and reacted to comes from studies of aversive conditioning, but most of what we know about how conditioned incentives are used to invigorate and guide behavior comes from studies of positive motivation. I'm therefore going to apply to negative motivation what is known about the role of the accumbens in positive motivation in an effort to move forward from our understanding of aversive conditioning into motivation circuits (the details may therefore need to be revised after

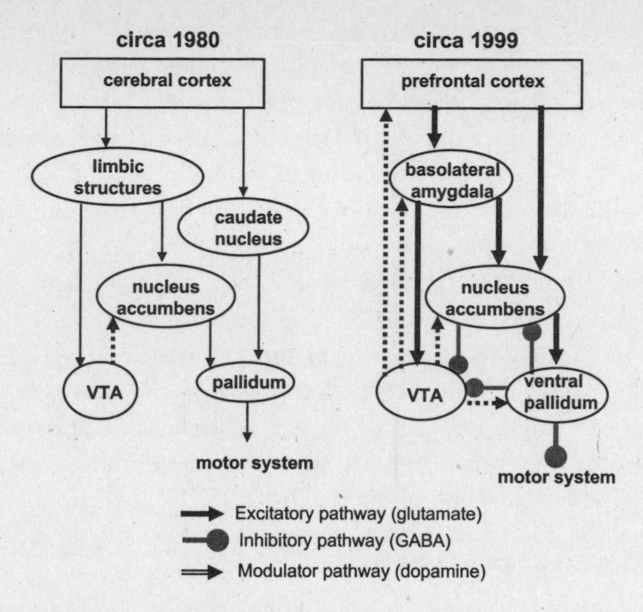

FIGURE 9.3 MOTIVE CIRCUITRY

Ideas about the nature of the motivational circuitry of the brain have been refined over the years. General terms like *cerebral cortex, limbic structures,* and *pallidum* have been replaced with more specific terms, such as *prefrontal cortex, basolateral amygdala,* and *ventral pallidum.* Further, different roles of excitation and inhibition in the motive circuits have emerged. However, the dopaminergic projection from the ventral tegmental area (VTA) to the nucleus accumbens remains a key feature of the circuitry. Based on Mogenson et al. 1980 and Kalivas and Nakamura 1999.

more research is performed on the role of the accumbens in negative motivation).

The processing begins in the sensory system that receives the stimulus (the auditory system, in the case of a tone that's been paired with an aversive shock) and then flows forward to the amygdala, especially the lateral nucleus, which, in turn, activates the central nucleus of the amygdala (chaps. 5 and 8). Outputs of the central nucleus initiate the expression of species-typical defense responses (like freezing and associated autonomic changes) as well as activate arousal systems in the brain stem, including the dopamine neurons in the ventral tegmental area. The tegmental cells then release dopamine from their axon terminals in the nucleus accumbens (as well as in other areas of the forebrain).

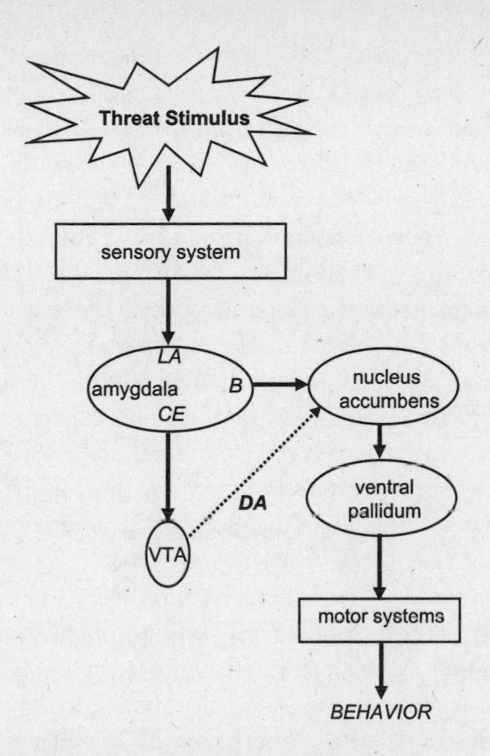

FIGURE 9.4. BEHAVIORAL INVIGORATION BY DOPAMINE

Key to modern notions of motivation is the role of dopamine (DA) in the invigoration or activation of behavior. One way this is believed to occur, at least in the presence of threatening stimuli, is illustrated. The lateral nucleus of the amygdala (LA) processes the sensory properties of the threatening stimulus. By way of connections to the basal nucleus (B), information about the threat is sent to the nucleus accumbens. By way of the central nucleus (CE), dopamine cells in the ventral tegmental area (VTA) are activated and these release dopamine in the accumbens. Dopamine facilitates the ability of accumbens cells to process the information sent from the amygdala. As a result, an amplified signal is sent to the ventral pallidum, which in turn activates motor systems that control instrumental (motivated) behavior.

As noted above, animals become active or invigorated when dopamine is injected into the accumbens.[49] This occurs because dopamine facilitates synaptic transmission in the pathway from the accumbens to the pallidum, which in turn connects with movement-control regions in the cortex and brain stem. With the pallidal output amplified, the motor regions are strongly activated, and movement is initiated (fig. 9.4). Behavior can potentially be invigorated by anything that activates tegmental cells and causes them to release dopamine in the accumbens. Novel stimuli, and conditioned and unconditioned incentives, are prime examples of invigorating stimuli.[50]

But invigoration alone is not sufficient: behavior also needs to be guided or directed.[51] Guidance of behavior is the job of conditioned incentives processed by the amygdala. The basal nucleus of the amygdala, as we've seen, receives information about a conditioned incentive from the lateral nucleus. It then transfers this conditioned incentive to the accumbens. When dopamine is elevated in the accumbens (as a result of the central nucleus's activation of dopamine neurons in the tegmentum that release dopamine in the accumbens), the arrival of an incentive stimulus in the accumbens from the basal nucleus will have a bigger effect on the activity of accumbens cells,[52] and presumably on neurons in the ventral pallidum that are downstream from the accumbens. The incentive thus leads to the release of dopamine, and dopamine facilitates the ability of the incentive to both invigorate and direct behavior.

Conditioned incentives are secondary reinforcers and can stamp in responses that lead to them. Animals will work to obtain conditioned incentives, even when they do not lead to primary incentives. As we've seen, after learning that a tone predicts shock, animals will take actions to stop the tone. Similarly, animals will perform a task (press a bar) to turn on a tone that was previously predictive of some tasty food, even if the tone no longer leads to food (this is comparable to looking at pictures of a loved one when he or she is away).

The manner in which conditioned incentives reinforce responses is not fully understood. However, LTP occurs in accumbens circuits, and dopamine is essential for the synaptic changes to occur.[53] A reasonable hypothesis is that dopamine facilitates Hebbian plasticity between active presynaptic and postsynaptic cells and thereby strengthens transmission between accumbens pathways that process the incentive and that control responses.

My conception of the role of the amygdala and accumbens in motivation borrows heavily from the work of Barry Everitt and Trevor Robbins at Cambridge University. These researchers have extensively studied how interactions between the amygdala and nucleus accumbens contribute to motivation, especially motivation by positive incentives.[54] They've shown that damage to the basal amygdala eliminates the ability of conditioned incentives to facilitate the learning of a new response by serving as a secondary reward. This disruption occurs for conditioned incentives created by pairing neutral stimuli (like tones) with a variety of primary rewards, including tasty food items, drugs like cocaine and amphetamine, and exposure of a male rat to a sexually receptive female. Through a clever combination of lesions, where the basal amygdala is lesioned on one side of the brain and the nucleus accumbens on the other, Everitt and Robbins have shown that connections between the amygdala and accumbens are key to the ability of a conditioned incentive to motivate new learning. Their work is part of a long tradition of research on the role of the amygdala in stimulus-reward learning. Some of the other key researchers in this area are Mort Mishkin, Edmund Rolls, Taketoshi Ono, Norman White, Michela Gallagher, and David Gaffan.[55]

Once an emotional habit is well learned, the brain systems involved in expressing it become simpler. The amygdala, for example, drops out of the circuit. After you know how to successfully avoid a specific danger, you no longer need the amygdala, because fear is no longer aroused. A dog needs its amygdala to learn that playing around in the road is dangerous, but once the learning has occurred, he can happily play in the yard next to the road. (In this case, avoidance of danger doesn't arouse fear, it prevents fear.) Characteristically, amygdala-dependent signs of emotional arousal, such as elevations of heart rate, occur during the initial phase of avoidance learning but disappear as the avoidance response is learned.[56] The accumbens likewise drops out once the response is learned; while it's needed to do the learning, it is not necessary to perform well-learned responses.[57] The exact systems that take over in learned defensive habits are not fully understood. Perhaps the accumbens, like other areas of the striatum, trains prefrontal cortical circuits, especially areas of the motor cortex, how to respond.[58] The learning, in other words, might be transferred from the striatum (accumbens) to the cortex. This would be similar to what occurs in declarative or explicit learning: initially, both the hippocampus and neocortex are involved, but once the hippocampus has slowly taught the neocortex the memory, the memory persists without the aid of the hippocampus (chap. 5).

Interactions between the accumbens and amygdala thus go a long way toward accounting, in neural terms, for some of the key aspects of motivation. But the motive circuit consists of other areas as well.[59] The hippocampus participates by way of its connections to the amygdala and accumbens, and may well be involved in the guidance of behavior on the basis of spatial and other kinds of relational cues in the environment: in order to find good things and avoid bad ones, you have to know where you are, where you need to go, how to navigate from where you are to where you need to go, and what sorts of stimuli along the way will be useful in guiding you. This information, drawn from long-term memory, is essential. It is also important to keep the overall goal in mind, and to constantly update your progress toward the goal. Working-memory functions of the prefrontal cortex probably are important for this ability. The prefrontal cortex receives dopamine inputs, and is connected with the accumbens, amygdala, and hippocampus. Certainly, when motivation is based on decisions, the prefrontal cortex, as we will now see, will often be involved.

DECISIONS UNDER FIRE

The study that Nader, Amorapanth, and I conducted, in which rats learned to escape from a sound that had been associated with a shock, was designed to address the question of how emotional arousal motivates action. The approach we took involved habit learning. But in real-life situations, we don't always have habits to fall back on, or time to learn a new one. We aren't only impelled by drives, pulled by incentives, and shaped by reinforcers. We live in a complex world where the physical and social environment changes from moment to moment and we often integrate immediate needs and past learning with predictions about the best course of action to take. We use our capacity to think, reason, and evaluate. We make decisions.

Decision-making compresses trial-and-error learning experiences into an instantaneous mental evaluation about what the consequence of a particular action will be for a given situation. It requires the on-line integration of information from diverse sources: perceptual information about the stimulus and situation, relevant facts and experiences stored in memory, feedback from emotional systems and the physiological consequences of emotional arousal, expectations about the consequences of different courses of action, and the like. This sort of integrative processing, as we've seen, is the business of working memory circuits in the prefrontal cortex. In chapters 7 and 8, we discussed the role of the prefrontal cortex in working memory and considered the contribution of the lateral and medial prefrontal cortex. Here, we will focus on two of the subareas of the medial prefrontal cortex in light of their relation to the motive circuits outlined above.

The anterior cingulate cortex receives inputs from the dopamine cells in the tegmentum, as well as from the basal amygdala, ventral pallidum, and hippocampus. In addition, it sends outputs to the accumbens and to the motor cortex.[60] It is thus in a position to receive information about behavioral arousal (dopamine connections from the tegmentum) and about conditioned incentives and their amplification by dopamine (connections from the amygdala and pallidum). It can then integrate this information with data from long-term memory (connections with the hippocampus) and with the temporary contents of working memory (connections with other prefrontal areas) in the process of controlling movement (connections to the motor cortex). That the anterior cingulate is also involved in processing unconditioned incentives is suggested by studies showing activation of this region in humans receiving painful stimulation.[61]

The orbital cortex, an area of the ventral prefrontal cortex, located at the bottom of the frontal lobe just above the eye sockets, is also important in motivation and decision-making. Studies of monkeys by David Gaffan and Edmund Rolls have implicated this region in emotional processing—in the processing of incentives (rewards and punishments) and in the temporary storage of incentive information.[62] This region is connected with the anterior cingulate, and, like the anterior cingulate, it also receives information from the amygdala and hippocampus.[63]

The work of Antonio Damasio has been especially significant in drawing attention to the critical nature of emotional information in human decision-making, and especially the role of the orbital cortex in this process. Much of Damasio's recent work on this topic is summarized in his two books, *Descartes' Error* and *The Feeling of What Happens*.[64] He and his colleagues have shown that patients with damage to the orbital prefrontal region have poor judgment and often make decisions that lead to socially inappropriate courses of action. Using a clever experimental task, called the gambling task, they demonstrated that patients with orbital cortex lesions are insensitive to changes in the incentive value of stimuli, despite the fact that they have relatively normal working memory, as assessed in purely cognitive ways.[65] Damasio and colleagues interpreted their results in terms of an inability of patients with orbital cortex damage to use emotional information to guide action, and have proposed that emotional information or knowledge normally biases reasoning ability by influencing attention and working memory processes.[66] Although they argued that their results dissociated decision-making from working memory processes, in light of the suggestion above that the orbital cortex is part of the working memory circuitry, and is especially involved in working memory about emotional information, an alternate conclusion is that the results show a dissociation between cognitive and emotional aspects of working memory.

The lateral, anterior cingulate, and orbital prefrontal regions are synaptically interconnected in various ways and should be thought of not as separate, independent modules, but as components of an integrated working memory system (chaps. 7 and 8). Although individual regions may make a relatively greater contribution to different aspects of working memory, especially in controlled laboratory experiments, in more natural settings, where decisions involve the integration of cognitive and emotional information, it is likely that interactions rather than dissociations between the areas will be commonplace.

But we have to be careful not to jump to the conclusion that decision-making is simply something carried out by the prefrontal cortex on its own. Recent studies by Paul Glimcher at NYU have pointed to the contribution of a region in the parietal cortex that appears to be a component of the decision-making process by which eye movements are controlled.[67] This region is part of the "where" pathway (chap. 7), is strongly connected with the lateral prefrontal cortex,[68] and has long been implicated in eye-movement control.[69] But Glimcher has upped the ante in this field by performing a sophisticated mathematical analysis that suggests that parietal neurons participate in decision-making. Basically, he's approached the problem of how populations of neurons in the brain make decisions in a manner similar to the way that economists approach the behavior of populations of people.[70] His analysis suggests that these cells are able to integrate information about a given stimulus and its implications for the amount of reward that a subject animal can expect on a particular trial, given what has been experienced in the past. It will be interesting to see whether this kind of analysis also works for cells in the prefrontal cortex and whether there are differences between parietal and prefrontal involvement. Also important to examine is whether there might be differences within prefrontal circuits. It might be expected that the medial or orbital areas would be more involved when emotional information is guiding decisions and the lateral prefrontal areas when emotional information is less important than cognitive information.

Many in the field have taken notice of Glimcher's work. Particularly interesting is his claim that his approach can put Descartes's dualism to rest—there's no mental stuff living in the neurons and making decisions, it's just the neurons carrying out mathematical calculations. While many have argued that a process of this type must underlie the mind-brain relation, Glimcher has actually eavesdropped on neurons while they do the calculations that we call decisions.

MOTIVATED MAN

Much of the work on motivation that we've considered here has been based on studies of animals. How far, you might ask, can this approach take us in understanding motivation in the daily life of human beings? The answer is, I think, fairly far. To understand why this is so, we need to consider two broad approaches to human motivation within psychology today.

The first, and most prominent, is the cognitive approach. In the 1950s, the social psychologist Leon Festinger fused the newly emerging cognitive view of the mind with the classic notion of drive in an attempt to account for the motivation of human behavior, especially in conflict situations. The theory, called cognitive dissonance, assumed that having inconsistent or conflicting thoughts (for example, "Wanting more money is greedy" and "Having more money is good") puts one in an uncomfortable state, a drive state, which demands reduction by changing one of the thoughts (for example, "Being greedy is not so bad" or "Having lots of money is not so good"). This theory gave rise to a tremendous amount of research in social psychology and spawned many other related theories.[71]

Social psychologists were one of the few groups of psychologists who remained interested in motivation during the cognitive revolution. But, over the years, their views lost their link to traditional notions of motivation and went farther and farther in the cognitive direction. Knowledge, beliefs, expectations, and self-awareness replaced drives and emotions in explanations of how behavior is aroused and directed.

For example, Nancy Cantor and Hazel Markus, leading contemporary cognitive motivational theorists, view motivation as a product of self-knowledge.[72] Self-knowledge obviously includes knowledge about one's emotions and their motivational consequences, but for Cantor and Markus motives are products of the self as much as contributors to it. Key to their theory is the notion of a working self, an on-the-fly construction about who we are that reflects who we've been (past selves), and who we want and don't want to be (future selves).[73] In contrast to earlier self theorists, such as Alfred Adler,[74] who viewed the self as a static and enduring entity, Cantor and Markus view the self as a dynamic and mutable construction that changes in different situations—we have different goals when at home than when on the job, for example, and the working self in each situation reflects such differences. One's working self is thus a subset of the universe of possible self-concepts that can occur at any one time—it is the subset that is available to the thinking conscious person at a particular moment, and is determined in part by memory and expectation, and in part by the immediate situation. These features of the working self explain how one can have both stable and mutable motives, and how motives can be conflicting or dissonant. The working self is a central part of one's mental apparatus. It influences perception, attention, thinking, memory retrieval, and storage, and guides action.

Building on William James, the cognitive psychologist John Kihlstrom has argued that a critical component of any conscious experience is the relation, in working memory, of the object of the experience to one's sense of self. Combining James and Kihlstrom with Cantor and Markus leads to the conclusion that the working self is constructed in working memory and the construction of the moment contributes in significant ways to on-line processing, decision-making, and behavioral control.

The second major approach to human motivation also originated in the 1950s when David McClelland proposed his need-achievement theory.[75] According to McClelland, humans have a limited number of inborn motives (drives or needs) that are sensitive to natural incentives (hunger, thirst, and sex, for example), and motivation occurs when a person does something to make these goal objects available, usually by performing some behavior that has been learned in the past. When the behavior leads to the incentive, emotion is aroused, and the behavioral sequence that led to the incentive is reinforced. Through experience, people also learn to recognize cues that signal the availability of natural incentives. These cues also arouse emotions. Because positive emotional states are reinforcing, people seek out situations in which learned or natural incentives are present. Motives, in McClelland's view, are thus emotionally charged states that anticipate goal objects. Upon this quasi-biological foundation of inborn motives and learned incentives, McClelland has built a theory of human behavior focused on positive motivational states like affiliation and achievement. McClelland does not deny the importance of negative motivation, but places this topic outside the area covered by his theory. Research on humans by McClelland and others accumulated over several decades has tended to support major aspects of achievement theory.

So how do we reconcile the two views of human motivation? McClelland's theory emphasizes nonverbal motivational systems, some of which are biologically organized. These work implicitly and function more or less similarly in humans and other mammals, though each species can have its own specialized motive systems as well. McClelland's views fit well with the drive/incentive/reinforcement tradition. But this approach does not cover the full range of human motivation, as McClelland himself has noted.[76] Consciously accessible, verbally encoded, explicit motives acquired through the use of language are also an important part of human mental life, especially in social situations. The self-consciousness view of motivation thus complements rather than contradicts the implicit view.[77]

Anthropologists Claudia Strauss and Naomi Quinn have used McClelland's rapprochement between implicit biological and explicit psychological motives to account for motivational similarities and differences between cultures.[78] They point out that some motives are universal and vary only in intensity between cultures, whereas others are unique to specific cultures. An obvious way to reconcile these theories is to place McClelland's biologically based motives in the former category and the explicit, self-conscious, individually constructed motives of Cantor and Markus into the latter. However, not all learned motives are related to explicit knowledge. Most motivational theories we've discussed, from Thorndike to Hull and Miller to McClelland, emphasize the learning of motivational habits. Habit learning generally is viewed as an implicit form of learning.[79] Strauss points out that habits are learned not just through encounters with reinforcers (primary or secondary) but also in common social situations in which we observe the successes and failures of others. Indeed, in spite of the past emphasis of social psychology on self-conscious motives, more and more research on social behavior is beginning to emphasize implicit or unconscious aspects of human motivation. Particularly noteworthy is John Bargh's work on *automotives* and Tim Wilson's theory of hidden selves.[80] It also needs to be pointed out, though, that while habit systems may indeed learn implicitly, when motivational habits are learned under conditions in which working memory attends to the present circumstances, the consequences of habit learning can be explicitly represented and consciously knowable, and only later, once routinized, sent to the depths of the mind. Because habits can be both useful and pathological (habitual avoidance of the outside world by a patient with panic disorder, for example), they are an especially important form of behavioral learning.

Both implicit and explicit systems thus contribute to motivation. Working memory is important in the guidance of behavior toward goals that are explicitly represented there and can be carried out under executive control functions. But, at the same time, we also have brain systems that work implicitly in the processing of incentives and the guidance of behavior toward goals. Sometimes, implicit and explicit motivation will be in sync, in which case working memory and implicit systems guide behavior with a common purpose. For example, following the explosion of a bomb, it is likely that both explicit and implicit systems will initiate behaviors that take you away from rather than toward the source of the explosion, all other things being equal. But other things are not always equal. If your spouse had gone to the concession stand to get a snack, and the stand is in the direction of the explosion, you may, through an executive decision, override the tendency to flee and instead run toward the explosion. Indeed, functional-imaging studies by Jonathan Cohen suggest that the prefrontal cortex, especially the anterior cingulate, is involved in resolving motivational conflict.[81] This region and its synaptic connections may therefore play a crucial role in overcoming fear or other emotional states when we need to take an action that goes against our innate or learned compulsions. (Might this circuitry also underlie cognitive dissonance?)

An informed conception about the brain regions and circuits underlying a particular function is the best way to begin to determine how the brain actually performs that function. Therefore, once we've translated basic motivational concepts into plausible neural circuits, the task of relating the enormously complex topic of human motivation to the brain becomes less daunting.

THE MENTAL TRILOGY IN ACTION

Through the topic of motivation, we begin to see the mental trilogy in action. A mind is not, as cognitive science has traditionally suggested, just a thinking device.[82] It's an integrated system that includes, in the broadest possible terms, synaptic networks devoted to cognitive, emotional, and motivational functions. More important, it involves interactions between networks involved in different aspects of mental life.

Often the things we attend to and remember are the things that are important to us. In such situations, cognitive processing will be accompanied by emotional arousal. And emotional arousal does not stop with a simple reaction, for we often use it to guide our behavior toward or away from the situation that the emotionally arousing stimulus signifies. In the process of doing so, we sometimes have to make decisions about what to do in order to keep working toward our goal. The goal has to be kept in mind, even when detours have to be taken. But the system also needs to be reset if a more important goal emerges along the way. This requires reallocation of cognitive, emotional, and motivational resources, and adjustments within and between the component processing systems.

Self-knowledge is certainly a significant aspect of human motivation, but even animals that are not self-aware, or at least not robustly aware of who they are the way a human is, are motivated to do things—they seek food and shelter and avoid predators and injury. Much of what we humans do is also influenced by processes that percolate along outside of awareness. Consciousness is important, but so are the underlying cognitive, emotional, and motivational processes that work unconsciously.

CHAPTER TEN

SYNAPTIC SICKNESS

CANST THOU NOT MINISTER TO A MIND DISEASED,
PLUCK FROM THE MEMORY A ROOTED SORROW,
RAZE OUT THE WRITTEN TROUBLES OF THE BRAIN,
AND WITH SOME SWEET OBLIVIOUS ANTIDOTE
CLEANSE THE STUFF'D BOSOM OF THAT PERILOUS STUFF
WHICH WEIGHS UPON THE HEART?

—Shakespeare, *Macbeth,* act 5, scene 3

In spite of Shakespeare's insight that mental problems are "troubles of the brain," hundreds of years later, proponents of a physical basis for mental illness still had little evidence with which to make their case. By the late nineteenth century, it was widely accepted that actual destruction of the brain (the result of diseases like syphilis) could change mental function drastically,[1] but the alterations of mood and thought that had come to be known as neuroses and psychoses were still resistant to physical explanation. Young Sigmund Freud sought a neurological account of mental illness, but realizing that such a goal was unattainable in his lifetime, turned to psychological explanations and psychological treatments instead. There was no "sweet oblivious antidote," to use Shakespeare's phrase, for hysteria, melancholia, or anxiety.

Obviously, the situation has changed considerably in the past few decades. Mental maladies have, in the eyes of many, come to be recognized as "troubles of the brain," and antidotes have emerged as the treatment of choice more often than not. Regardless of how one feels about the biological orientation of psychiatry today (and it has many critics),[2] two facts must be acknowledged. The essence of who we are is encoded in our brains, and brain changes account for the alterations of thought, mood, and behavior that occur in mental illness. The key issue is not whether mental illness is really neural in nature. It is instead the nature of the neural changes that underlie mental problems, and the manner in which treatment should proceed.

Biological psychiatry was founded on, and still largely adheres to, the assumption that mental disorders are due to chemical imbalances in the brain. The brain, in this view, is like a delicate soup. To maintain its characteristic flavor, just the right mix of key ingredients is required. Too little or too much of one or another item, and its character is altered. But the normal character can sometimes be restored with the right mix of additional ingredients that either replace a missing ingredient or dampen the effects of an overly abundant one. Like a master chef's, the biological psychiatrist's job is to adjust the blend, to balance the chemistry, so that the desired character is restored.

The soup model evolved in psychiatry for two reasons. One was practical—a number of mental disorders were found, sometimes accidentally, to be helped by altering brain chemistry. Another was conceptual—neuroscientists had come to think of chemicals as an important contributor to the way in which mental life is created by the brain. The practical justification remains, but, as we'll see now, the conceptual one has changed.

Although it seemed in the early 1960s that there might be a particular chemical code for different mental states[3]—molecules of pleasure, fear, and aggression were proposed—this notion is now viewed by many as naive. Mental states are not represented by molecules alone, or even by a mix of molecules. As we've seen, they are instead accounted for by intricate patterns of information processing within and between synaptically connected neural circuits. Chemicals participate in synaptic transmission, and in the regulation or modulation of transmission, but it is the pattern of transmission in circuits, more than the particular chemicals involved, that determines the mental state. The old battle cry "No twisted thought without a twisted molecule"[4] needs replacing. Synaptic changes, not molecules, underlie mental illness. The importance of circuits has been illustrated repeatedly in the previous chapters. In this chapter, I want to emphasize the implications of a circuit point of view for understanding the nature of psychiatric disorders.

Biological psychiatrists do realize that circuits are significant, and it's not out of ignorance that they adhere to the soup model. They would love to have smart drugs, akin to smart missiles, that, when taken orally, would go straight for the circuits that underlie a particular disorder and avoid all others. With this sort of drug, dreaded side effects would be eliminated.[5] But achieving such precision would require that the exact circuits altered in particular mental disorders be known, and that drugs be available to adjust selectively the chemistry of the relevant circuits. Although not all of the pieces of the puzzle are in place today, research efforts are under way to obtain the information that would make this possible. As Steve Hyman, director of the National Institute of Mental Health, has noted, psychiatry, arm in arm with neuroscience, is poised to answer many of its central questions.[6] Hyman, in fact, refers to the time ahead as "the millennium of mind, brain and behavior." To facilitate the achievement of his goal, the National Institute of Mental Health recently has established several Research Centers for the Neuroscience of Mental Disorders (I serve as the director of one of these, the Center for the Neuroscience of Fear and Anxiety, which I'll describe later in the chapter).

Most critics of biological psychiatry will not be satisfied by the shift in paradigm from soup to circuits. They will argue that the important question is not whether we should focus on connections instead of chemicals, but whether we should focus on the brain at all in our efforts to understand and treat mental illness. Psychological problems, such critics insist, are rooted in life experiences rather than in malfunctioning brain hardware, and should be treated by helping the patient come to terms with the underlying problem rather than by trying to alter brain circuits. But as the arguments presented in this book make clear, life's experiences leave lasting effects on us only by being stored as memories in synaptic circuits. Because therapy is itself a learning experience, it, too, involves changes in synaptic connections. Brain circuits and psychological experiences are not different things, but rather, different ways of describing the same thing. Still, the manner in which brain changes are effected by psychotherapy is not necessarily the same as the way in which they are by drugs, which is why therapy works in some cases where drugs don't and vice versa, or why a combination of drugs and psychotherapy can sometimes be more productive than either alone.

Enough generalities—let's examine what is known about the biology of mental illness. I can't cover all categories of mental illness here, and so will focus on schizophrenia, depression, and anxiety disorders.[7] In assembling a coherent account of the history of drug therapy for these disorders, I've borrowed from books by Samuel Barondes (*Molecules and Mental Illness*) and Elliot Valenstein (*Blaming the Brain*).[8] To find out more about various forms of mental illness and their treatments, consult the Websites listed in the following box entitled "Mental Health Organizations."

MENTAL HEALTH ORGANIZATIONS

Anxiety Disorders Association of America: *www.adaa.org.*

American Psychiatric Association: *www.psych.org.*

American Psychological Association: *www.apa.org.*

National Alliance for the Mentally Ill: *www.nami.org.*

National Alliance for Research on Schizophrenia and Affective Disorders: *www.narsad.org.*

National Depressive and Manic-Depressive Association: *www.ndmda.org.*

National Institute of Mental Health: *www.nimh.nih.gov.*

National Mental Health Association: *www.nmha.org.*

World Fellowship for Schizophrenia and Allied Disorders: *www.world-schizophrenia.org.*

TRIPPIN' UTERUSES

The mid-twentieth century was an exciting time in neuroscience research. With the end of World War II, national scientific resources were shifted to peacetime efforts, and one area that boomed was brain studies. The neuron doctrine had long since prevailed, and synaptic transmission between neurons was widely accepted as the currency of the brain. But the nature of synaptic transmission was still the topic of much investigation, the interpretation of which was hotly debated. Sir John Eccles pushed the notion of electrical transmission, while Sir Henry Dale emphasized the chemical nature of synaptic communication.[9] In the end, both were right—some synapses are chemical and others electrical.[10] Nevertheless, chemical transmission is believed to be the dominant form of synaptic communication, a finding to which even Eccles came around. Still, it is a long leap from the idea that tiny amounts of neurotransmitter are squirted across individual synapses to the notion that mental states and mental disorders are chemically coded. Another set of events, culminating around the same time, played an important role in closing the gap.

Ingestion of a variety of substances from nature, like peyote or certain mushrooms, had long been known to induce hallucinations, and by 1937, it had been proposed that the hallucinations induced by peyote resembled the symptoms of schizophrenia. But it was the accidental discovery of a novel hallucinogen, d-lysergic acid, or LSD, that helped jump-start the psychopharmacological industry in the 1950s.[11]

Chemists had determined in the 1930s that ergonovine was the active ingredient in the fungus ergot, which had been used by midwives to induce uterine contractions and control bleeding during childbirth. In the process of trying to formulate a new drug that would be useful for obstetricians, Albert Hoffman was accidentally exposed to one of the chemical compounds with which he was working and entered a "dream-like state" in which he experienced a "kaleidoscopic play of colors." Suspecting that the culprit was LSD-25 (the twenty-fifth modification of an acidic extract of ergonovine), he decided to take a small amount to see if the experience would recur. Again, he had powerful hallucinations. Furniture in the room "assumed grotesque, threatening forms," and the woman next door became a "malevolent, insidious witch with a coloured mask." His feeling that a demon had invaded him lasted for about fourteen hours. Others at the company tried the substance as well, all with the same result. News got out, and so did the drug. The key event for our story was not the popularization of LSD by figures like Ken Kesey and Timothy Leary, but rather the discovery in 1953 by John Gaddum, a Scottish scientist, that LSD blocked uterine contractions induced by the chemical 5-hydroxytryptamine, otherwise known as serotonin. Serotonin subsequently was found in the brain, and shown to be a neurotransmitter. These facts gave Gaddum an idea. If LSD produced psychotic states (hallucinations) by altering neurotransmission in the brain, sanity might require a certain level of brain neurotransmitters, and changing transmitter levels might be a way of treating mental illness.

That drugs might be useful in treating mental illness was not a novel idea. What was new was the notion that abnormal mental function might result from alterations in synaptic transmission, and that drugs that affected synaptic transmission might therefore be used to treat mental illness. Gaddum's hypothesis thus marks the beginning of the modern approach to biological psychiatry.

FROM SEDATION TO CORRECTION IN PSYCHOSIS

As many psychiatrists are aware, the neo-Freudian Harry Stack Sullivan encouraged his patients to partake of alcohol to loosen them up before starting psychoanalysis. This was not drug therapy per se, as it was not intended as a cure. In fact, during the first half of the twentieth century, the noncurative use of drugs was common, not so much to facilitate therapy as to sedate agitated or manic patients and make them more manageable, especially in institutional settings.[12]

Insulin was one of the first drugs conceived of as helping to cure a mental disorder. Like many other drugs, its therapeutic effects were discovered by chance. In 1933, a Viennese physician named Manfred Sakel gave a small dose of insulin, which lowers blood sugar levels and thereby stimulates appetite, to some of his agitated schizophrenic patients who had not been eating. Surprisingly, their psychotic symptoms were somewhat relieved. When Sakel tried a larger dose, the extreme lowering of blood sugar put the patients in a coma for several days, but when they recovered, their psychotic symptoms were much improved. Although the reason this treatment worked was a complete mystery, insulin coma therapy came to be widely used, in spite of its dangers, since there was no other viable treatment for schizophrenia.

In the 1950s, though, two new psychotic treatments emerged, both of which were explained in terms of new discoveries about neurotransmission. One of these was reserpine, a drug isolated from a plant called *Rauwolfia serpentina. Rauwolfia* had been used to treat insomnia and insanity in India since ancient Hindu times. When Indian doctors in the 1930s found that *Rauwolfia* could reduce high blood pressure, drug companies were motivated to isolate reserpine, the active substance in the plant. In the meantime, Nathan Kline, a New York psychiatrist who knew of the use of *Rauwolfia* to treat insanity in India, obtained reserpine and administered it to schizophrenic patients.[13] He noted that the drug made the patients less suspicious and more cooperative. Soon thereafter, Kline convinced the governor of New York that reserpine should be given to the 94,000 patients in the massive New York State Psychiatric Hospital system. Klein's report came right around the time that Gaddum had made the connection between serotonin and sanity, and it wasn't long afterward that scientists determined that reserpine reduced the level of monoamines (including serotonin, norepinephrine, and dopamine) in the brains of experimental animals. By the mid-1950s, it seemed that psychosis was indeed related to alterations in monoamine levels in the brain, and that restoration of monoamine function might therefore be a way of treating this condition.

This conclusion was supported and refined by the near-simultaneous discovery of the beneficial effects of chlorpromazine, a phenothiazine drug marketed as Thorazine. This drug was isolated because its chemical structure was similar to that of antihistamines. A French surgeon tried administering chlorpromazine to some of his patients to test whether it might be more effective than other antihistamines in sedating patients and reducing respiratory complications in surgery. The drug turned out to have a powerful calming or tranquilizing effect, which led the surgeon to propose using it with psychotics. Two French psychiatrists, Jean Delay and Pierre Deniker, tried administering chlorpromazine to a few schizophrenic patients who were resistant to all other forms of treatment.[14] Not only were they tranquilized, their psychotic symptoms (paranoia and hallucinations) were also reduced.

One of the side effects of both reserpine and chlorpromazine, however, was the development of a movement disorder called tardive dyskinesia, involving muscle rigidity similar to that typical of Parkinson's disease. Since Parkinsonism had been found to be associated with a reduction in dopamine in the brain,[15] the suggestion that antipsychotic drugs were useful in treating schizophrenia because they altered dopamine transmission arose, as did the related notion that schizophrenia was due to excess dopamine. In fact, it came to be believed that the potency of antipsychotic drugs was directly related to their ability to produce tardive dyskinesia.[16] The view that schizophrenia involved dopamine transmission was strengthened by the complementary observation that psychotic symptoms are brought on by too high a dose of L-dopa, which increases dopamine levels and is used to treat Parkinson's patients, or by amphetamine (speed), which artificially stimulates dopamine receptors.

An experiment performed by a Swedish scientist, Arvid Carlsson, helped to relate the effects of reserpine and chlorpromazine to dopamine.[17] He examined the effects of each drug on the level of the various monoamines in rats. While reserpine caused a reduction, chlorpromazine had no effect. This was a troubling finding for the emerging dopamine theory of schizophrenia. But in a follow-up experiment, he found that the breakdown products of dopamine, and not the other monoamines, were increased. This implied that more dopamine was being released, possibly to compensate for the block of dopamine receptors by chlorpromazine. The logic is complex, but the important point here is that this study suggested that both antipsychotic drugs acted by reducing dopamine transmission in the brain. While reserpine worked by reducing dopamine release from presynaptic terminals, chlorpromazine's effect was due to a blockade of postsynaptic dopamine receptors. Carlsson shared the Nobel Prize in 2000 for this and related work.

The idea that schizophrenia might be explained at the level of dopamine receptors was strongly supported by research carried out by the Dutch scientist J. M. van Roussum.[18] Rats given amphetamine displayed increased locomotor activity. Van Roussum showed that both the antipsychotic drugs chlorpromazine and reserpine blocked this increased activity. Although these two drugs affect all monoamines, a third potent antipsychotic drug, haloperidol (Haldol), which only affected dopamine, was particularly powerful in reducing the locomotion. Consequently, van Roussum proposed that antipsychotic drugs do their work by blocking dopamine receptors, and that overstimulation of dopamine receptors was at the heart of schizophrenia.

Still, the evidence linking dopamine neurotransmission to schizophrenia was circumstantial. An additional discovery strengthened the connection considerably. By the mid-1970s, a number of antipsychotic drugs had been developed. Studies by Philip Seeman in Toronto and by Ian Creese and Soloman Snyder in Baltimore rank-ordered the drugs by their clinical effectiveness and then examined the ability of each to block dopamine receptors.[19] Both studies found that the correlation was nearly perfect: the more potent its dopamine-receptor-blocking ability, the more effective the drug was in treating schizophrenia. The demonstration that the same relation did not hold between clinical effectiveness and blockade of serotonin or norepinephrine bolstered the view that it was dopamine, and not the other monoamines, that was involved.[20] Dopamine receptors come in two groups: D1-like and D2-like. The fact that all of the classic dopamine-receptor-blocking drugs were found to produce their antipsychotic effects by acting on the D2 class added further specificity to the role of dopamine in schizophrenia.[21]

The bottom-line conclusion from the early work on the relation of monoamines to schizophrenia was this: Too much dopamine induces a psychological disorder, and too little, a movement disorder similar to Parkinson's disease. A balance must be maintained. When it is tipped, the brain does not function normally. Subsequent work, as we'll see in the following section, has drawn a more complex picture of the neural basis of schizophrenia. Dopamine is still believed to be involved, but not quite in so simple a way as the original imbalance hypothesis suggested.

RETHINKING PSYCHOSIS

Although blocking the action of dopamine at its receptors in the brain is an effective way to treat psychotic symptoms, and although artificial stimulation of dopamine receptors (with L-dopa, amphetamine, or cocaine) can bring on psychotic symptoms, the dopamine receptor hypothesis of psychosis ran into trouble for several reasons. First, the therapeutic effects of dopamine drugs are far too slow to be accounted for by a simple receptor process. The drugs bind to receptors as soon as they are released from the presynaptic terminal, and have their physiological effects moments thereafter. But actual therapeutic effects of these substances take a week or two (or longer) to begin. The therapeutic effect is therefore more like a long-term adjustment to the receptor blockade than a direct and immediate effect of the blockade. Second, attempts to demonstrate that dopamine levels are elevated in schizophrenia proved unfruitful, in contrast to the relative ease with which decreases in dopamine were found in Parkinson's disease. The search for changes in dopamine receptors, rather than an increase in dopamine itself, was more successful,[22] but the evidence was controversial, since most of the research was performed on patients who had been on antipsychotic medication, which makes it difficult to know whether the change in receptors was due to the disorder itself or was a consequence of the treatment.[23] Third, standard manifestations of schizophrenia include positive or added symptoms (hallucinations, abnormal thought patterns, paranoia, delusions, agitation, hostility, bizarre out-of-context behavior) and negative symptoms or deficiencies (blunted emotions, cognitive deficits in attention and working memory, poor hygiene, poverty of speech, social isolation, loss of motivation).[24] The classic or typical antipsychotic medications, D2 receptor blockers, are mainly useful in treating positive symptoms, and are less helpful in treating negative ones,[25] suggesting that the dopamine theory is at best incomplete. Finally, recent studies have shown that schizophrenia can also be helped by drugs that target other systems, including serotonin and norepinephrine.[26] These so-called atypical antipsychotic medications turn out to be more useful for negative symptoms, further challenging the sufficiency of the original dopamine theory.

Recent efforts to explain schizophrenia and other mental disorders have turned to more complex conceptions that focus on alterations of function in specific brain regions and circuits rather than global changes in the level of monoamines. For example, a number of studies have demonstrated structural differences between the brains of schizophrenics and normal control subjects.[27] Included are changes in size or volume of certain brain regions, and the number of cells and their shape and arrangement in different regions. Some of the key areas in which structural changes have been discovered include the prefrontal cortex and medial temporal lobe (hippocampus and amygdala). Changes in the number of dopamine receptors have also been found in the prefrontal cortex and basal ganglia.[28] The prefrontal cortex, hippocampus, and amygdala also exhibit functional anomalies in blood flow and/or neural activity, as determined by PET and functional MRI scanning techniques.[29]

The structural and functional differences in the prefrontal cortex have attracted considerable attention and have led to a revised dopamine theory of schizophrenia. Classic or typical antipsychotic drugs are believed to achieve their beneficial effects on positive symptoms by blocking D2 receptors in the basal ganglia. Indeed, a number of studies have found that D2 receptors are elevated in the basal ganglia of schizophrenics,[30] including patients not previously medicated.[31] However, as we've seen, D2 blockade has little effect on negative symptoms. Some of the most prominent negative symptoms include cognitive changes, and the degree of cognitive deficit is directly related to treatment prognosis.[32] It has been proposed that deficits in working memory underlie the cognitive changes in schizophrenia.[33] Working memory, you'll recall, crucially involves the prefrontal cortex. Given that there are relatively few D2 receptors in the prefrontal cortex,[34] it is not surprising that D2-related drugs have little effect on cognitive negative symptoms.

At the same time, the D1 class of receptors is abundant in the prefrontal cortex of normal persons but is reduced in schizophrenics.[35] Further, working memory performance in schizophrenics is related to the number of D1 receptors in the prefrontal cortex—the fewer the number of receptors, the poorer a patient performs.[36] Moreover, cognitive tests that require temporary storage and executive function typically result in increased functional activity in the prefrontal cortex in normal humans, but not in schizophrenics.[37] Consistent with these results is the fact that blockade of D1, but not D2, receptors in the prefrontal cortex of monkeys impairs delayed response performance (a cognitive test of the temporary storage aspect of working memory).[38]

The revised version of the dopamine theory thus proposes that schizophrenia involves overactivity of the D2 class of receptors in the basal ganglia and underactivity of D1 receptors in the prefrontal cortex.[39] The hyperactivity of D2 receptors in the basal ganglia accounts for positive symptoms, and the underactivity of D1 receptors in the prefrontal cortex accounts for negative symptoms. This explains why drugs that block D2 receptors only treat positive symptoms, but leaves open the question of how atypical antipsychotic drugs, which mainly target serotonin or norepinephrine systems, treat negative symptoms.

The improvement in negative symptoms is believed to come about because the atypical medications cause an elevation in the amount of dopamine released from terminals in the prefrontal cortex and thereby help overcome the deficits caused by D1 understimulation. The site of the drugs' action is not in the prefrontal cortex itself, but in the ventral tegmental area, which is the home of the neurons whose terminals release dopamine in the the prefrontal cortex. For example, some very effective atypical antipsychotic drugs block the serotonin 2A receptors, which are located on ventral tegmental dopamine neurons. When the serotonin receptors are stimulated, they inhibit the ability of the tegmental cells to release dopamine in the prefrontal cortex, and other areas. By blocking serotonin receptors, atypical antipsychotic drugs block the inhibition normally produced by serotonin. This leads to an increased firing of the tegmental cells and thus more dopamine release in the cortex. The extra dopamine released helps compensate for the reduced number of D1 receptors in the schizophrenic brain.

Research on antipsychotic drugs is currently advancing in many directions. In addition to work on drugs that target monoamines, much work is attempting to determine whether alterations in glutamate and/or its receptors, especially NMDA receptors, may help treat schizophrenia.[40] Other studies are examining the role of GABA transmission.[41]

In general, there is a growing interest in the idea that alterations in synaptic connectivity in neural circuits, rather than just levels of neurotransmitters or receptors, are important.[42] Neurotransmitters and receptors still figure prominently in this approach, but in the context of connections within and between areas. For example, Francine Benes, noting that the number of GABA cells is decreased in the hippocampus of schizophrenics, has proposed that the disorder involves a shift of dopamine connections from excitatory glutamate cells to GABA cells in the hippocampus to compensate for a loss of inhibitory cells, and David Lewis has proposed that a specific class of GABA cells located in a specific layer of the prefrontal cortex plays a critical role in cognitive regulation and is altered in schizophrenia.[43] In imaging studies, researchers such as David Silbersweig and Emily Stern are beginning to examine changes in neural activity in networks involving interconnections of the prefrontal cortex with other cortical and subcortical areas.[44]

Perhaps the most sophisticated circuit theory of schizophrenia to date has been the work of Anthony Grace.[45] Based on his studies of dopamine in various forebrain circuits, he has proposed a complex theory that attempts to explain how synaptic interactions between the prefrontal cortex, hippocampus, and amygdala account for certain aspects of schizophrenia. Recall from our discussion of motivation circuits in chapter 9 that the prefrontal cortex sends connections to the nucleus accumbens, which connects back with prefrontal areas by way of the ventral pallidum. This is potentially a significant set of connections, since it includes the key areas involved in both negative and positive symptoms. The prefrontal cortex, as we've seen, has abundant D1 receptors that seem to be involved in negative symptoms. Similarly, the nucleus accumbens is one of the key regions of the basal ganglia that receive dopamine inputs from the ventral tegmental area and is loaded with D2 receptors, which, as we've also seen, are involved in positive symptoms.

Grace showed that the ability of the prefrontal cortex to activate the accumbens was regulated by the hippocampus and amygdala. That is, when there was synaptic activity from the hippocampus or the amygdala to the accumbens, the prefrontal cortex could fire accumbens cells; otherwise, it could not. The hippocampus and amygdala thus gate or regulate the prefrontal activation of accumbens. Given the role of the hippocampus in contextual processing of the environment, and the role of the amygdala in emotional processing, Grace proposed that these connections allow prefrontal activity to be adjusted in response to changes in the global environmental situation as well as to specific emotional stimuli. In schizophrenia, alterations in the hippocampus and amygdala (both of which have been found) could affect prefrontal processing in the accumbens, reducing the ability of the patient to produce responses appropriate to the immediate cognitive and emotional situation.

In Grace's theory, dopamine modulates the synaptic interactions in all of the regions. Specifically, Grace has proposed that there are two different pools of dopamine, which have different roles. The most obvious pool is the dopamine that is released by axon terminals in the striatum or prefrontal cortex when dopamine cells in the brain stem fire action potentials. This is called the phasic pool of dopamine. Some of the dopamine released phasically does not get broken down or sucked back up but instead diffuses away from the immediate area of the synapses and collects over time. This chronic pool of dopamine will affect dopamine receptors that are in its vicinity even if they are not at the moment receiving phasic dopamine from the brain stem cells. Grace argues that the chronic dopamine is what is altered over the course of long-term drug treatment, which is why it takes so long for such treatment to be effective. These alterations, in turn, adjust the phasic actions of dopamine and influence processing in the circuits just described.

Clearly, research in this area has made much progress. Although the situation has grown more complicated now that schizophrenia is no longer thought of as simply a problem of too much dopamine, the puzzle is more likely to be solved by accepting complexities than by ignoring them. Circuit

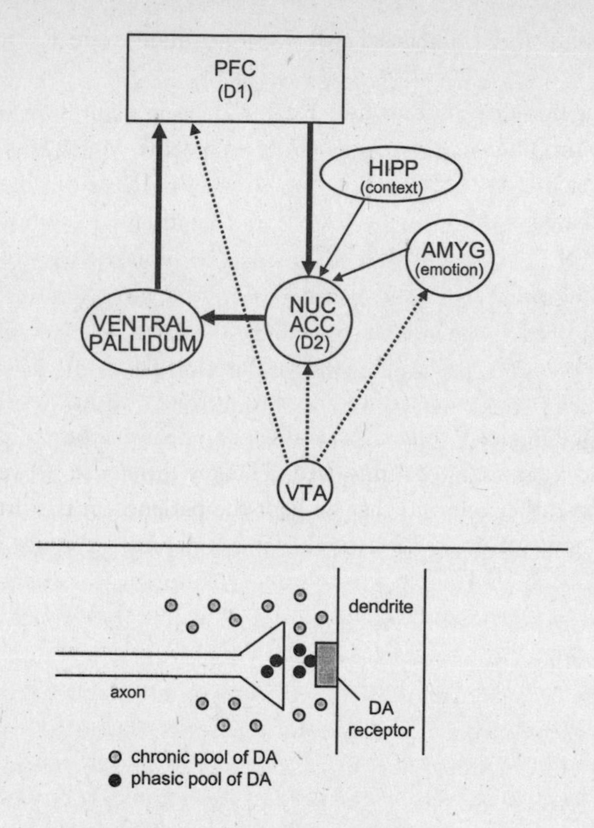

FIGURE 10.1 A CIRCUIT THEORY OF SCHIZOPHRENIA

Based on the work of Anthony Grace (see text). Abbreviations: AMYG, amygdala; D1, D2, dopamine receptors; HIPP, hippocampus; NUC ACC, nucleus accumbens; PFC, prefrontal cortex; VTA, ventral tegmental area.

models, for example, are more consistent with the fact that schizophrenia is not a single disorder but a family of conditions that can be manifested in different ways in different people.[46] The particular circuits affected, including transmitter alterations, may well determine the extent to which positive and negative symptoms appear and the manner in which they will be expressed. Key to future breakthroughs may be the ability to relate specific symptoms to specific circuits, and the ability to develop drugs that target those circuits.

BALANCING DEPRESSION

Around the same time that drugs were being developed to treat schizophrenia in the 1950s, depression, too, was proving susceptible to chemical amelioration. Although it was not realized at first, monoamines were also involved in this illness.[47]

One of the first drugs successfully used in treating depression was iproniazid. This compound was developed from hydrazine, which was originally used as rocket fuel by the Germans in World War II. After the war, other compounds based on hydrazine were developed and tested for medical applications. One of these, iproniazid, was found to inhibit the bacteria that give rise to tuberculosis. When TB patients were treated with the drug, they were also observed to become euphoric. On the basis of this effect, iproniazid was given to depressed patients. Although the results were not convincing, Nathan Kline, the man who gave reserpine to schizophrenics in the New York State system, decided to give iproniazid a try. He put a diverse group of patients on the drug, and depressed ones improved. The key difference between Kline's study and the earlier one was that he kept the patients on the drug for five weeks; it is now commonly accepted that antidepressant drugs take some time to kick in, so Kline's decision to extend treatment was crucial. According to Kline, within a year of reporting his results, 400,000 depressed patients had been treated with iproniazid.[48]

The mechanism by which iproniazid worked was unknown, but studies of animal brains soon revealed that monoamines were involved. Normally, the enzyme monoamine oxidase causes monoamines to be degraded after they are released from axon terminals, helping to end the action of the monoamines at postsynaptic receptors (stopping transmission is as important as starting it). Iproniazid was found to prevent the breakdown of monoamines after they are released. As a result, the monoamines remain around the synapse longer, and have a greater effect. Because iproniazid achieves its effects by inhibiting monoamine oxidase, it is called a monoamine oxidase inhibitor (MAO inhibitor).

The MAO inhibitors quickly became a common form of treatment for depression. But the treatment, successful or not, had a cost. MAO is a player in many brain areas, not just those that are altered in depression, and also participates in various functions outside the brain. When a drug is taken orally, it affects all relevant functions and not just those the patient wants corrected. One of the roles of MAO is the breakdown of the amino acid tyramine, which enters the body from ingested food substances (for example, certain cheeses). Unchecked, a rise in tyramine in the body can lead to a sudden and life-threatening rise in blood pressure.

The other major treatment for depression that emerged in the 1950s involved a class of drugs known as tricyclics (due to the three rings of their chemical structure). One of these substances, imipramine, was initially tried as a treatment for schizophrenia, due to some similarity to phenothiazines. However, because the drug seemed not to calm the patients but to energize them and elevate their mood, it was tested as a treatment for depression, and positive results were obtained. This class of drugs had fewer side effects than the MAO inhibitors, and tricyclics with brand names like Tofranil and Elavil became widely used to treat depression.

Like the MAO inhibitors, the tricyclics work by maintaining elevated levels of monoamines at postsynaptic sites. But the mechanism of their activity is different. The action of monoamines, once released, is terminated not just by the substances' being broken down, but also by being transported from the synaptic space back into the presynaptic terminal. Tricyclics block the transportation step, thereby allowing the monoamines to remain available to bind to postsynaptic receptors for longer periods of time. Their effect was discovered by Julius Axelrod, who received a Nobel Prize for this and other work on a monoamine chemistry. Because the side effects of tricyclics were less severe than those of MAO inhibitors, they soon came to be the first line of attack against depression.

By the mid-1970s, it seemed that depression was due to too little monoamine transmission, just as schizophrenia was due to too much. This conclusion implied a rather simple, in fact simplistic, picture of the illness. We now know that its mechanism is much more complicated. When considered from the point of view of whether patients were helped, rather than whether the underlying theory was correct, the monoamine hypothesis was a success. But the success came not from the fact that the monoamine drugs that were developed in the 1950s, 1960s, and 1970s were the ultimate answer. Instead it was because they helped pave the way for a logical approach for future drug development—by making effective drugs more specific, side effects might be reduced. As in the case of schizophrenia treatment, recent attempts to develop drugs for depression are more promiscuous in their flirtations with transmitter systems in an attempt not just to reduce side effects but to create more effective treatments. This has severely compromised the monoamine theory, but has facilitated the search for new and more effective approaches to treating depression.

DRUGS THAT TALK

In the late 1980s, a new chapter in the treatment of depression began with the development of drugs that selectively enhanced the availability of serotonin, in contrast to the earlier generation of drugs, which augmented both serotonin and norepinephrine. Offering the relief of depression with fewer side effects, and possibly even making "normal" people happier, the selective serotonin reuptake inhibitors (SSRIs) took the world by storm. Cover stories in magazines like *Newsweek* and a best-selling book, Peter Kramer's *Listening to Prozac,* helped make them a popular phenomenon.[49] SSRIs have been prescribed to millions and millions of people around the world, not just for debilitating depression but also for various shades of unhappiness, back pain, anxiety, stress, PMS, and a variety of other discontents. Prozac came to be seen as the ticket to a happier life, free of stress and filled with friends and success. As *Newsweek* put it, "Shy? Forgetful? Anxious? Fearful? Obsessed? How Science Will Let You Change Your Personality with a Pill."[50]

Over the past decade or so, a number of different SSRIs have been marketed, including Prozac, Zoloft, Paxil, Luvox, and Celexa. These drugs are as effective as tricyclics in helping depressed people, but probably not more so.[51] This is to be expected, since both classes of drug basically do the same thing—make more serotonin available at synapses. But the SSRIs do the job with fewer side effects—they are selective for serotonin, so the side effects caused by enhancing norepinephrine are eliminated.

More people have taken SSRIs than any other prescription medication. About 3.5 million prescriptions were written in one year alone.[52] Because the side effects are less severe than those of MAO inhibitors or tricyclics, patients don't have to be monitored so closely, and many prescriptions are given out by general practitioners rather than psychiatrists.

There is, however, a growing movement against the use of SSRIs and other psychiatric medications. In books like *Talking Back to Prozac* and *Toxic Psychiatry,* Peter Breggin has mounted an attack against the widespread use of drugs, arguing for an approach to psychiatry based on love, trust, understanding, and traditional psychotherapy.[53] His Website claims that the side effects of SSRIs include violent and suicidal tendencies, and cites all sorts of literature on the topic, including articles with titles like "Prozac and Xanax Found by Court to Cause Criminal Conduct" and "Was School Shooter Eric Harris Taking Luvox?" Side effects of SSRIs, including facial tics and sexual dysfunction, are also emphasized in *Prozac Backlash,* by Joseph Glenmullen.[54] While these are legitimate concerns, it is important that they be evaluated in the same way as the positive effects of the drugs, that is, through controlled studies rather than on the basis of anecdotes.

Yet another group of books makes a personal case for the use of such drugs.

It's hard not to be moved by the autobiographical accounts of people who, after years of suffering with depression, have been given a new lease on life by drug therapy. Kay Redfield Jamison's *An Unquiet Mind*[55] is especially noteworthy, but there are others.[56] Again, though, careful research is the best way to evaluate drug efficacy.

What, then, are the chances that these drugs actually will help an individual? Scientific studies show that many clinically depressed patients (about 50 to 60 percent) improve when given SSRIs or tricyclics, but some people given placebos (sometimes as many as 30 percent of the sample) get better as well.[57] The most cynical view of such data is that drugs don't accomplish much more than placebos. The more positive view—that some people really *are* helped by drugs—is suggested by the fact that in contrast to SSRIs, placebo effects tend to be smaller and shorter-lasting and hard to interpret.[58] Further, a recent report has challenged the validity of placebo effects.[59]

But what about the notion, promoted by critics, that many of the people taking SSRIs are not really clinically depressed but just out for a little mood boost or personality makeover? I asked Greg Sullivan, a psychiatrist friend of mine from Columbia University, what he thought of this criticism, and he responded as follows:

> In my experience I do not know of anyone for whom this is true. In general, these drugs are not fun to take (it's bothersome to remember a daily pill, they often cause sexual dysfunction, they can lead to slow weight gain, and friends and family sometimes imply that taking drugs for depression is a sign of mental weakness). If someone is pleased with the effects of an SSRI, that's usually an indicator that the drug has had a significant impact on serious symptoms, including those caused by a chronic low-level depression (dysthymia). Such people are amazed once they learn that low energy, chronic sleep disturbance, and a pessimistic approach to just about everything doesn't have to be the norm for the rest of their life. But SSRIs are not "happy pills," and people without significant mood or anxiety disturbance do not generally get anything beneficial from them, certainly nothing that would make them sustain the use. This is a sharp contrast to psychiatric drugs that are sometimes abused such as Ritalin, which makes people high if they administer enough, or benzodiazepines, which some people use in order to become numb to their difficulties and to life in general.

The bottom line seems to be that, if you are seriously depressed, drug treatment, especially in conjunction with psychotherapy, is probably your best hope, especially if psychotherapy alone has been unhelpful. There are likely to be side effects, and you have to decide whether the side effects are going to be worse than living with depression. But this is a question that has to be posed about almost any medication, since there are few magic bullets in medicine.

By now, SSRIs have been in the spotlight for over a decade and have been very profitable for drug companies. Some scientists have suggested, with reason, that the success of these drugs has taken some of the impetus out of developing others.[60] At the same time, the best way to make money in the drug business is by creating better drugs. Much of the effort in this area in recent years has gone into making serotonin-related drugs more specific, focusing on selective receptors rather than on presynaptic uptake in general. There are numerous serotonin receptors, and the more specific the drug can be made and still achieve a therapeutic effect, the fewer the side effects should be.

Because depression is not a single clinical condition, the same treatment doesn't work in all depressed people. One strategy being pursued is to find new treatments that involve novel combinations of drugs.[61] For example, in some cases resistant to either tricyclics or SSRIs, a cocktail involving a mixture of an SSRI and a tricyclic is useful.[62] The reason for this is not clear, but this is not unusual in psychiatric drug therapy, since so little is known about the underlying brain malfunction. In fact, one of the obstacles to the development of better drugs for psychiatric disorders has been that the main insights into the underlying brain malfunctions have come from findings, often accidental ones, about effective treatments.[63] This is how the monoamine approach emerged, and why it continues to be a major focus of treatment strategies. However, there is growing recognition that, while monoamines are involved, the fundamental problem underlying depression is not in the monoamine system per se.[64] Drug development is, as a result, moving in other directions, targeting other transmitters, modulators, hormones, second messenger systems, and neurotrophic factors. Some of these efforts are described below.[65] As molecular biological approaches become better integrated with circuit-level approaches to brain function, new and better ideas about the nature of depression and its treatment are likely to emerge.

STRESS IS DEPRESSING

One of the most consistent biological findings about depression is that the adrenal cortex secretes more of the stress-related hormone cortisol in depressed people.[66] This simple fact, which can be determined from a cotton swab containing saliva, links depression to the biology of stress. Over the years, a great deal of research has gone into understanding the relation between stress and depression, both through studies of depressed patients[67] and experimental animal models of depression.[68] However, recent findings about the adverse consequences of stress on the brain have led to a powerful new hypothesis about the origin of depression.

As we saw in chapter 8, in stressful situations, the concentration of cortisol rises in the bloodstream. This occurs because during stress, the amygdala and other brain regions alert hypothalamic neurons to release a peptide from their terminals in the pituitary gland. The peptide, called corticotrophin-releasing factor (CRF), causes the pituitary gland to release the hormone ACTH (adrenocorticotrophin hormone), which travels in the bloodstream and leads to the release of cortisol from the adrenal gland. The cortisol is then transported through the blood to various bodily organs and tissues (it eventually reaches the salivary glands, which is why it can be measured in saliva). In the brain, cortisol binds to receptors in the hippocampus, among other regions. When a sufficient number of receptors are occupied in the hippocampus, signals are sent to the hypothalamus telling it to stop releasing CRF. In this way, the hippocampus regulates the stress response triggered by the amygdala, keeping the release of cortisol within a normal safe range.

In the short run, stress responses are useful in mobilizing bodily resources to cope with danger. But if the stress is severe and continuous, the consequences can be serious. Your cardiovascular system can be compromised, your muscles can weaken, and you can develop ulcers and become more susceptible to certain kinds of infections.[69] But none of these should happen if the hippocampus is working properly to shut down the stress reaction. As we discussed in chapter 8, during prolonged and severe stress, the ability of the hippocampus to do its stress-control job falters.

Studies by Robert Sapolsky, Bruce McEwen, and others have shown that stress damages the hippocampus, leading to a shrinkage of dendrites and ultimately to cell death.[70] Not surprisingly, functions that depend on the hippocampus, like explicit or declarative memory, become severely compromised. Stress hormones do not damage the hippocampus directly, but instead deplete its neurons of glucose, their main source of energy, and make them less capable of performing their job in the face of stringent demands. As a result, they become especially sensitive to elevations of excitatory transmitters like glutamate during periods of increased neural activity, such as occurs in stress.

Specifically, hippocampal cells have a toxic reaction to synaptically released glutamate in the presence of depleted glucose.

Cell shrinkage and cell death mainly take place in one area of the hippocampus, the CA3 region. In another area, the dentate gyrus, cell death is far less likely to occur, but stress also takes its toll there. This is one of the few regions of the brain that is known to undergo neurogenesis, the production of new neurons, in adult organisms.[71] Neurogenesis increases when animals learn new things, and decreases when they are under stress, or when they are given steroid hormones that mimic the elevations of cortisol that occur in stress. Together, the cell shrinkage and cell death in the CA3 area and the loss of new neuron production in the dentate gyrus probably account for the fact that the volume of the human hippocampus is smaller in people who have elevated levels of cortisol due to stress or other conditions.

For example, as we've seen, cortisol is elevated in depressed patients, who often have a smaller hippocampus and who accordingly suffer memory problems.[72] Cortisol is also elevated in elderly people, especially those with memory problems and depression.[73] Further, there's a condition known as Cushing's disease in which the adrenal cortex secretes excess cortisol.[74] Cushing's patients also have a smaller hippocampus and memory problems, and many develop depression. Reduction of cortisol levels with drugs often reverses these structural and functional consequences. Similarly, people who receive prolonged steroid treatment for inflammation develop memory problems and depression, conditions that are reversed when the treatment is discontinued. One of the current trends in the treatment of depression is the development of drugs that alter the ability of CRF to lead to a rise in cortisol. Since cortisol is believed to be a major chemical culprit in stress (though perhaps not the only one), preventing cortisol from rising can sometimes help decrease stress and reverse depression.

The body system involved in controlling stress-hormone release is called the hypothalamic-pituitary-adrenal axis (HPA axis). This system did not evolve its ability to secrete cortisol for the purpose of making us depressed, but came about to help organisms deal with short-term physical dangers—situations that disrupt the normal physiological balance of the body, such as food or water deprivation, injury, or encounters with a predator or an enemy of your own species.[75] The job of cortisol is to mobilize body resources in the short run, after which physiological normality, or homeostasis, is restored. As stress guru Robert Sapolsky notes, animals in the wild do not have to deal with thirty-year-long escaping sprints from lions, but humans have thirty-year mortgages and other long-lasting problems. It is the relentlessness of human stress, the long-term disruption of homeostasis, that increases our risk of disease, including mental disorders like depression.[76]

Ron Duman, a Yale researcher, has proposed a theory that explains how stress can predispose one to develop depression.[77] According to Duman, depression results from an inability to make the appropriate adaptive response to stress. Duman's theory borrows heavily from the basic science findings of Sapolsky, McEwen, and others and emphasizes the effects of stress on the initiation of cell death in the CA3 area and on the suppression of neurogenesis in the dentate area. In the most general sense, he proposes that stress elevates adrenal steroids and decreases neurotrophic factors, like BDNF, a molecule that is essential for cell growth and survival (chap. 4) and for the maintenance of synaptic connectivity (chap. 6). In the presence of increased neural activity and glutamate release, CA3 and dentate cells depleted of glucose and lacking BDNF are endangered. In CA3, they atrophy and die. In the dentate, new cell growth and connectivity of new cells with existing networks is suppressed. Following antidepressant treatment, though, these changes can be overcome—dendrites of CA3 cells expand and the cells survive, and dentate cells continue to be generated and inserted into networks.

How, then, does antidepressant treatment achieve these goals? It has long been known that antidepressant drugs do not treat depression by simply altering synaptic levels of monoamines, since, as we saw earlier, their effects on synaptic levels are rapid but the treatment effects themselves take weeks. Duman proposes that the key to the drug effects are to be found in the second messenger cascades activated by the drugs. For example, when an SSRI increases the amount of serotonin available at serotonin receptors, these receptors are stimulated for longer periods of time. As a result, a stronger intracellular response is generated and a more intense activation of second messenger systems occurs, leading to enhanced gene activation and protein synthesis. The long-lasting effects of drugs, then, may be due to new growth processes, which take time to be put in place and to alter synaptic networks.

Let's look a little more closely at how this might work. Recall from earlier chapters that during memory formation, calcium elevation inside cells raises cAMP levels, which activates protein kinases (PKA and MAP kinase) that phosphorylate the transcription factor CREB. CREB then induces genes to make proteins that become new tools of synaptic transmission (new receptors, new channels that pass calcium or other ions, new transcription factors or kinases, and so on). Antidepressant treatment skirts real-world learning experience and directly increases the amount of calcium inside the cell, triggering other second messengers that induce genes to make proteins. A brain on antidepressants can be brought back from its state of isolation from the outside world and encouraged, even forced, to learn. The brain, in other words, is duped into being plastic by these treatments.

Antidepressant medication thus enhances plasticity and thereby makes the brain more adaptive, better able to overcome the endangered state into which it was placed by excess cortisol. While the treatment itself does not directly substitute for direct experience (it doesn't give you new memories), by placing the brain in a state in which new memory formation is facilitated, depressed persons may be able to learn new mental states and behaviors that override the modes that they had been locked into by depression. A therapist who understands the patient and the underlying effects of the drugs may be able to serve as a guide to the recovery process, helping the patient turn to new, positive live experiences, or to turn back to old ones, at just the right time of antidepressant treatment.

Before leaving the topic of stress and depression, it's important to point out that the hippocampus is not the only region of the brain that is altered in depressed people. Imaging studies in humans show changes in size and in functional activity in the prefrontal regions as well.[78] Further, animal studies show that this brain region has abundant receptors for adrenal steroid hormones and, like the hippocampus, is implicated in the regulation of HPA axis function.[79] The elevated cortisol that occurs in depression therefore also attacks the prefrontal cortex, which might help account for other cognitive changes associated with depression, including poor short-term (working) memory, distractibility, and altered decision-making and executive functions.[80] Changes also occur in the amygdala.[81] More studies are needed to determine the extent to which antidepressant treatment reverses structural and functional changes associated with depression in these areas.

Depression, like schizophrenia, is thus no longer simply viewed as a monoamine imbalance. It is instead believed to involve altered circuits that lock one into a state of neural and psychological withdrawal in which the brain's ability to attend to, engage, and learn about the world is reduced. Any treatment that can reengage a person with the world is likely to help. Whether psychotherapy alone can achieve this will likely depend on how far the maladaptive changes have progressed, as well as on factors peculiar to the individual, such as genetic makeup and past learning experiences.

MOTHER'S LITTLE HELPER

Anxiety disorders are the most prevalent form of mental illness. Schizophrenia afflicts about 1 percent of the population and depression about 15 percent, but roughly 25 percent of all adults are said to suffer from some manifestation of clinical anxiety.[82] Everyone is a little anxious at times, but an anxiety disorder is said to exist when day-to-day psychological functioning and interpersonal relationships are disrupted by symptoms such as worry, tension, sleep disturbances, irritability, and somatic complaints. Anxiety is not a single disorder but a spectrum of conditions that includes nonspecific or generalized anxiety, phobias, panic attacks, posttraumatic stress disorder, and obsessive-compulsive disorder.

Psychotherapy is more effective in treating anxiety disorders than any other group of mental problems.[83] This is true both of generalized anxiety and of the more specific conditions. The most effective forms of psychotherapy for anxiety involve behavioral and cognitive-behavioral approaches, which help the patient cope with and eliminate the unpleasant symptoms that accompany and perpetuate anxiety.

Medicinal approaches to anxiety have also been valuable.[84] Alcohol is the oldest and most widely used anxiety-reducing drug. Many unwind from the tensions of the working world by stopping off for "happy hour" at their favorite watering hole, or with a "cocktail hour" at home. But because alcohol is intoxicating and addictive, it is not a practical solution for severe and chronic anxiety.

The first widely used medical treatment for anxiety involved barbiturates, long-acting drugs that depress brain activity. At low doses, they reduce tension and anxiety without producing intoxication, and at higher doses, they induce sleep. However, these drugs are highly addictive, and the difference between a dose that induces sleep and one that is fatal (by suppressing breathing) is quite small. The development of a new class of anxiety drugs in the 1950s was thus welcomed.

Meprobamate was identified in the process of screening variations of mephenesin, a drug that relieved anxiety but had a very short-lasting effect. One method that drug companies used to test the value of antianxiety drugs was to give them to monkeys, which tend to be wild and aggressive in captivity. Meprobamate made them calm, less vicious. When given to humans, it reduced anxiety without producing drowsiness. This compound was then made available by prescription under the brand names Equanil and Miltown, and came to be known, in the popular press, as "happy pills." Though initially believed to be a safe, nonaddicting replacement for barbiturates, later reports were less favorable. Addiction did result, and sometimes medical complications occurred when the drug was stopped.

The next generation of drugs for anxiety emerged when a chemist who had worked developing dyes in Poland moved to the United States and joined a drug company. He tested a number of the compounds he had worked with in his dye research in Poland and found that one of them, and only one, was effective in calming wild monkeys. Upon further investigation, it turned out that this compound was different from all the rest. Chlordiazepoxide, as the compound was called, became the first of a new class of drugs, the benzodiazepines.

The effects of chlordiazepoxide were further tested in rats using the conflict test. The rats were made hungry, and then were allowed to obtain food by pressing a bar. However, if they pressed while a light was on, they were shocked. The light itself came to be a fear-arousing stimulus and led to outward signs of nervousness (freezing, defecation, etc.). When treated with chlordiazepoxide, though, they pressed more and were less visibly afraid even while the light was on. The drug allowed them to earn their daily bread and to be less fearful in doing so—perfect effects for a treatment of human anxiety. The conflict test in rats has, in fact, become one of the key methods for determining whether new compounds are likely to be effective in relieving anxiety in humans.

Chlordiazepoxide was sold as Librium. Then came diazepam, better known as Valium, and alprazolam, marketed as Xanax. By 1975, 15 percent of the U.S. population was said to have taken at some point a version of "mother's little helper." Though highly effective in relieving anxiety, and far less problematic than barbiturates and meprobamate, benzodiazepines are likewise addictive, and can be dangerous when mixed with alcohol. Judy Garland, for example, is believed to have died from this combination.

In contrast to most other drugs used to treat mental illness, all the antianxiety drugs work immediately. That is why a drink can calm your nerves, and why Valium and other benzodiazepines are useful in acute anxiety. In *Starting Over,* a romantic comedy from the late 1970s, the character played by Burt Reynolds is having a severe anxiety attack in Bloomingdale's while a crowd gathers around him. Someone calls out for a Valium, and essentially the entire crowd reaches into their purses or pockets and pulls out a vial. Valium jokes are, in fact, still popular in comedies. Hugh Grant's character in

Woody Allen's recent film *Small Time Crooks* kept running to the bathroom for a Valium when things didn't go his way.

While benzodiazepines are also useful as sleeping pills on an as-needed basis, these drugs can produce temporary memory loss after waking. There are infamous stories of businessmen taking a benzodiazepine to sleep on an overnight trip to Europe and then waking up and not being able to remember why they had flown overseas. A new class of drugs, imidazopyridines, are more useful for inducing sleep. Ambien is a prime example. Though their chemical structure is somewhat different from benzodiazepines, they work similarly (though more specifically) at the chemical level. Because these drugs have a shorter life span in the blood, and because they are not muscle relaxants, they have fewer lingering effects the next day on cognitive and behavioral functioning.

The classic antianxiety drugs work primarily by facilitating GABA inhibitory transmission in the brain, making it harder for glutamate to elicit excitation at its postsynaptic receptors. This is true for alcohol, barbiturates, and benzodiazepines (and imidazopyridines). However, each achieves its effects differently. Recall from chapter 3 that when a neurotransmitter binds to its receptor, the receptor opens, allowing electrically charged chemical ions to flow from the extracellular space into the cell. In the case of GABA receptors, the flow of chloride ions makes the inside of the cell more negative, which means that more positively charged ions have to flow into the cell through glutamate receptors to initiate an action potential. Barbiturates enhance inhibition by acting directly on GABA receptors, keeping the chloride channel open longer, allowing more negative ions to enter the cell. Alcohol has a similar effect, but at a different GABA receptor from the one affected by barbiturates. Benzodiazepines, however, work differently, as they have their own receptors, which are linked to GABA receptors. So when benzodiazepine receptors are occupied, linked GABA receptors bind GABA more easily. As a result, the same amount of GABA released from a GABA terminal will have a greater inhibitory effect on the postsynaptic cell. The benzodiazepines therefore work only at those sites where GABA is being released naturally. Imidazopyridines work like benzodiazepines, but at a more selective part of the GABA receptor, which is probably why they have fewer side effects. (Parenthetically, the fact that benzodiazepine receptors exist in the brain means that the brain also has a natural supply of benzodiazepines. It may be that calm, worry-free people have more of this chemical, and anxious people less.)

Because GABA and benzodiazepine receptors are dispersed throughout the brain, drugs that bind to them do not affect anxiety exclusively. Anxiety may

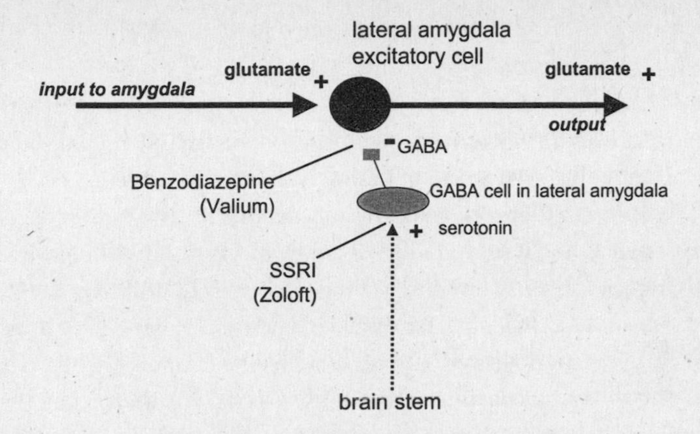

FIGURE 10.2 ANTIANXIETY DRUGS AND THE AMYGDALA

Two of the major types of drugs used to treat anxiety are benzodiazepines (like Valium) and selective serotonin reuptake inhibitors (SSRIs). Both may achieve anxiety relief, at least in part, through a common mechanism in the amygdala—enhancement of the inhibitory actions of the neurotransmitter GABA. Benzodiazepines directly enhance the inhibitory effects of GABA. By increasing inhibition, benzodiazepines weaken the ability of external or internal stimuli to activate the amygdala and produce fear and anxiety. SSRIs may also increase inhibition in the amygdala, though indirectly. It is known that increasing serotonin levels in the amygdala leads to an inhibition of amygdala activity, but this is achieved by exciting GABA cells, which then do the inhibiting. Since SSRIs make more serotonin available at the synapse by preventing uptake and thus breakdown, they would presumably also lead to GABA excitation and thus inhibition. This explanation works better for benzodiazepines than for SSRIs, since the former but not the latter are effective immediately in treating anxiety. Additional mechanisms, such as alteration in second messengers, are required to more fully account for the antianxiety effects of SSRIs (see earlier discussion of SSRIs and depression).

well be generated by specific networks, but drugs that treat it affect the entire brain. This accounts for some of the side effects of antianxiety drugs, such as drowsiness and muscle relaxation. Drowsiness, of course, is not considered a side effect when the drug is used for sleep induction rather than for anxiety reduction while awake.

Newer treatments for generalized anxiety are also available. Though these drugs produce fewer side effects, they have the disadvantage of not working immediately. Most of them enhance serotonin function. BuSpar (buspirone), for example, stimulates serotonin receptors (specifically serotonin 1A receptors). SSRIs, which as we've seen make more serotonin available at receptors, also appear to be helpful with these disorders.

WORRY CIRCUITS

Where does anxiety come from? Jeffrey Gray has long believed that the antianxiety drugs are the key to understanding the nature of generalized anxiety, and specifically the circuits in the brain that generate anxiety.[85] As he argues, alcohol, barbiturates, and benzodiazepines are different drugs from a chemical point of view, and each has different side effects, but each relieves anxiety. By studying the effects of the three drugs on a variety of different behavioral tasks in rats, and identifying a set of tasks that are affected by all three drugs, a behavioral portrait of anxiety can be drawn. And if brain regions can be discovered that when damaged lead to the same effects as the antianxiety drugs on these "anxiety" tasks, then the neural home of anxiety might be isolated.

This research program led to a neural theory of anxiety published in 1982. Because the common effects of the antianxiety drugs were similar to the effects of damage to the septum and/or hippocampus, these areas were postulated to be at the core of anxiety. We've encountered the hippocampus frequently, but not the septum. This is a region of the forebrain—actually part of the so-called limbic system—that is intimately associated with the hippocampus and regulates some of its activity.

From a psychological point of view, the septum and hippocampus were thought of as constituting the brain's behavioral inhibition system, a network that detects and responds to aversive stimuli, those that produce pain, punishment, failure, or loss of reward, or that elicit novelty and uncertainty. When the behavioral inhibition system is activated, ongoing behavior is inhibited (resulting, for example, in freezing), and the organism becomes aroused, attentive, and vigilant. Administering antianxiety drugs prevents aversive stimuli from eliciting septohippocampal arousal and vigilance, and damaging these areas reduces anxiety by eliminating the brain regions where arousal and vigilance would be elicited by aversive stimuli.

As we saw above, the classic antianxiety drugs work by enhancing GABA transmission. Although the drugs each affect GABA transmission by different means, they all work through GABA. A key part of Gray's theory was that GABA neurons were the common denominator in anxiety. However, because classic antianxiety drugs, when administered to the hippocampus itself, proved not very effective in relieving anxiety, Gray proposed that these substances alter GABA transmission in another brain region that subsequently influences the septohippocampal system. The key area of action was proposed to be the monoamine systems in the brain stem, specifically the serotonin and norepinephrine systems. These areas were selected on the basis of earlier studies showing that, when stimulated in rats or monkeys, they elicited anxiety-like behaviors, and that antianxiety drugs injected in the same areas inhibited cell activity and relieved anxiety-like behaviors.[86]

Before considering Gray's theory of how anitianxiety drugs work, let's briefly look more closely at his conception of what anxiety is at the level of neurons. During threats, serotonin and norepinephrine cells in the brain stem are activated, and serotonin and norepinephrine are released from the terminals of these cells. Although the terminals are located in many regions of the brain, because serotonin and norepinephrine are modulators, their main effect is to alter transmission at active synapses rather than to elicit activity themselves (chap. 3). Since the septum and hippocampus are involved in processing threats, these regions would be active in the presence of threats, and serotonin and norepinephrine would enhance their synaptic processing, leading to arousal and vigilance, and anxiety.

Now for the antianxiety effects. The antianxiety drugs, as we've seen, enhance GABA transmission. The enhancement mainly occurs at synapses where GABA transmission is taking place, rather than at inactive synapses. Because GABA activity is often triggered in response to excitation in nearby cells, the enhancement of GABA by antianxiety drugs in the presence of threatening stimuli will be concentrated at those synapses processing the threat. In addition to activating cells in the septum and hippocampus, threats activate serotonin and norepinephrine cells. Gray therefore proposed that antianxiety drugs enhance GABA transmission at these brain stem cells, and thereby reduce the release of serotonin and norepinephrine in the forebrain. The enhanced processing that these chemicals were engendering in the septum and hippocampus will in turn be eliminated, decreasing arousal and vigilance, thereby reducing anxiety.

Although Gray's theory was a tour de force of neuropsychological logic, it ran into trouble. First of all, it focused exclusively on GABA transmission, since this was a point of convergence between all the classic antianxiety drugs. The emergence of drugs that successfully treated anxiety without affecting GABA transmission therefore challenged the theory. Buspirone and SSRIs, for example, are both effective in relieving anxiety and both work at the level of receptors postsynaptic to serotonin terminals rather than at cell bodies in the brain stem. Indeed, these drugs are somewhat effective when injected directly into the hippocampus, where they selectively enhance serotonin transmission.[87] While this helps salvage part of the theory, it weakens the core concept that facilitation of GABA transmission in the brain stem is key. Second, the theory did not allow a role for the amygdala, which as we've seen plays an important role in processing danger and threats.

In a recent revision of the theory, Gray and his longtime collaborator Neil McNaughton attempted to address both of these shortcomings.[88] Their new formulation conceives of anxiety as a psychological state emerging out of a synaptically connected network involving, among other regions, the septo-hippocampal area, amygdala, and prefrontal cortex, all modulated by brain stem monoamine systems. Although the new theory is broader in its coverage, in my opinion, it still gives the septum and hippocampus too prominent a role, at the expense of the amygdala and prefrontal cortex.

Anxiety, in my view, is a cognitive state in which working memory is monopolized by fretful, worrying thoughts. The difference between an ordinary state of mind (of working memory) and an anxious one is that, in the latter case, systems involved in emotional processing, such as the amygdala, have detected a threatening situation, and are influencing what working memory attends to and processes. This in turn will affect the manner in which executive functions select information from other cortical networks and from memory systems and make decisions about the course of action to take.

I believe that the hippocampus is involved in anxiety not because it processes threats, as Gray suggests, but instead because it supplies working memory with information about stimulus relations in the current environmental context, and about past relations stored in explicit memory. When the organism, through working memory, conceives that it is facing a threatening situation and is uncertain about what is going to happen or about the best course of action to take, anxiety occurs.

In my formulation, the amygdala plays a greater role in threat processing than the hippocampus. I have described the role of the amygdala in processing and responding to threats in earlier chapters and in *The Emotional Brain*,[89] and will only briefly recount it here. When sensory information about a threatening stimulus is detected by the amygdala, output connections to response-control systems in the brain stem initiate the expression of defense responses (freezing) and supporting physiological changes in the body (rises in blood pressure and heart rate, stress hormone release, and so on), some of which give rise to signals that are fed back to the brain and influence ongoing processing. Although Gray's theory did not address the question of how monoamine systems are activated by threat, one of the key ways this occurs is through direct connections from the amygdala to the monoamine cells. Thus, during threats, serotonin and norepinephrine (and dopamine) are released in widespread forebrain areas (including the prefrontal cortex, hippocampus, and amygdala, as well as other areas). Direct connections from the amygdala to prefrontal areas (including anterior cingulate and orbital cortex) allow the detection of threats by the amygdala to directly influence working memory processing. But the prefrontal cortex and its working memory functions are also influenced through other routes, including the amygdala-initiated release of monoamines and the feedback from hormonal and other bodily responses. When the amygdala detects a threat, it triggers consequences that ultimately place working memory in a vigilant processing state, causing it to continue to attend to whatever it is occupied with at the moment, biasing thoughts, decisions, and actions. Further, output connections of the amygdala to the nucleus accumbens, as we saw in the last chapter, allow threatening stimuli to motivate the organism to avoid the source of the threat. This is an important factor, given that pathological avoidance of possible threat sources is a paramount behavioral symptom of anxiety disorders.

Up to this point in the discussion, I've not made a clear distinction between fear and anxiety. Classically, though, fear is viewed as a reaction to a specific and immediately present stimulus, whereas anxiety is a concern about what *might* happen. One possible resolution of the Gray-McNaughton theory with the amygdala theory would be if amygdala networks took care of fear and hippocampal networks took care of anxiety. This might have been an acceptable thesis some years ago, but there is now strong evidence showing that injection of benzodiazepines directly into the lateral and basal amygdala, the input stages of the amygdala, reduces anxiety behaviors in several (though not all) of the classic tasks used to test the efficacy of antianxiety drugs.[90] That finding is consistent with the fact that benzodiazepine receptors are concentrated in these input regions of the amygdala.[91] Further, atypical antianxiety drugs that target serotonin receptors also relieve anxiety in animal models when injected into the amygdala.[92] Both classical and atypical antianxiety drugs may therefore achieve their anxiety-reducing effects, at least in part, at the input stages of amygdala processing, making it harder for threatening stimuli to initiate activity in the amygdala, thereby preventing amygdala activation from arousing the rest of the brain.

But another theory has been proposed that distinguishes fear and anxiety in a different way. Michael Davis has performed experiments suggesting that anxiety might be a function of the bed nucleus of the stria terminalis,[93] a brain region that is considered an extension of the amygdala and whose output connections are remarkably similar to those of the amygdala.[94] Because of their similar outputs, the amygdala and bed nucleus can affect many of the same target brain areas (prefrontal cortex, monoamine systems, and so on) in comparable ways and produce the same kinds of bodily responses (muscle tension, fast-beating heart, sweaty palms, tight stomach, and so on). But because the inputs to the two structures are different, they might be activated under different conditions—the amygdala in response to immediately present threats, the bed nucleus to anticipated ones. This distinction between the role of the amygdala in fear and the bed nucleus in anxiety is appealing because it potentially explains why people on antianxiety medication can be generally less worried (that is, less anxious), and still be capable of responding to an immediate threat. However, much work remains to be done to clarify the role of these two regions and their relation to the broader circuits involved in anxiety.

In summary, generalized anxiety is an aroused state of mind initiated and maintained by emotional processing. As a result, it requires, at a minimum, networks involved in arousal (monoamine systems), emotional (amygdala, perhaps including the extended amygdala), and cognitive (prefrontal cortex, hippocampus) functions. And while individual brain regions and networks make distinct contributions to the processes that together constitute anxiety, anxiety itself is best thought of as a property of the overall circuitry rather than of specific brain regions.

THE SPECTRUM OF ANXIETY

So far, this overview has focused on free-floating or generalized anxiety. But the anxiety disorders also include phobias, panic disorder, posttraumatic stress disorders, and obsessive-compulsive disorder. These are distinct conditions with different causes and symptoms, and the treatments that work for one do not necessarily work for the others.[95]

Benzodiazepines are of little help in treating the so-called simple phobias (pathologic fear of snakes, spiders, heights, and so forth). In fact, the most effective treatment for these conditions is psychotherapy, especially cognitive behavioral therapy, in which the symptoms are reduced by exposure to the feared stimulus under various conditions (see the box entitled "Cognitive Behavioral Therapy"). Another kind of phobia, social phobia, centers around a fear of being negatively evaluated by others—for example, by people with whom one works or shares interests. Some forms of social phobia involve a

COGNITIVE BEHAVIORAL THERAPY

Therapy is a process of changing the way a patient thinks, feels, and/or acts. There are many different forms of therapy, but for the purpose of explaining what cognitive therapy is, we can contrast it with other approaches. Psychoanalysis attempts to get at the cause of the maladaptive condition, often by making unconscious (repressed) memories conscious. This is a long, slow process. Behavioral therapy, by contrast, seeks to alter maladaptive conditions (basically, bad habits) through the learning principles of behaviorist psychology, namely, reinforcement, extinction, counterconditioning, and so forth. Behavioral therapy, also known popularly as behavioral modification, focuses on observable symptoms and teaches the person to act in new ways. While fairly effective in rapidly altering certain pathologic tendencies, behavioral therapy was criticized for dealing with symptoms rather than causes, and for failing to acknowledge the importance of mental life in the initiation and maintenance of pathological conditions. Cognitive therapy, by contrast, takes as its starting point the notion that dysfunctional mental states (beliefs, attitudes, ideas) contribute significantly to psychopathology, and that the pathological conditions can be altered by helping the patient to identify and correct the beliefs. Rarely, though, does the simple realization that one has been thinking irrationally lead to improved mental health. The person has to learn new ways of thinking and acting. Many contemporary therapists blend cognitive and behavioral approaches and practice cognitive behavioral therapy (CBT), a process in which bad habits (mental and behavioral ones) supported by dysfunctional cognitions are changed. CBT has proven to be fairly effective as a treatment for a wide variety of nonpsychotic conditions, including the various anxiety disorders and depression, sometimes used alone and sometimes in conjunction with medication.

SELECTED READINGS

Beck, A. T. 1991. Cognitive therapy: a 30-year retrospective. *Am. Psychol.* 46: 368–75.

Gorman, J. M. 1996. *The New Psychiatry*. New York: St. Martin's Press.

Hollon, S. 1999. What is cognitive behavioral therapy and does it work? *Curr. Opin. Neurobiol.* 8: 289–92.

Zinbarg, R. E., D. H. Barlow, T. A. Brown, and R. M. Hertz. 1992. Cognitive-behavioral approaches to the nature and treatment of anxiety disorders. *Ann. Rev. Psychol.* 43: 235–67.

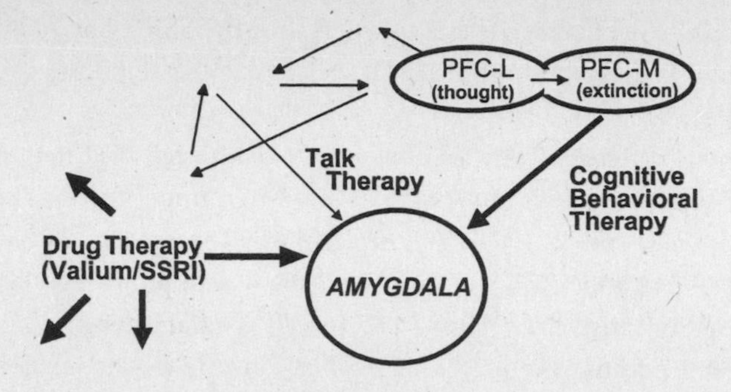

FIGURE 10.3 THERAPY AND THE AMYGDALA

Possible modes of action of three broad classes of therapy on amygdala function and dysfunction are illustrated. Classic psychotherapy (talk or insight therapy and all variations that require some conscious understanding in order to achieve a therapeutic effect) is likely to directly require the working memory functions of the prefrontal cortex While the lateral prefrontal cortex (PFC-L) is the classic working memory area, other more medial regions (PFC-M) also appear to be involved in working memory. The absence of direct connections from the PFC-L to the amygdala may be related to why talk therapy for psychiatric conditions that involve amygdala-related conditions is relatively inefficient (in terms of the amount of time required to achieve a therapeutic effect). Behavioral therapy (including cognitive behavioral therapy) is less dependent on conscious insight and more dependent on extinction processes and on the development of new associations, skills, and habits (that is, on implicit learning). Some of these processes (especially extinction) involve the PFC-M. The direct connection of the PFC-M with the amygdala may explain why cognitive behavioral therapy is more efficient for certain fear/anxiety-related problems. Drugs can go straight to the amygdala, but they will also affect other brain regions, possibly accounting for some of the psychiatric side effects of drug therapy.

fear of being the center of attention in public. Examples of the latter include an extreme debilitating fear of public speaking or performance, and fear of eating in public or using public bathrooms. Antidepressant medications, including certain SSRIs and monoamine oxidase inhibitors, are often used in treating these conditions. Cognitive behavioral therapy is also useful. Drugs like propranolol, which reduce the bodily signs of fear (high heart rate and so forth) but that do not affect the brain directly, are sometimes used to aid in public speaking and stage fright. As with social phobia, the preferred treatment for panic disorder and posttraumatic stress disorder today is the use of an SSRI in combination with some form of psychotherapy involving anxiety management and cognitive behavioral therapy. A variety of other drugs (including monoamine oxidase inhibitors, tricyclics, and benzodiazepines) also have some positive effects in these disorders, but the SSRIs have a better side-effect profile. Obsessive-compulsive disorder likewise responds to SSRIs to some degree.

Because the different forms of anxiety are manifested in different mental, behavioral, and physiological symptoms and often under distinct conditions (having to give a speech may have little effect on a spider phobic but will be a source of fear and anxiety to one who has a fear of public speaking), there is something unsatisfactory about the use of similar medications to treat them all. And because the preferred medications are antidepressants, rather than drugs developed for specific anxiety disorders, one can't help but feel that from the point of view of drug development and treatment, we have hardly touched these mental problems, some of which are widespread and devastating. Certainly, following the strategy of using information about effective treatments to give clues to the underlying biological disturbance is not going to reveal differences between various anxiety disorders, since the same treatment (SSRIs) is used for different anxiety disorders, as well as for depression.

Earlier, I mentioned that I am the director of the Center for the Neuroscience of Fear and Anxiety (*http://www.cns.nyu.edu/CNFA/*). This center brings together scientists at all of the major biomedical research institutions in Manhattan, including New York University (Elizabeth Phelps, Mony de Leon, and me), Rockefeller University (Bruce McEwen), Cornell University Medical College (David Silbersweig and Emily Stern), Mt. Sinai Medical School (John Morrison and Patrick Hof), and Columbia University (Jack Gorman). The research involves studies of animals and of patients with anxiety disorders. Our immediate goal is to elucidate the ways in which the brain is changed in anxiety disorders, in the hope of facilitating, over the long run, the development of more selective and more effective treatments with fewer side effects. The patient work involves the use of state-of-the-art functional MRI imaging techniques, and the animal work takes advantage of the vast arsenal of neuroscience tools that has been discussed throughout this book.

Because anxiety disorders involve alterations in the processing and/or reaction to threat and danger, and fear mechanisms are to a large extent conserved in the brains of humans and other mammals (chap. 8), research on the psychology and neuroscience of fear can inform ideas about the underlying disturbance in different anxiety disorders.[96] Although there have been several imaging studies of patients with anxiety disorders of various types,[97] the aim of our center is to use the enormous progress that has been made in elucidating the neural pathways underlying fear conditioning in animals to try and understand the nature of the brain changes that take place in anxiety disorders, especially panic disorder and posttraumatic stress disorder (PTSD).

Panic disorder and PTSD both involve alterations in the processing of threats, but the causes and symptoms are distinct.[98] In panic disorder, fear is expressed in the form of discrete and sudden panic attacks that arise without an obvious environmental stimulus and lead to chronic anticipatory anxiety and various degrees of avoidance behavior. Fear in panic patients has no apparent relationship to any actual threat and often involves abnormal sensitivity to uncomfortable somatic sensations. In PTSD, fear is expressed as increased sensitivity to stimuli reminiscent of an original, life-threatening traumatic event and includes flashbacks and increased startle response. Different neural mechanisms should therefore be altered in these patients.

As we've seen repeatedly, the amygdala is a key brain region in the acquisition and expression of fear elicited by stimuli associated with threatening experiences. And several studies of normal humans have now shown amygdala activation during fear conditioning, even when the conditioned stimulus is presented subliminally, that is, unconsciously.[99] We therefore can use fear conditioning to activate the amygdala and determine whether the pattern of activation there is altered in patients with anxiety disorders. Our expectation is that because changes in the way fear-arousing stimuli are perceived and responded to characterize all anxiety disorders, we should see increased levels of activity in the amygdala in both panic and PTSD patients.

But the amygdala is not the only area implicated in fear conditioning. As we've seen, the hippocampus is involved in processing fear-arousing situations or contexts, and the medial prefrontal cortex in the adjustment of fear reactions in response to changing environmental conditions, as in extinction (the process whereby a conditioned stimulus loses its fear-arousing properties when it no longer predicts danger). Recent studies by Liz Phelps have confirmed the role of the human hippocampus in processing contextual fear stimuli,[100] and we are currently testing whether activation of the human medial prefrontal cortex will be altered during extinction.

The use of fear conditioning allows us to assess whether patterns of neural activity within and between the amygdala, hippocampus, and medial prefrontal cortex differ in panic and PTSD patients from normal controls. Further, we can determine whether successful treatment (e.g., the use of SSRIs and cognitive behavioral therapy) will cause the brains of these patients to look more like those of normal controls. By testing patients after treatment with either SSRIs or cognitive behavioral therapy alone, we should be able to determine whether the two kinds of therapy achieve their results by altering the same or different components of the circuits.

Once we have some sense of how the brains of panic and PTSD patients differ from each other and from those of normal persons, we can turn to animal studies and explore what kinds of life experiences and biological conditions might predispose the brain to change in these ways. For example, suppose we find that the amygdala is hyperactive in both panic and PTSD, but that the hippocampus and prefrontal cortex change in different ways in each condition. We could look for stressful experiences that produce these patterns of change in the rat brain. If such patterns can be identified, then the changes can be pursued at the level of synaptic alterations and explained in biological terms: as neurotransmitter actions on receptors that open channels and allow calcium to flow inside the cell and trigger kinases that phosphorylate transcription factors that induce genes to make proteins that stabilize synaptic changes. To the extent that any one of these molecular steps differs between anxiety disorders, clues about future drug development would be available.

By now you've surely noticed a trend. Neuroscientists have proposed that the prefrontal cortex, hippocampus, and amygdala are altered in some way in all forms of mental illness we've considered so far. It seems that mental disorders are giving us a message, telling us that these three brain regions are especially critical to understanding who we are and why we are that way. But we should remember that the same brain areas can be involved in different disorders for different reasons. At a fairly gross level of explanation, neural activity in a particular area might be up in one disorder and down in another. But in order for neural activity to change in this way, more basic changes have to be taking place within and between individual cells in the area in question. Since these deeper changes can be revealed only by exacting neurobiological studies, which can be done only in experimental animals, it is essential that research on psychiatric disorders in humans and basic science investigations of brain function in animals be coordinated.

We have just begun this research program, and it is too early to know just what we'll turn up. However, because of the systematic nature of the research program, we will almost certainly be able to point to differences in the brains of panic and PTSD patients, even if they are not the ones we expected. But, in many ways, unexpected findings are the most interesting, as they force thinking about problems in new ways.

BRAINS, GENES, AND THERAPY

We've gotten all the way to the end of this chapter on biological approaches to mental illness without discussing genes. To many, especially to critics, a biological approach to mental disorders implies an emphasis on genetics. But I hope I've presented a credible alternative view. As in all other aspects of mental life, genes can predispose people in certain directions without altogether predetermining their destiny. Genes contribute to, rather than solely dictate, synaptic connectivity.

Critics such as Colin Ross and Alvin Pam are concerned that a biological approach to mental disorders implies that "the individual with behavior problems must suffer in some way from defective protoplasm, a constitutional predisposition to mental illness," and that the body or brain is responsible for mental illness rather than the family or society.[101] They strongly equate biological psychiatry with genetics. However, while a genetic approach to mental disorders is by definition biological in orientation, the reverse is not necessarily true: a biological approach is not exclusively or even primarily genetic. Let's first take a look at the genetic approach, and then consider the broader picture.

A massive effort is now under way to determine the extent to which genes contribute to mental illness. For example, years of research have gone into pedigree studies of families or groups of people with a tendency toward a mental disorder, and into concordance studies of identical twins, which determine the likelihood that the presence of the disorder in one predicts the disorder in the other. Schizophrenia, depression, anxiety disorders, alcoholism, and a variety of other psychiatric and neurologic conditions are being studied this way.

For illustrative purposes, consider the work on schizophrenia.[102] The research shows a strong correlation between the number of genes shared and the likelihood of developing the disorder.[103] Schizophrenia occurs in about 1 percent of the general population. In contrast, in identical twins (100 percent gene overlap), if one child has schizophrenia then there's roughly a 50 percent chance the other will develop it at some point. But in fraternal twins (50 percent gene overlap), the likelihood drops to 17 percent. In siblings (25 percent gene overlap), the figure falls to 9 percent, and in first cousins (12.5 percent gene overlap) to 2 percent. The overall picture that emerges is that schizophrenia is strongly tied to genetic factors. That is, in genetic terms, there's 50 percent concordance among identical twins. But one can view this cup as half full or half empty—there's also 50 percent discordance. If genes fully "explained" schizophrenia, concordance would be 100 percent.

The discordance between identical twins is due to several factors.[104] One is that gene expression is an *epigenetic* phenomenon—it involves the interaction of the gene with environmental factors. An extreme example of epigenesis is the disease phenylketonuria (PKU), a genetic form of mental retardation that is not expressed unless the essential amino acid phenylalanine is ingested. (It is present in dairy products, avocado, certain nuts, and in the sugar substitute aspartame.) In this case, early detection and treatment can prevent the symptoms of the disease. In most genetic conditions, though, the environmental co-conspirators are not known. Another factor working against concordance is the polygenetic nature of the genetic contribution. In the days when psychiatric disorders were thought to be due to a simple imbalance in a particular brain chemical, it was plausible that the disorder might be caused by a single gene (one that makes transmitter or receptor molecules). But as we've seen, psychiatric disorders involve complex circuits in the brain, and therefore any genetic contribution is likely to involve interactions among multiple genes. Because the expression of every gene also involves interactions with environmental factors, polygenetic inheritance greatly expands the manner in which the environment can influence genetic predispositions. Further, in some cases, people with the gene for a disease show little or no evidence of having the symptoms of the disease. This is called *nonpenetrance.* In other instances, genetic diseases are variably expressed, where some family members may have the full-blown disease and others only mild symptoms. Finally, mental disorders can probably have multiple unrelated causes. It is believed, for example, that in some instances, schizophrenia can be induced by environmental factors, like brain injury or infection, in persons with little or no genetic leaning toward the disorder. So discordance between identical twins might sometimes be due to the fact that the afflicted twin acquired schizophrenia for nongenetic reasons.

Biological psychiatrists are interested in the possibility that genes contribute to mental disorders because genetic problems might in the future be fixable. A gene working overtime and thereby causing a malfunction might, with a pill, be turned down or off, and one that is sluggish might be encouraged to work harder. Whether these feats will ever be successfully accomplished is unknowable at the present. But if they are, many people are likely to be helped.

It should be noted that success or failure for one psychiatric disorder will not predict the success or failure in treating others. Each disorder will have its own peculiar advantages and disadvantages for a genetic approach, and each has to be attacked individually. At the same time, research strategies developed for one problem may help in the pursuit of the others. With the human genome recently pronounced as "knowable,"[105] this kind of work is likely to expand in the coming years.

But perhaps I'm not being completely fair to the critics. One of their concerns is that a focus on genes or other aspects of biology puts too much blame on the body, discouraging mental professionals from using nondrug forms of therapy, insurance companies from paying for nondrug therapy, and patients from taking responsibility for their own healing. But these concerns, I believe, reflect a misunderstanding of the aims of biological psychiatry.

In spite of the current enthusiasm for genetic analyses of mental disorders, most biological psychiatrists view genes in a realistic way. Raymond DePaulo, a biological psychiatrist at Johns Hopkins University, expresses what is probably the most common view: "I don't think depression is all genetic. If you can nail down one end of the tent, however, it's easier to figure out by stretching out the rest of the tent what it should look like."[106] In other words, once you know the genetic component of a disorder, it should be easier to figure out which kinds of environmental experiences interact with it to trigger the disorder and make it worse. Genes are part of the story, and if that part of the story can be interpreted, we should be better off than if we ignore it.

Experience is often considered the counterpoint to genes. But *experience* is a complex notion with many possible meanings and endless implications. As a result, the role of experience is probably going to be even more difficult to understand than the role of genes. But the more we understand about how experience changes the brain, the better off we'll be in discovering how it contributes to mental disorders. As we've seen, experiences affect the brain when they are stored as synaptic changes in one or more systems during learning, which is why research on synaptic plasticity, and on learning and memory, is so important.

Where does a biological approach to mental health and mental disorders leave psychotherapy? Eric Kandel, the renowned neuroscientist, started his career as a psychiatrist. Recently, he boldly proposed a "new intellectual framework for psychiatry," one founded on biological principles.[107] He argued that we are on the threshold of understanding memory and emotion systems in the brain, and that this information, combined with continued advances in molecular neuroscience, offers a wealth of opportunities for psychiatry.

As I pointed out at the beginning of this chapter, psychotherapy is fundamentally a learning process for its patients, and as such is a way to rewire the brain. In this sense, psychotherapy ultimately uses biological mechanisms to treat mental illness. This does not mean, however, that psychotherapy involves learning while drug therapy involves something else, like correction of genetically dictated chemical imbalances. Even if chemical imbalances were to account for mental disorders, the imbalance could result purely from environmental factors, like some intensely stressful experience, or from environmental events that trigger and amplify a genetic predisposition. The therapist's job, whether using drugs or not, is to restore mental well-being. If the patient's problem involves neural changes that are locking the brain in an aroused state, relief will come only when the aroused state is reduced, either by psychotherapy or drug therapy, or by some combination of the two. Regardless of whether the initial cause is social stress or a genetic time bomb, unless the changes in the brain that accompany the disorder can be reversed or circumvented, the problem is unlikely to dissipate.

Contemporary approaches to the treatment of mental disorders are not distinguished from older ones so much by an emphasis on drugs as by an emphasis on *effective* treatment. Jack Gorman of Columbia University, a leading biological psychiatrist, puts it this way: "In the Old Psychiatry, each clinician learned a set of techniques from one school of thought and insisted on applying them to all comers. In the New Psychiatry clinicians may still feel most comfortable with the tools of one school, but they readily admit when they don't have the right skill for a particular patient."[108] Today, psychoanalysts are willing to refer patients who need medication to a psychiatrist with skills in psychopharmacology, just as those trained in psychopharmacology increasingly send patients to a psychotherapeutic specialist. "The New Psychiatry," according to Gorman, "is only interested in providing safe and effective treatment. If that means borrowing from different points of view to offer a patient a combination of therapies, then that is what must be done."

Once a disorder exists, and the brain has changed, the changes have to be dealt with in some way in order for a patient to recover. Drugs can induce adaptive changes in neural circuits, or put neural circuits in a state where adaptation and learning are promoted. But there's no guarantee that, left to its own devices, the brain will learn the right things. Patients, in other words, are likely to benefit most from drug therapy when the drug-induced adaptivity of their brains is directed in a meaningful way. This is probably best achieved by traveling down the pharmacological road to recovery with someone who understands not just the drug or the person, but the drug, the person, and the life situation the person is experiencing. HMOs may not like it, but the drug, therapist, and patient are partners in the synaptic adjustment process called therapy, with drugs attacking the problem from the bottom up, the therapist from the outside in, and with the patient reaching up and down from his or her own synaptic self.

CHAPTER ELEVEN

WHO ARE YOU?

WHO ARE YOU—WHO WHO, WHO WHO?

—The Who

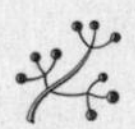

Figuring out how the brain works is a daunting task. For that reason, neuroscientists usually work only on pieces of the puzzle—like aspects of cognition, emotion, or motivation—rather than on the whole organ and its systems at once. But if we want to understand how our brains make us who we are, we have to figure out how these individual processes blend together to cause a person to emerge effortlessly from the electrochemical activities of the protoplasmic mass that is his or her brain. It's time for me to bite the bullet and explain how I think the brain, specifically its synapses, makes us who we are.

BRAINS AND OTHER PARALLEL COMPUTERS

I once had a friend who supported a bohemian life in downtown Manhattan with a job as a computer programmer in a neuroscience lab. He was extremely talented, too talented for such work, and eventually got a position more suited to his abilities. He was hired by a company in Boston that made powerful computers called "connection machines." It was through him that I first heard about parallel computers.

Parallel computers work differently from the standard model with which we're familiar. Rather than doing computations one at a time in sequence (in other words, carrying out the steps of a program line by line in serial order), they process many steps simultaneously. Parallel computers can function this way because, in contrast to your desktop PC or Mac, they have many processing units that can be devoted to the execution of a given task. By distributing the workload across the various processors, they can perform the task much faster than serial computers. (Don't plan on getting one for your home or office, however: they cost millions of dollars.)

The brain is also sometimes described as a parallel computer, but it actually functions differently from an off-the-shelf connection machine. The brain is organized into processors[1] (neural systems) that function independently of one another (at least to some extent). Since each of these systems has a specific job assignment, several types of tasks can be done by the brain simultaneously, that is, in parallel. This architecture enables you to chew gum and walk down the street, guiding yourself toward your destination while feeling happy and rehearsing the phone number your friend gave you a block back, all at the same time as your posture is maintained upright, your blood pressure is kept at a safe level, and your rate of breathing is paced to the oxygen needs imposed by all the activities in which you're engaged.

Connection machines can, like brains, be divided up in such a way that different groups of processors are responsible for particular tasks.[2] Although each task is then performed less efficiently than it would be if all the processors were devoted to it, overall this can be a more efficient use of the machine since multiple tasks can be worked on at the same time. Reversing the logic, if we had fewer neural systems using up the same overall computing power in our brains the systems would each be more powerful. However, because we have to do lots of different things each day to stay alive and well (eat, sleep, walk, avoid danger and pain, hear, see, smell, taste, talk, and think, to name some), with fewer brain systems we would almost certainly be less capable, even if the remaining systems were each more proficient at their particular tasks.

Over the long, slow evolutionary process of building the brains of vertebrates, and then of mammals, and eventually of our own species, the neural systems we have were specifically designed to take care of important jobs. We don't have extra ones that we can easily give up and continue to lead life as before. Similarly, new systems aren't easily acquired. For example, the addition of language and related aspects of cognition to the primate brain was not a trivial process—the brain was by that point in its evolution already fully booked. Adding a whole new set of functions, therefore, required either a loss of space devoted to some functions or an increase in brain size. In fact, both seem to have occurred. The human brain is bigger than that of other animals (relative to our body size)[3] and also seems to have undergone some reorganization. For example, the neural mechanism underlying the perception of spatial relations is present in both hemispheres of other primates; it is mainly on the right side in humans. This implies that spatial perception was forced from the left during the language invasion of human synaptic territory.[4]

Life requires many brain functions, functions require systems, and systems are made of synaptically connected neurons. We all have the same brain systems, and the number of neurons in each brain system is more or less the same in each of us as well. However, the particular way those neurons are connected is distinct, and that uniqueness, in short, is what makes us who we are.

THE PARADOX OF PARALLEL PLASTICITY

We've met a number of the brain's neural systems throughout this book. Included are networks involved in sensory function, motor control, emotion, motivation, arousal, visceral regulation, and thinking, reasoning, and decision-making. What is remarkable is that synapses in all of these systems are capable of being modified by experience. Consider a few examples.

Emotion systems, as we've seen, are programmed by evolution to respond to some stimuli, so-called innate or unconditioned stimuli, like predators or pain. However, many of the things that elicit emotions in us or motivate us to act in certain ways are not preprogrammed into our brains as part of our species heritage but have to be learned by each of us. Emotion systems learn by association—when an emotionally arousing stimulus is present, other stimuli that are also present acquire emotion-arousing qualities (classical conditioning), and actions that bring you in contact with emotionally desirable stimuli or protect you from harmful or unpleasant ones are learned (instrumental conditioning). As in all other types of learning, emotional associations are formed by synaptic changes in the brain system involved in processing the stimuli. Some of the brain's plastic emotional processors include systems involved in detecting and responding to danger, finding and consuming food, identifying potential mates and having sex.

Sensory systems are likewise plastic. Until very recently, it was believed that perceptions are stable from day to day and year to year because sensory systems are immutable after childhood. But studies by Norman Weinberger, Charles Gilbert, Mike Merzenich, and others have shown that these systems are remarkably susceptible to modification in response to external stimulation.[5] Merzenich, for example, has shown in a variety of ways how experience with a particular stimulus alters the net area, and thus the underlying synaptic territory, devoted to processing that stimulus. Following amputation of a finger, for example, the area in the somatosensory cortex devoted to that finger shrinks, while extensive stimulation of a particular finger expands the area of cortex devoted to it.

Motor systems are also plastic. This is obvious from the fact that we can learn skills and can improve our ability to perform certain movements with practice. As we saw in chapter 5, changes in synaptic transmission in the cerebellum are important for some forms of motor skill learning. Synaptic plasticity also occurs in the motor cortex, basal ganglia, and other brain regions involved in motor control.[6]

Because synaptic plasticity occurs in most if not all brain systems, one might be tempted to conclude that the majority of brain systems are memory systems. But as I argued in chapter 5, a better way of thinking about this is that the ability to be modified by experience is a characteristic of many brain systems, regardless of their specific function. Brain systems, in other words, were for the most part not designed as storage devices—plasticity is not their main job assignment. They were instead designed to perform particular tasks, like processing sounds or sights, detecting food or danger or mates, controlling actions, and so on. Plasticity is simply a feature that helps them do their job better.

The fact that plasticity does occur in so many brain systems, however, raises interesting questions. How does a person with a coherent personality—a fairly stable set of thoughts, emotions, and motivations—ever emerge? Why don't the systems learn different things and pull our thoughts, emotions, and motivations in different directions? What makes them work together, rather than as an unruly mob?[7]

LESSONS FROM DISCONNECTION

Before we examine what holds the self together, let's consider how fragile a patch job it is. The bottom line is simple: Functions depend on connections; break the connections, and you lose the functions. This is true of the function of a single system (a lesion of the visual thalamus, for example, will prevent information from the eyes from reaching the cortex, and thus will prevent the cortex from being able to perceive the visual world) as well as of interactions

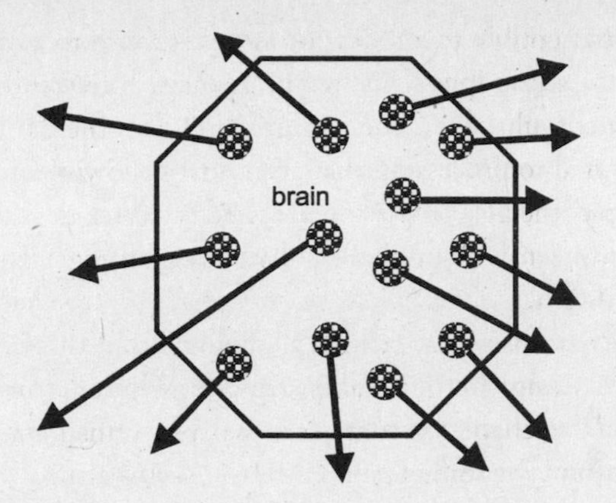

FIGURE 11.1 BRAIN SYSTEMS AS AN UNRULY MOB

Given that we have so many brain systems capable of learning and storing information about who we are, how is it that we develop distinct personalities? How can we function as an individual, a person, with goals and aspirations, with an identity, rather than as an unruly collection of systems that learn and store information on their own?

between systems (a lesion in a certain part of the temporal lobe will prevent information about visual objects from reaching the prefrontal cortex and thus will prevent that stimulus from being held in working memory, and hence from being used as the basis for thinking and decision-making).

The most striking instance of a broken connection I've ever personally witnessed involved a teenage boy who underwent split-brain surgery for the control of epilepsy.[8] In this operation, the nerves between the two sides of the brain are severed to prevent the seizures from moving between the left and right hemispheres. One of its unwanted consequences is that the two hemispheres become somewhat independent. Several days after the surgery, the boy was observed pulling his pants down with his right hand and up with his left. Since the right hand is under the control of the left hemisphere, and the left hand of the right hemisphere, the normal integration of motor control that exists so effortlessly in each of us broke down in this boy because the brain systems that had final say over what the two hands do were no longer connected. (This sort of conflicted behavior has been described for many such patients but was amazing to see firsthand.)

Another example involves conduction aphasia. Patients who suffer from this condition can speak without trouble and can understand spoken words (for example, they can point to the picture of an object named by someone else), but cannot, upon hearing a spoken word or sentence, repeat it, or answer a question posed to them. The reason for this, according to the late Norman Geschwind, one of the great neurologists of the twentieth century, is that the neural pathway that transmits information between the areas of the brain responsible for the comprehension and production of speech is cut.[9] Conduction between the brain areas is thus disrupted.

Geschwind used the phrase *disconnection syndromes* to characterize conditions in which specific behavioral or mental consequences result from disrupting communication between brain regions. What is striking about these disorders is that the deficit results not from the loss of a particular function, but from the inability to exchange information between brain areas. Disconnection syndromes illustrate how critical internal coordination between brain systems is in maintaining the unity of mind and behavior.

While the effects of brain lesions are not invariably interpreted as disconnection syndromes, disconnections always occur when the brain is damaged. For example, lesions of the prefrontal cortex, especially the ventral or orbital frontal cortex, have, since the nineteenth century, been known to drastically alter personality.[10] After an iron rod passed through the head of Phineas Gage in a railroad accident, this upstanding citizen suddenly became fitful, irreverent, and unrestrained. As one contemporary observer put it, "The equilibrium between his intellectual faculties and animal urges seems to have been destroyed."[11] Building on this case and numerous others, Antonio Damasio has suggested that damage to the ventral prefrontal cortex can lead to a loss of social control, and in the extreme can cause sociopathic behavior.[12] But it would be wrong to think of the ventral prefrontal region as the center of social grace. According to Damasio, the consequences of damage to the ventral prefrontal cortex can be conceived of as a breakdown in the ability to use emotional information to guide thoughts and actions. This seems reasonable given that some of the key connections of the ventral prefrontal cortex include areas involved in higher cognitive processes (other prefrontal regions such as the lateral prefrontal cortex and anterior cingulate cortex, as well as the hippocampus) and emotional and motivational functions (the amygdala and nucleus accumbens). Ventral prefrontal damage thus does more than just create a hole in this part of the brain. It removes that brain region from the circuits in which it participated. Regardless of whether it is obvious or not, brain lesions always produce disconnections.

Connectivity changes also take place in psychiatric disorders, but they are typically more subtle than those in neurological patients with overt brain lesions. As a result, psychiatric disorders might be best thought of as malconnection rather than disconnection syndromes. For example, as described in the previous chapter, a growing body of evidence suggests that certain forms of depression appear to involve alterations in the way circuits in the hippocampus, as well as in the prefrontal cortex and amygdala, adapt to the consequences of long-term elevations of stress hormones. And just as a brain lesion in one area can affect the functions mediated by other regions or systems with which it is connected, so, too, can alterations in the synaptic operation of a region. The only thing one brain area knows about another is the state of its synapses. Change the synapses in one area, and like dominoes in a line, synapses in others will be altered as well.

Most of the time the brain holds the self together pretty well. But when connections change, personality, too, can change. That the self is so fragile an entity is disconcerting. At the same time, if the self can be disassembled by experiences that alter connections, presumably it also can be reassembled by experiences that establish, change, or renew connections. An important challenge for the field of neuroscience is to figure out how to manipulate the brain in a way that patients with mental disorders can, either alone or with the help of a therapist, try to put the self's synapses back together.

SELF-ASSEMBLY

As described in chapter 4, your brain was assembled during childhood by a combination of genetic and environmental influences. Genes dictated that your brain was a human one and that your synaptic connections, though more similar to those of members of your family than to those of members of other families, were nevertheless distinct. Then, through experiences with the world, your synaptic connections were adjusted (by selection and/or instruction and construction), further distinguishing you from everyone else.

Synaptic connections are adjusted by environmentally driven neural activity in specific neural systems. When these changes occur during early life, they are said to involve developmental plasticity; when they occur later, they are considered as learning. But the line between developmental plasticity and learning is a fine one and perhaps nonexistent. I will therefore ignore this distinction, and plunge right into the question of how synaptic plasticity occurring in multiple neural systems is coordinated in the process of assembling, and maintaining, the self. The manner in which this occurs can, I think, be understood in terms of seven principles.

PRINCIPLE 1 DIFFERENT SYSTEMS EXPERIENCE THE SAME WORLD.

Although the different neural systems have different functions, because they are part of the same brain they will be involved in encoding the same life events. One system processes the sights, another the sounds, and still another the smells in a given scene. Additional systems will determine whether those sights, sounds, or smells indicate that danger is present, or that there might be something tasty to eat out there. From the point of view of the organism, and the world with which it is interacting, these are not different experiences, but rather different aspects of a single experience. And although each system is plastic, and can thus learn and store information, each is learning and storing information about the same experience. Just as people living in different towns of a country who never meet face-to-face can share a culture because they have similar environmental influences (similar climate, similar geography, similar myths and legends, similar political histories, similar current political situation, similar social institutions), within the brain, a kind of shared culture develops between the various systems because they are exposed to similar environmental circumstances.

To make this principle more explicit, I've put it in pictorial form in figure 11.2. Part A shows a hypothetical brain with three neural systems. Each system receives inputs from the outside world. In this model brain, the systems do not communicate directly with one another. But because they have the same inputs, they come to know exactly the same things about the world. They are thus completely redundant processors.

Part B adds a bit of complexity. Now the three systems are distinct and nonredundant—each encodes the world differently (as represented by the different fills in the shapes). These can be thought of, for example, as different sensory systems—visual, auditory, and olfactory, if you like. In spite of the fact that each processes qualitatively different information, because they experience the same events, there is still a strong degree of overlap in what they represent about the world. The sounds, sights, and smells that occur during an experience are different pieces of information from the point of view of each of the three systems, but are all part of the same experience from the point of view of the brain and person.

Now, imagine that our hypothetical brain visits three different places where it has three different experiences. Each experience will be encoded in parallel by the three systems—as sights, sounds, and smells. Contrast this to what would take place if, during each of the three experiences, only one of the neural systems was active, and the one that was active differed in each situation.

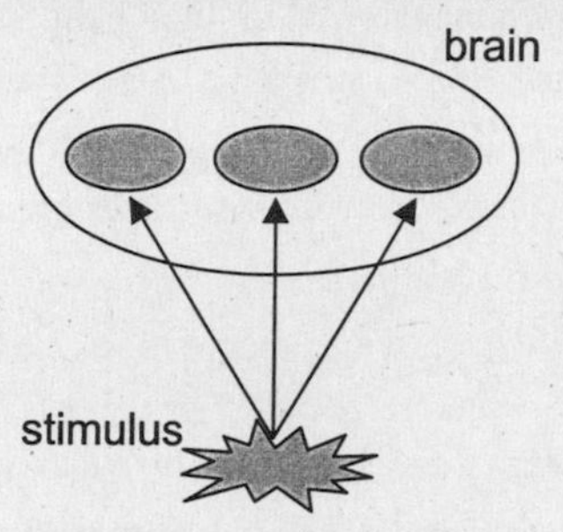

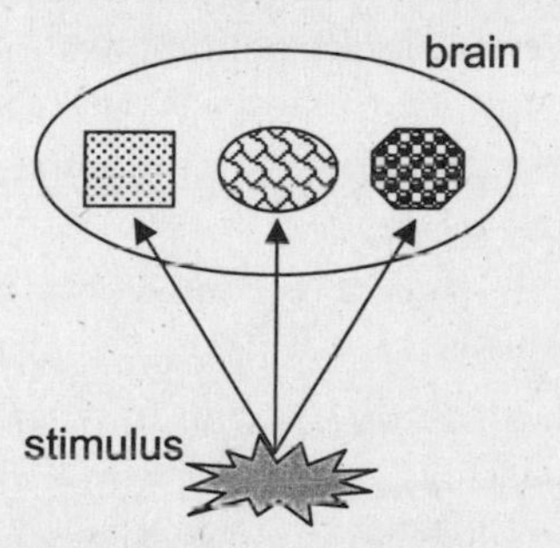

FIGURE 11.2 SHARED INPUTS TO DISTINCT SYSTEMS COORDINATE PARALLEL PLASTICITY

Although we have many brain systems that are capable of learning and storing information, because they experience the same events, they all learn and store information about the same things rather than about different things. However, each system is different. As a result, although they learn and store information about the same events, they process different aspects of these events.

In this case, although the organism and its brain had three unique experiences, each system had only one. As a result, this brain has information about one environment in the form of sights, the second in sounds, and the third in smells, but because no parallel encoding took place, there is no shared information about the three experiences across the three systems.

Normally, this condition does not occur. The various systems of your brain share the same experiences. They encode them differently, but they encode the same external events. They will not always focus on the same details, and each may not always participate in every experience. But to the extent that a neural system encodes an experience, it is likely that some other systems of the brain are encoding the same experience. As a result of parallel encoding by, and parallel plasticity within, neural systems, a shared culture develops and persists among the systems, even if they never communicate directly.

PRINCIPLE 2 SYNCHRONY COORDINATES PARALLEL PLASTICITY.

In real brains neural networks do not exist in isolation. They communicate with other networks by way of synaptic transmission. For example, in order to see an apple, instead of a roundish, reddish blob, the various features of the stimulus, each processed by different visual subsystems, have to be integrated. As we saw in chapter 7, the problem of understanding the manner in which this occurs is called *the binding problem.* One popular solution to this problem is based on the notion of neuronal synchrony.[13] Synchronous (simultaneous) firing, and thus binding, has been proposed as an explanation of consciousness (chap. 7), but our interest here is more in the ability of synchronous firing between cells in different interconnected regions to coordinate plasticity across the regions.

Wolf Singer is one of the major proponents of synchrony as a means of integrating plasticity across regions, especially within the visual system.[14] In brief, his basic idea is that information processing across different interconnected regions is coordinated when cells in the individual regions fire action potentials synchronously—that is, at the same time. Form and color, for example, are brought together for an immediately present object by the fact that the cells processing the particular form and particular color are active at the same time. By way of the synaptic interconnections between cells in the color and form regions, Hebbian plasticity occurs (since the cells will be activating each other at the same time they are being activated by the external visual stimulus). Hebbian plasticity thus binds simultaneously active cells together so that the next time the same or similar stimulus occurs, the same cells and connections will be activated (fig. 11.3). That synchronous firing can lead to Hebbian plasticity (as opposed to conscious perception) is incontrovertible. In fact, simultaneous (or near simultaneous) activation of inputs is what accounts for Hebbian plasticity (chap. 6).

Unfortunately, little is known about whether changes of this type actually take place between networks in the brain (as opposed to within individual

Co-active inputs create associations in each area

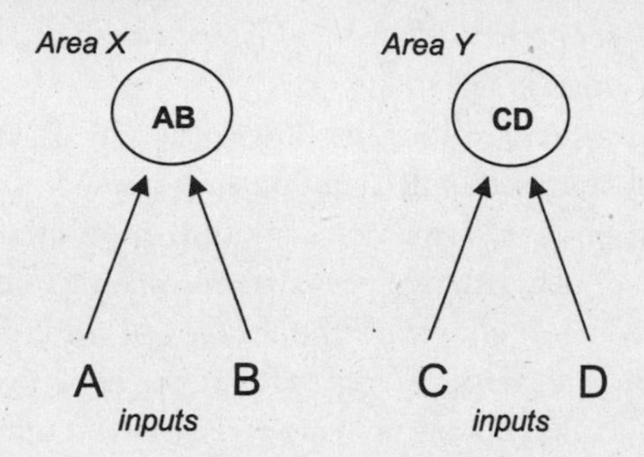

Interactions between areas also create associations

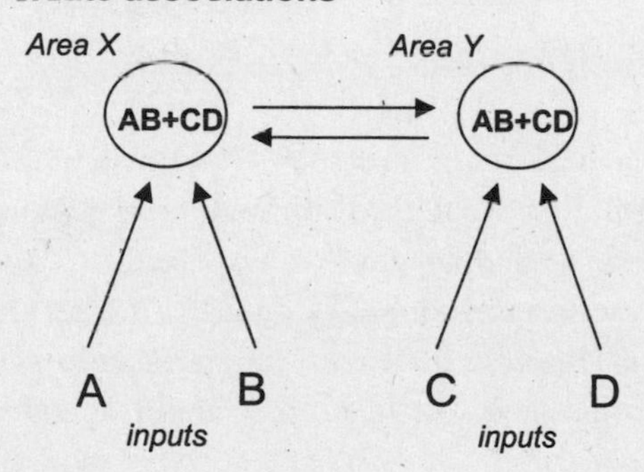

FIGURE 11.3 PROPAGATION OF PLASTICITY ACROSS NETWORKS COORDINATES PARALLEL PLASTICITY

In order for Hebbian plasticity to be useful in integration across networks, as opposed to simply inducing plasticity within a region, the plasticity would need to propagate between networks. Two areas (X and Y) that receive nonoverlapping inputs (X receives A and B, and Y receives C and D) are shown. If inputs A, B, C, and D are all active at the same time, the co-activity of A and B will induce Hebbian plasticity in X, and co-activity of C and D will produce plasticity in Y. But because X and Y are connected, activation of X (produced by activity in A and B) will lead to activation of Y at roughly the same time that Y is being activated by C and D. As a result, the CD association in Y wili be associated with X (that is, with the AB association). Similarly, activation of Y (produced by activity in C and D) will lead to activation of X at about the same time X is being activated by A and B, and the AB association in X will be associated with Y (that is, with the CD association).

networks). However, recent studies using computer simulations have begun to explore how plasticity occurring in individual systems alters information processing in interconnected systems.[15] This work provides a foundation for pursuing similar studies in real brains.

Integration across brain regions (binding) is usually discussed in the context of perception. However, understanding long-distance communication in the brain is also important for memory, emotion, motivation, and other systems. Exploration of interactions across systems is going to be especially crucial as we try to come to grips with the relation of the self to the brain. A ridiculously simple example is the fact that our perception of an apple is not just based on the integration of the shape and form and other visual features of the object, but also on the integration of these features with information stored in memory about the object and our experiences with it, and its significance for us at the moment, in the past, and in the future.

PRINCIPLE 3 PARALLEL PLASTICITY IS ALSO COORDINATED BY MODULATORY SYSTEMS.

Parallel processing in different brain systems is further coordinated by modulators. As we've seen, these are released throughout the brain in the presence of significant stimuli, including novel, unexpected, or painful stimuli, or stimuli that otherwise signal emotional arousal. In the last chapter, we examined the role of one class of modulators, the monoamines, in mental disorders. Here we examine their role in normal brain function. These are two sides of the same coin.

In the middle of the twentieth century, researchers discovered a region of the brain stem that was required for alertness and arousal.[16] Damage to this region put animals and people into a comatose state. Stimulation of the same region could awaken an animal from a deep sleep, and if an animal was already awake, alertness and attention were enhanced by the stimulation. This area came to be called the "reticular activating system" or the "reticular formation." Subsequently, it was discovered that arousal functions were largely accounted for not by a single integrated system in the brain stem but by the activities of several different groups of neurons in the vicinity, with each group having a unique chemical signature (different neurotransmitter molecules are present in the different groups).[17] The chemicals in question are all amines, and specifically include the monoamines—dopamine, norepinephrine, epinephrine, serotonin—and acetylcholine.

The cells that produce modulators are located primarily in the brain stem, but their axons are distributed throughout the brain. Consequently, when these cells are activated, many brain areas are affected. The widespread action of modulators makes them especially useful in broadcasting that something significant has happened, but they are less suited to identifying exactly what it is that's happening. The modulatory system functions somewhat like an alarm sounded by the firehouse in the center of a small town. The alarm is very effective in alerting all the town's firemen to the fire, and in summoning them to the station, but it doesn't tell them whose house is on fire. This they have to learn by other means, just as brain areas have to determine precisely what it is that's causing the arousal by other means.

The main job of modulators is to regulate neurotransmission between neurons, but they don't work at all the many synapses that they bathe. They are mainly effective in modulating transmission at synapses that are already active when the modulator arrives (fig. 11.4).

Because modulatory systems are activated during significant experiences, modulators can selectively facilitate transmission at the synapses actively processing information about such experiences across widely distributed neural systems. Emotional or otherwise significant experiences are the ones we tend to form memories about, and as described in chapter 8, it is well established that modulators like norepinephrine are involved in the enhancement of memory that occurs during emotional events. Norepinephrine has also been implicated in the induction of long-term potentiation (LTP),[18] which, as we've seen, is a laboratory procedure for studying synaptic plasticity. Thus, LTP is facilitated when this chemical is present, and disrupted when it is absent. So not only can modulators produce a momentary facilitation of transmission in circuits actively involved in processing significant events, they can also promote synaptic plasticity, and thus learning and memory, in those circuits. Recall that one of the most prominent current theories supporting monoamine treatment of mental disorders is that these drugs make more serotonin and/or norepinephrine available at synapses and thereby trigger intracellular molecular cascades that promote synaptic plasticity. Modulation of plasticity across brain systems is thus important in both normal and pathological mental conditions.

One of the most important features of modulators is that, once released, they have a prolonged action, at least with respect to transmitters like glutamate or GABA. The primary action of glutamate or GABA is typically concluded within a matter of milliseconds, whereas modulators can have effects that last for seconds. Given that not all brain systems operate at exactly the same rate (some involve more distant pathways with more connections, and,

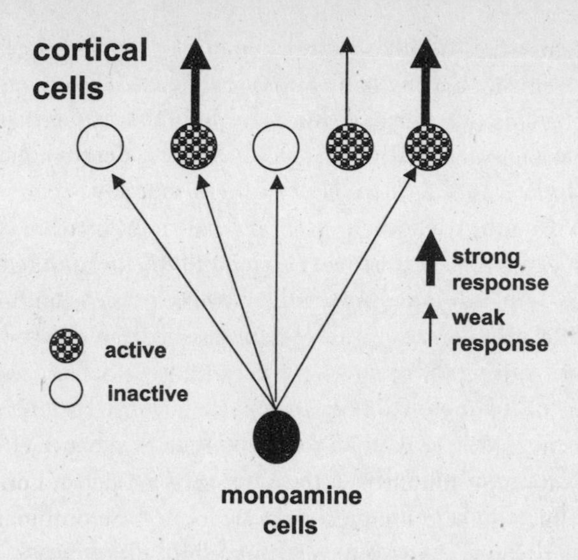

FIGURE 11.4 MODULATORY CHEMICALS COORDINATE PARALLEL PLASTICITY
Monoamine cells in the brain stem send connections to widespread brain regions (cells in different regions of the cortex are illustrated) and release monoamines during significant events. Although cells in many regions will simultaneously be bathed by monoamine release, only active cells (cells actively involved in processing current events) will be affected. For example, three active cortical cells are shown. The two active cells that receive monoamine inputs produce a stronger response than the one cell that does not receive the monoamine inputs. One effect of monoamines is to facilitate plasticity. Thus, learning is facilitated in those cells in areas actively processing the event. In this way, plasticity is coordinated across widespread regions during significant events, increasing the likelihood that cells actively engaged in processing the event will store information about the event. Because different brain regions store different aspects of an experience, such coordination is important to the unity of our memories (explicit and implicit) of an experience.

in general, the more complex the process, the longer it takes to occur because more connections are involved), the slow recovery time of modulators allows them to affect a wide range of processes, from the earliest and simplest in an episode to the last and most complex, promoting learning independent of information extracted during different components of an experience.

Although different systems learn about different aspects of an experience, the widespread action of modulators increases the likelihood that when something significant takes place, plasticity will occur in parallel at active synapses in all these systems. As a result, the learning of multiple elements of an experience (its sights, sounds, and smells, its emotional and motivational significance, its movement patterns, and so on) is facilitated, allowing the whole experience to be stored at once, albeit across multiple systems. Included, of course, would be systems that process information both implicitly and explicitly, but more about that later.

Not all modulators have the same effect (some inhibit rather than enhance plasticity), and the same modulator can have different effects depending on the particular postsynaptic receptor with which it interacts. Moreover, the same interaction between a modulator and its receptor can be different depending on the other cells in the circuit. For example, when serotonin interacts with one of its receptors, it produces inhibition, and when it interacts with another receptor, it produces excitation. But the net effect (that is, whether excitation or inhibition is the outcome) will depend on the kind of neurons on which the serotonin receptors are located. Serotonin, for instance, inhibits the activity of amygdala projection cells (excitatory cells that transfer information from one region of the amygdala to another). But it achieves this inhibition by way of an excitatory serotonin receptor located on GABA cells.[19] Thus, serotonin excites GABA inhibitory cells, and these inhibit amygdala projection cells. While the outcome of the interaction between serotonin and its receptor is excitatory, the outcome for the overall circuit is inhibition. Much more work is needed to better understand the contribution of modulators to routine transmission in specific circuits, and to the induction of plasticity (learning) and its maintenance (memory). But the gaps in our knowledge do not take away from the established role of some modulators in enhancing plasticity, and thus potentially serving as regulators of plasticity across neural systems.

PRINCIPLE 4 CONVERGENCE ZONES INTEGRATE PARALLEL PLASTICITY.

In the examples cited so far, brain systems learn in parallel. While parallel learning is surely an important part of the complex process by which the self is assembled, parallel learning, on its own (even when buttressed by synchrony and modulatory chemicals), is not sufficient to account for the coherent personality of a human being.

Another important mechanism in self-assembly, especially in humans and other primates, is the existence of convergence zones, regions where information from diverse systems can be integrated. Figure 11.5 shows two independent processing units, the outputs of which meet up in a third, a convergence

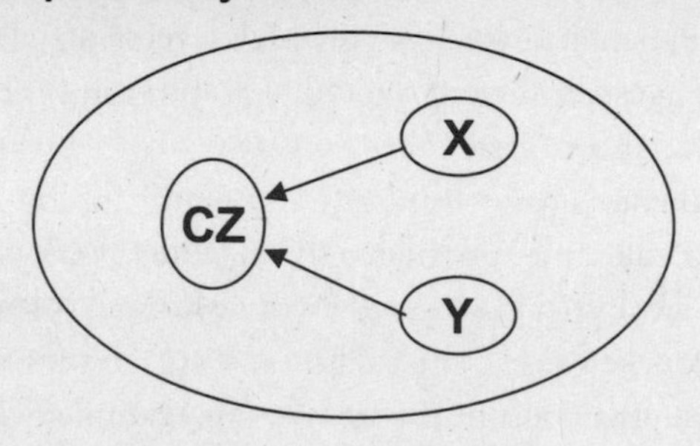

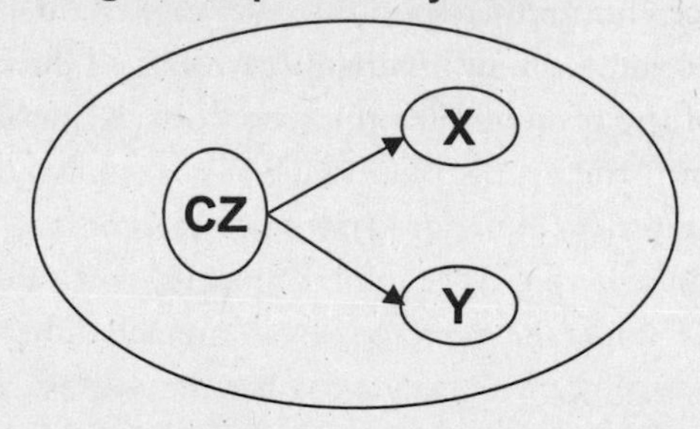

FIGURE 11.5 CONVERGENCE ZONES (CZ) INTEGRATE PARALLEL PLASTICITY

A convergence zone (CZ) is a region that receives inputs from other brain regions and that integrates the information separately processed by the other regions. Important convergence zones are located in the prefrontal cortex. Once information is integrated, it can then be used to influence the activity of the input regions. These are examples of bottom-up and top-down processing. The ability of working memory to integrate information from various systems and hold that information temporarily for the purpose of performing mental operations (comparing, contrasting, recognizing) is a typical bottom-up process, and the ability of working memory to use the outcome of this processing to regulate what we attend to is a typical top-down or executive function.

zone. Many kinds of animals have multiple independent learning systems that can be coerced into learning simultaneously by modulatory chemicals and synchronous firing, but only some animals have convergence zones in their cortex.[20] The cognitive sophistication of a mammalian species, in fact, is nicely predicted by the extent of convergence that occurs in its cortex—more is present in humans than in monkeys, for example, and more in monkeys than in rats. When plasticity occurs simultaneously in two regions that feed into a convergence zone, plasticity is also likely to occur in the convergence zone since it will be the recipient of the high level of activity that occurs when plasticity is being established in the individual regions. Obviously, synchrony and modulation also influence convergence zones, further increasing their potential to integrate information across systems.

Convergence takes place within systems before it takes place between systems. This can be illustrated by considering object recognition in the "what" stream of the visual cortex (chap. 7). The object recognition system, like other cortical processing systems, is hierarchically organized.[21] The later stages depend on the earlier ones, and information representation becomes increasingly complex as the stages are passed through. In the earliest stage, for example, each cell primarily responds to the orientation of a contour or edge of a small part of the stimulus. Across many cells, all the contours that make up the shape of the stimulus are represented. In the next stage, cells receive convergent inputs from the earlier-stage cells. As a result of getting inputs from cells that represent different parts of that object, each cell in the second stage can represent a larger part of the object. This sort of convergence continues through the hierarchy until at the final stage individual cells represent much of the entire object. These latter cells were once affectionately known as "grandmother cells," since they were supposed to be able to receive all the information needed to represent a stimulus as complex as your grandmother's face. While it is no longer believed that grandmother cells exist, many scientists do believe that small sets of synaptically connected cells, called ensembles, receive convergent inputs from lower levels in their processing hierarchy, and represent faces, complex scenes, and other objects of perception.[22] This difference is sometimes described as one between "pontifical" cells, which would make final decisions alone about the way things are, and "cardinal" cells, which do things as a small group.[23] Although single cells have been shown to have remarkable capacities,[24] most researchers accept that ensembles rather than single cells underlie mental and behavioral functions.[25]

Once convergence is completed within systems, it begins to occur across systems. In 1970, E. G. Jones and T. P. S. Powell published a landmark study in which they identified several regions in the monkey cortex that receive convergent inputs from the last processing stage of two or more sensory systems in the cortex.[26] Some of the key convergence zones identified were the posterior parietal cortex, the parahippocampal region,[27] and areas of the prefrontal cortex. The ability of these regions to integrate diverse kinds of information explains why they are involved in the most sophisticated cognitive functions of the brain. As we've seen, areas of the prefrontal cortex are involved in working memory functions, which underlie many aspects of thinking, planning, and decision-making. The posterior parietal area plays an important role in the cognitive control of movement in space in nonhuman primates,[28] and in humans is crucially involved in language comprehension in the left hemisphere and spatial cognition in the right.[29] Rhinal cortical areas are part of the medial temporal lobe memory system (chap. 5). They establish critical links between sensory areas of the cortex and the hippocampus, and thus provide the hippocampus with the raw materials needed to form relations between external stimuli in the process of establishing long-term explicit memories. The hippocampus, too, is a convergence zone; rather than integrating inputs from different sensory systems per se, it receives inputs from other convergence zones, and is thus something of a super-convergence zone.[30]

In convergence zones like the hippocampus, it is possible for completely independent sensory representations to be synthesized into memory representations that transcend the individual systems involved in the initial processing. Thus, while different systems may form independent memories of separate aspects of an experience, memories formed in, or by way of, a convergence zone are multifaceted—they include information extracted from different systems. Such memories reflect the whole experience of the organism, rather than bits and pieces of an experience recorded by other systems. But because the bits and pieces are the raw materials, there is a kind of unity of experience between the memory established by a convergence zone and by its lower connections. And because the hippocampus and other convergence zones receive inputs from modulatory systems, during significant states of arousal plasticity in these networks is coordinated with the plasticity occurring in other systems throughout the brain.

One important consequence of this arrangement is that though memories are formed by systems that function both implicitly and explicitly during significant experiences, the memories are coordinated to some degree.[31] That is, the elements you are able to consciously remember about an experience overlap often with some of the elements that were also being separately stored implicitly in other systems. Convergence zones such as those in the medial temporal lobe make possible the creation of consciously accessible memories that integrate elements being encoded separately and implicitly in the other systems. But, remember, while the medial temporal lobe system forms memories in a way that allows them to be consciously accessible, these memories only enter consciousness when they are placed in working memory. And once in working memory, memories and thoughts can, as we see next, influence activity back down the processing hierarchy.

PRINCIPLE 5 DOWNWARDLY MOBILE THOUGHTS COORDINATE PARALLEL PLASTICITY.

So far, I've placed much of the burden of assembling the self on processes that work more or less automatically, from the bottom up. But this is only part of the story. Convergent representations built from the bottom up are also used to direct activity back down the processing hierarchies. Thoughts and memories placed in working memory, for example, can influence what we attend to, the way we see things, and the way we act. These executive control functions of working memory, which were discussed in chapter 7, are possible because the prefrontal cortex, like other convergence zones, reciprocates projections. That is, connections are sent back to the regions that provide the convergent inputs. By pulling the right strings (activating the right axons), working memory can direct traffic in the areas with which it is connected, enhancing the processing of stimuli that are relevant to the task on which it is engaged and suppressing the processing of other stimuli.[32]

The process by which a thought can cause the brain to issue certain orders is known as downward causation.[33] We prove that downward causation exists every time we carry out an intention. Downward causation is only mind-boggling if you believe that thoughts are one phenomenon and brain activities another. It's still a difficult problem even if you view thinking as a form of brain activity, but the nature of the solution is far more obvious in this case.

If a thought is embodied as a pattern of synaptic transmission within a network of brain cells, as must be the case, then it stands to reason that the brain activity that is a thought can influence activity in other brain systems involved in perception, motivation, movement, and the like. But there's one more connection to make. If a thought is a pattern of neural activity in a network, not only can it cause another network to be active, it can also cause another network to change, to be plastic.

All that is required to induce plasticity at a synapse is the right kind of synaptic activity. If cells processing sensory events can undergo plasticity as a result of the kind of activity those events trigger in sensory systems, then why can't cells processing a thought change the connections of the cells with which they communicate? Obviously, they do; we simply need to learn more about precisely how this happens.

The downward mobility of thought provides a powerful means by which parallel plasticity in neural systems is coordinated. The more elaborate the convergence zones present in a species, the more elaborate will be the cognitive capacity of the species and the more sophisticated will be the ability of information convergence to coordinate plasticity in that species. With thoughts empowered this way, we can begin to see how the way we think about ourselves can have powerful influences on the way we are, and who we become. One's self-image is self-perpetuating.

PRINCIPLE 6 EMOTIONAL STATES MONOPOLIZE BRAIN RESOURCES.

Emotions, too, play a key role in organizing brain activity.[34] We've already had a hint of this when we discussed the function of modulators in coordinating parallel plasticity, since emotional stimuli are some of the most potent activators of modulatory systems. But the influence of emotions is much broader than simply activating modulatory systems, and can be illustrated by reconsidering the various ways that the amygdala affects other brain systems when it detects danger (fig. 11.6).

In the presence of a threatening stimulus, the amygdala sends direct feedback by way of neural connections to sensory areas of the cortex, encouraging these areas to stay focused on those aspects of the stimulus world that are critical. Amygdala feedback also reaches other cortical areas engaged in thinking and explicit memory formation, encouraging them to think certain thoughts and to form certain memories about the current situation. In addition, the amygdala sends connections to arousal networks, causing them to release their modulatory chemicals throughout the brain. The synapses that are actively involved in processing the external world, in thinking about the world, in forming memories about it, and in receiving the amygdala's feedback will thus be enhanced. In addition, plasticity will be facilitated at these active synapses. And interconnections between active cells in different regions and systems that fire synchronously will be linked by the plasticity that is induced. At the same time, bodily responses controlled by the amygdala will be expressed, and these will provide additional feedback to the brain, not only in the form of bodily sensations that are part of the "felt" response of the emotion, but also in the form of hormones that further affect synaptic activity, over a longer time scale than even modulators. The net result is that emotional arousal penetrates the brain widely, and perpetuates itself.

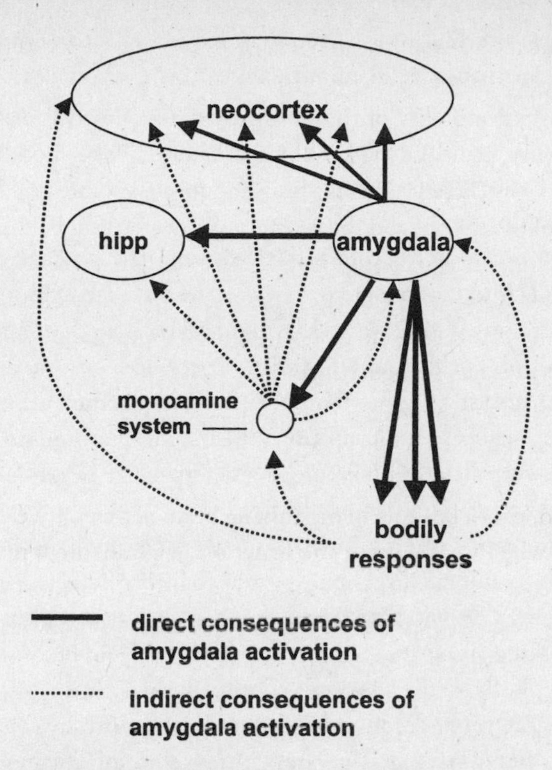

FIGURE 11.6 EMOTIONAL MONOPOLY

Once the amygdala determines that danger is present, it activates a variety of other brain networks. The net result is that the various systems affected are coordinated in their response to the threatening situation. Thus, while many brain systems are active during danger, they are activated in a coordinated way because of the extensive connectivity of the amygdala. Abbreviation: hipp, hippocampus.

As we've seen, the brain has a number of emotion systems, including networks involved in identifying sexual partners and food sources, as well as detecting and defending against danger. When one of these is active, the others tend to be inhibited. For example, other things being equal, animals will tend to spend time in areas where they are safe. So when it comes time to search for food, their fear of certain places, like wide-open spaces or places where they've previously encountered a predator, has to be overcome if that's where food is likely to be found. The hungrier the animal is, the more it will tolerate fear and anxiety and take risks to obtain nourishment. Similarly, both eating and sexual arousal are decreased by activation of systems involved in fear and stress.[35] But, once aroused, sexual desire can override many other brain systems—people risk all sorts of adverse consequences for an adulterous fling. Not only does the arousal of an emotional state bring many of the brain's cognitive resources to bear on that state, it also shuts down other emotion systems. As a result, learning is coordinated across systems in a very specific manner, ensuring that the learning that does occur is relevant to the current emotional situation.

Because emotion systems coordinate learning, the broader the range of emotions that a child experiences the broader will be the emotional range of the self that develops. This is why childhood abuse is so devastating. If a significant proportion of the early emotional experiences one has are due to activation of the fear system rather than positive systems, then the characteristic personality that begins to build up from the parallel learning processes coordinated by the emotional state is one characterized by negativity and hopelessness rather than affection and optimism.

The wide influence of emotional arousal results in many brain systems being activated simultaneously, many more than if one is engaged in quiet cognitive activity, like lying back musing about something, or even when vigorously thinking about the solution to a problem. And because more brain systems are typically active during emotional than during nonemotional states, and the intensity of arousal is greater, the opportunity for coordinated learning across brain systems is greater during emotional states. By coordinating parallel plasticity throughout the brain, emotional states promote the development and unification of the self.

PRINCIPLE 7 IMPLICIT AND EXPLICIT ASPECTS OF THE SELF OVERLAP, BUT NOT COMPLETELY.

In spite of the multiplicity of checks and balances that help keep the various systems of the brain on one track, learning about the same experiences and about the same things in each experience, the job performed is not always performed perfectly. Sometimes, the things learned explicitly are not the things that were focused on by the implicit systems, especially emotional systems—recall the ability of the amygdala to learn independent of the cortex (chap. 8). Although there are probably many reasons why this is so, the most obvious one is that there is an imperfect set of connections between cognitive and emotional systems in the current stage of evolution of the human brain. This state of affairs is part of the price we pay for having newly evolved cognitive capacities that are not yet fully integrated into our brains. Although this is also a problem for other primates, it is particularly acute for humans, since the brain of our species, especially our cortex, was extensively rewired in the process of acquiring natural language functions.

Language both required additional cognitive capacities and made new ones possible, and these changes took space and connections to achieve. The space problem was solved, as we saw earlier, by moving some things around in existing cortical space, and also by adding more space. But the connection problem was only partially solved. The part that was solved, connectivity within cortical processing networks, made the enhanced cognitive capacities of the hominid brain possible. But the part that hasn't been fully solved is connectivity between cognitive systems and other parts of the mental trilogy—emotional and motivational systems. This is why a brilliant mathematician or artist, or a successful entrepreneur, can like anyone else fall victim to sexual seduction, road rage, or jealousy, or be a child abuser or rapist, or can have crippling depression or anxiety. Our brain has not evolved to the point where the new systems that make complex thinking possible can easily control the old systems that give rise to our base needs and motives, and emotional reactions. This doesn't mean that we're simply victims of our brains and should just give in to our urges. It means that downward causation is sometimes hard work. *Doing* the right thing doesn't always flow naturally from *knowing* what the right thing to do is.

In the end, then, the self is maintained by systems that function both explicitly and implicitly. Through explicit systems, we try to willfully dictate who we are, and how we will behave. But we are only partially effective in doing so, since we have imperfect conscious access to emotional systems, which play such a crucial role in coordinating learning by other systems. In spite of their importance, though, emotion systems are not always active and have only episodic influence on what other brain systems learn and store. Furthermore, because there are multiple independent emotion systems, the episodic influence of any one system is itself but a component of the total impact of emotions on self-development.

YOU ARE YOUR SYNAPSES

Synaptic connections hold the self together in most of us most of the time. Sometimes, though, thoughts, emotions, and motivations come uncoupled. If the mental trilogy breaks down, the self is likely to begin to disintegrate and mental health to deteriorate. When thoughts are radically dissociated from emotions and motivations, as in schizophrenia, personality can, in fact, change drastically. When emotions run wild, as in anxiety disorders or depression, a person is no longer the person he or she once was. And when motivations are subjugated by drug addiction, the emotional and intellectual aspects of life suffer.

That the self is synaptic can be a curse—it doesn't take much to break it apart. But it is also a blessing, as there are always new connections waiting to be made. You are your synapses. They are who you are.

NOTES

CHAPTER ONE THE BIG ONE

1. Although "the self" and "personality" are not major topics in the field of neuroscience, some brain scientists and psychologists have discussed the relation of self and personality to the brain. Most discussions of the self and the brain have focused on conscious aspects of the self. My approach, in contrast, gives as much if not more weight to unconscious or implicit aspects. The attempts of others to relate personality to the brain have mostly treated personality as a set of fairly static traits. I'm attempting to construct a way of thinking about personality as a set of brain processes that are in constant flux due to their capacity to learn and remember. The relation of the terms *self* and *personality* is considered in chapter 2. For discussions of self and the brain by others see: Popper and Eccles 1977; Gazzaniga 1985; Gazzaniga 1998; Stuss 1991; Brothers 1997; Arbib 1999; Llinas 2001; Damasio 1999; Feinberg 2000; for personality (or temperament) and the brain see: Gray 1991; Schore 1994; Davidson 1992; Kagan 1994; Kagan 1998; Zuckerman 1991.
2. The two other major possible alternatives to the synaptic view are that the self is mediated by intrinsic properties of individual neurons (rather than connections between them) or that it is mediated by large aggregates of neurons that act globally as a field or Gestalt (rather than by particular connections between specific ones). The aggregate field theory has generally not been given much credence in recent years, in part due to experiments that argue against it (Sperry and Miner 1955) and in part due to the fact that the synaptic approach has been successful. In contrast, the notion that intrinsic properties of neurons are important is indisputable (Llinas 1988; Llinas 2001). However, in order for intrinsic properties of any cell to be expressed in psychological functions of the brain it is necessary for that cell to interact with others by way of synapses. This is discussed further at the end of chapter 3.
3. Pinker 1994; Pinker 1997; Dawkins 1996; Wilson 1999.
4. Tellegen et al. 1988.
5. Kagan 1999; Kagan 1998.
6. Pinker 1997; Harris 1998; Gazzaniga 1992.
7. Harris 1998. Also see the Nurture Assumption Website (*http://home.att.net/~xchar/tna/*). For rebuttals to Harris, see: Gardner 1998; Kagan 1999; LeDoux 1998.
8. O'Connor et al. 2000; O'Connor and Rutter 2000.
9. Blanchard and Blanchard 1972.
10. This does not mean that the rat's innate fear of cats is programmed solely by genes in the absence of any environmental influence. Fear circuits, like other circuits, get

wired by a combination of genetic programming of synaptic connections and various environmental influences. As a result, although the ability to respond to the cat doesn't require experience with the cat, it may require other kinds of experience to get the amygdala properly wired. Unfortunately, relatively little is known about amygdala development.

11. The actual sequence of events was startle and then freeze.
12. Explanations like this are the business of evolutionary psychology (Tooby and Cosmides 2000). See chapter 4 for a critique of this field.
13. Bolles and Fanselow 1980; Blanchard and Blanchard 1972.
14. Blanchard and Blanchard 1972; LeDoux 1996.
15. LeDoux 1996.
16. The bodily responses that occur during fear reactions include the so-called flight-fight responses. Actually, a better term might be the freeze-flight-fight response, since freezing often occurs first. Supporting physiological changes include a redistribution of blood away from the skin and gut and toward the brain and muscles, since the latter will need energy during the upcoming fight or flight. These changes in blood flow account for the alterations in blood pressure and heart rate that occur, as well as for the alterations in skin temperature. Hormones are also released from various organs that support these processes. For a more extensive discussion, see LeDoux 1987.
17. Rushdie 1990.
18. Sperry 1966; Sperry 1984; Gazzaniga 1970; Popper and Eccles 1977; Gazzaniga and LeDoux 1978; Gazzaniga 1988; Szentagothai 1984; Gazzaniga 1985; Gazzaniga 1992; Crick and Koch 1990; Stoerig 1996; Penrose 1989; Singer 1998; Edelman and Tononi 2000; Edelman 1993; Crick 1995; Damasio 1999; Llinas 2001; Zeki and Bartels 1999.
19. Horgan 1996.
20. By this statement, I'm not denying the existence of consciousness in other animals but only saying that the unique kind of consciousness we have is probably not present in other animals, owing to the fact that our brain is different from most others in terms of its size (relative to body weight) and complexity (especially in the frontal neocortex). I'll have more to say about these issues later, especially in chapters 7 and 8.
21. As pointed out in the previous note, I'm not denying that other animals have some kind of conscious awareness, and instead am only saying they don't have the kind of conscious awareness that comes from having a human brain. In particular, their capacity for self-reflectance is probably missing. They are not strictly speaking unconscious in the sense of being asleep or knocked out. They are instead unconscious in the sense of not being self-aware in the way humans are. I will avoid saying which animals are conscious and which are not and instead emphasize that only humans are conscious in the way humans are conscious. Animal consciousness is discussed in some detail in chapters 7 and 8.
22. Bargh 1990; Bargh and Barndollar 1996; Bargh and Chartrand 1999; Greenwald and Banaji 1995; Bowers and Meichenbaum 1984; Greenwald 1992; Jacoby and Woloshyn 1989; Kihlstrom 1987; Kihlstrom 1990; Meichenbaum and Gilmore 1984; Merikle 1992; Öhman and Soares 1994; Öhman 2000; Rozin 1976; Shevrin et al. 1992; Nisbett and Wilson 1977; Erdelyi 1985; Wilson et al. 2000; Wilson (in press).
23. Rozin 1976; Shevrin and Dickman 1980; Kihlstrom 1987; Kihlstrom 1990.

24. Popper and Eccles 1977; Stuss 1991; Sperry 1984; Gazzaniga 1985; Brothers 1997; Arbib 1999; Llinas 2001; Damasio 1999; Feinberg 2000.
25. Damasio 1999; Gazzaniga 1998.

CHAPTER TWO SEEKING THE SELF

1. James 1890.
2. Hall et al. 1998; Mischel 1993.
3. Title of a song by the saxophonist King Curtis.
4. The conference was sponsored by the Vatican Observatory and the Center for Theology and Natural Sciences. The proceedings were published, Russell et al. 2000.
5. The theological problem raised by this discussion, of course, is one of figuring out when God acts and when He doesn't.
6. Christian 1977.
7. Christian 1977.
8. Flew 1964.
9. Quoted in Walter 1953.
10. I'm grateful to Stephen Happell and Nancey Murphy for their helpful suggestions on the content of this paragraph. They were participants at the Vatican conference in Poland.
11. For a summary of how Descartes's views came to be so influential, see Rorty 1979.
12. Bremmer 1993; Snell 1960.
13. Flew 1972.
14. Plato, cited in Flew 1964.
15. Flew 1964; Flew 1972.
16. Happel 2000.
17. For contemporary discussions of the mind-body possibilities, see: McGinn 2000; Humphrey 1992, 2000; Metzinger 1995; Searle 1992, 2000; Dennett 1991; Churchland 1984; Block 1995; Chalmers 1996; Clark 1998. For a Website with a bibliography on the mind-body problem, see: *http://www.u.arizona.edu/~chalmers/biblio.html.*
18. Chalmers 1996.
19. Although I believe that my mind (and yours) is the product of a physical system, I don't outright reject other ways of thinking about the mind. Reductionism is a good approach to brain research, but isn't necessarily a good principle for guiding us through daily life, say, when we are wooing a partner, raising children, climbing, or descending, the corporate ladder, or hiring a plumber. These activities, of course, all depend on and are even potentially explainable in terms of brain mechanisms, but when scientists or lay persons do these things, they don't necessarily need to know about the neurobiological underpinnings involved. Of course, facts about how the brain operates can work their way into everyday activities (people freely take drugs to control anxiety or depression, eliminate aches and pains, or to manage epilepsy or Parkinson's disease). But there's nothing special about brain research in this regard, as our culture is constantly changing on the basis of developments in the humanities as well as the sciences. Literature, for example, offers ideas that are often useful to people in their lives, and may even be helpful in understanding how the mind, through the

brain, makes us who we are. Dostoevsky, for one, had lots of interesting ideas about the importance of unconscious processes in mental life. Nonscientific approaches (literature, poetry, psychoanalysis) and nonreductionist sciences (linguistics, sociology, anthropology) can, I believe, coexist with and complement neuroscience. For example, new facts about how the brain works may help anthropologists understand human evolution, and new discoveries in anthropology or other social sciences might lead neuroscientists toward novel experiments on the mind. In a similar vein, as I said above, a spiritual view of persons doesn't have to be mutually exclusive with a neural view. Though I'm not particularly religious, I know scientists who are, and even some with a mystical side. Reduction is often treated disparagingly by those outside science. This is partly because people like to think of themselves in terms of their own self-awareness, and they don't like the idea that the self might exist at some level other than at the level of conscious awareness. Reduction also has a bad name because carried to its logical extreme, it would require that we, for example, describe poetry in terms of subatomic particles. This is the so-called absurd kind of reduction that we have to avoid. But I'm looking for nonabsurd reductions, reductions that make sense, and I believe it is reasonable to begin to think of the self in terms of synapses.

20. Philosophers can and have helped in the area of mind and brain by analyzing the mental in ways that can be pursued by brain researchers. Jerry Fodor's philosophical analysis of what constitutes a mental module, a self-contained mental system, has been very useful in stimulating research and discussion in neuroscience (Fodor 1983). There have been proponents (Tooby and Cosmides 2000; Gazzaniga 1992) and detractors (Elman et al. 1997; Fuster 2000). Ned Block's view that the reason it is so hard to think about brain and consciousness is that different kinds of consciousness are often confused and mixed together is also helpful (Block 1995). His analysis gives rise to a distinction between phenomenal and access consciousness, with one being about subjective experience and the other about control processes that regulate mental and behavioral states. Although subjective experiences are difficult to investigate scientifically, control processes, like attention, are amenable to experimental study. Regardless of whether Block's distinction is ultimately right, it helps researchers think in concrete terms about which aspects of consciousness are most profitably pursued in the brain, given current understanding and research tools. Also important to keep in mind is a distinction made by John Searle and others between the search for the neural correlates of consciousness and the search for the mechanisms of consciousness (Searle 2000). That is, many brain events may occur during a conscious experience, but not all of these will be related to the generation of that experience. Pat Churchland has written philosophy for neuroscientists on several occasions, sometimes in collaboration with neuroscientist Terry Sejnowski (Churchland 1986; Churchland and Sejnowski 1992). And Nick Humphrey, a neuroscientist turned philosopher, has made the interesting point that maybe clever thinking about the way the brain works, not just about how the mind works, may be a key to progress (Humphrey 2000).
21. Referenced in Dennett 1976.
22. Strawson 1959.
23. Both quotes appeared in Strawson 1959.
24. Dennett 1976.

25. Rawls is quoted in Dennett 1976.
26. Nagel is quoted in Dennett 1976.
27. Gallagher 2000; Sorabji 2001.
28. Gallagher 2000.
29. Dennett 1991; Dennett 1988; Neisser and Fivush 1994.
30. Gallagher 2000.
31. Foucault 1978; Gergen 1990; Butler 1990; Lutz 1988; for review, see Strauss and Quinn 1997.
32. Social constructivists emphasize the relativistic nature of reality and assume that there is no one underlying reality waiting to be discovered by scientists. Some even reject the notion that people exist as psychological beings, and thus argue for an elimination of psychology. For a sampling of writings about this topic, see: Gross et al. 1996; Martin and Sugarman 2000; Gergen 1997; Sass 1992.
33. Kolm 1985.
34. James 1890; Elster 1985; Neisser 1988.
35. Gallagher 1996; Rochat 1995; Damasio 1999; Bermudez 1996.
36. Neisser 1988.
37. Nagle 1974.
38. Others might disagree. For example, Leslie Brothers, a neuroscientist turned philosopher, has embraced Strawson's idea that persons are defined by their conscious states and has combined it with the social theories of George Herbert Mead and Rom Harré, and added a bit of evolutionary psychology to it (Brothers 1997). Like Strawson, Brothers says that a person is a "being with a mental life, an 'owner' of conscious subjective experience." Following Mead and Harré, she proposes that "Self consciousness arises in the process of social experience" and *"Only brains in a social context can generate the kind of consciousness that includes 'I.'"* In the tradition of evolutionary psychology, she notes, "Human beings are biologically prepared to subscribe to the concept of a person just as we are biologically prepared to learn a language." For reasons that should be obvious by now, I disagree with Brothers's idea that consciousness is the main key to persons, but more about that later. On the other hand, I agree with Brothers that it is important to understand the role of brain mechanisms in social interactions, but I'd like to try to reach the social level by climbing up to it from the neurobiology of specific brain networks rather than start at the social level and try to find brain correlates. I also agree with the idea that we should look, to the extent possible, for evolutionary mechanisms, though I'm less enthusiastic about evolutionary psychology (as opposed to evolutionary biology) than she is. That is, I believe the brain rather than the mind per se evolves.
39. See the discussion of Brothers, Harré, and Mead in the previous note.
40. Boring 1950; Gardner 1987.
41. Boring 1950; Gardner 1987.
42. Watson 1925.
43. Ryle 1949.
44. Gardner 1987.
45. Bruner et al. 1956.
46. Miller 1956.

47. Gardner 1987.
48. See Shevrin and Dickman 1980; Kihlstrom 1987; Erdelyi 1985; LeDoux 1996; Wilson et al. 2000; Wilson (in press); Bargh 1990; Bargh and Chartrand 1999; Greenwald and Banaji 1995; Zajonc 1984; Loftus and Klinger 1992; Bowers 1984; Bowers and Meichenbaum 1984; Öhman 2000; Debner and Jacoby 1994.
49. Gardner 1987.
50. Gazzaniga 1995.
51. LeDoux 1984; LeDoux 1996; Zajonc 1984; Ekman and Davidson 1994.
52. Hilgard 1980.
53. Hall et al. 1998; Boring 1950.
54. Hall et al. 1998.
55. Kagan 1994; Hall et al. 1998.
56. Freud 1915.
57. Hall et al. 1998.
58. Hall et al. 1998.
59. *Personality* is usually a broader term since *self* usually means "conscious self." But in my scheme, *self* is the broader term since only people are persons but all organisms have selves.
60. Rogers is quoted in Hall et al. 1998, p. 463.
61. Markus and Kitayama 1991.
62. Munroe 1955.
63. Bargh 1990; Greenwald and Banaji 1995; Bargh and Chartrand 1999; Wilson et al. 2000; Wilson (in press).
64. See Squire et al. 1993; Schacter 1987; Cohen and Eichenbaum 1992.
65. The effects of emotion and stress on memory are described in some detail in LeDoux 1996 and will also be discussed in later chapters.
66. Popular trait theories are those of Raymond Cattell and Hans Eysenck. See chapters 7 and 8 of Hall et al. 1998 for a summary.
67. Tellegen et al. 1988.
68. See Zuckerman 1991; Gray 1982; Gray 1991; Kagan 1994; Kagan 1992; Kagan 1998; Eysenck and Eysenck 1985; Davidson 1992.
69. Schwartz et al. 1999.
70. Kagan 1994; Kagan 1992; Kagan 1998.
71. Mischel 1993; Mischel 1990.
72. Carlson 1993; Zuckerman 1991.
73. Interview with Bob Dylan, *Newsweek,* October 13, 1997.
74. Roth 1986.
75. Epstein 1995.
76. James 1890.
77. Virginia Woolf, *Orlando,* chapter 6.
78. Klee 1957.
79. The main exception is biological trait theory, especially Eysenck's trait theory. It has been proposed that the trait called neuroticism is related to overactivity in the brain's fear/anxiety system, whereas extraversion is due to overactivity in the pleasure or reward system (see Gray 1982, 1991; Zuckerman 1991). For a critique of trait theory, see Mischel 1993.

80. For other views of the self and the brain, see the list of citations below. For the most part, these have tended to emphasize the conscious aspects of the self. My view, in contrast, includes unconscious as well as conscious aspects. Others that include a role for unconscious aspects of the self include Antonio Damasio (Damasio 1999) and Michael Gazzaniga (Gazzaniga 1985, 1992, 1998; Gazzaniga is working on a book, *The Last to Know,* which emphasizes unconscious processing in the construction of consciousness). For ideas about the brain and the conscious self, see: Popper and Eccles 1977; Stuss 1991; Sperry 1984; Gazzaniga 1985, 1992, 1998; Brothers 1997; Arbib 1999; Llinas 2000; Damasio 2000; Feinberg 2000.

CHAPTER THREE THE MOST UNACCOUNTABLE OF MACHINERY

1. For a summary, see LeDoux 1987.
2. The divisions of the forebrain include: the thalamus, hypothalamus, basal ganglia, limbic system, old cortex, and neocortex.
3. Ariëns Kappers 1909; Papez 1937; MacLean 1949; MacLean 1952; Nauta and Karten 1970.
4. Nauta and Karten 1970; Northcutt and Kaas 1995; Karten and Shimizu 1991.
5. Lettvin et al. 1959; Camhi 1984.
6. Camhi 1984; Suga 1990; Gould 1982.
7. For an alternative view that emphasizes evolutionary pressures on the whole brain rather than on specific systems, see Finlay and Darlington 1995.
8. Killackey 1990; Preuss 1995.
9. Brodmann 1909; Economo and Koskinas 1925; Campbell 1905.
10. Gazzaniga et al. 1996; Feinberg and Farah 1998; Ramachandran and Blakeslee 1998.
11. Nonneuronal cells communicate with each other but not the way neurons do. The electrochemical process of synaptic transmission is unique to nervous tissue.
12. Cell theory discussion based on Shepherd 1998, Jacobson 1993, and Microsoft Encarta 2000.
13. Based on chapter 3 in Shepherd 1988.
14. Shepherd 1988, p. 41.
15. Jones 1961, p. 32.
16. Jones 1961, p. 34.
17. Freud 1887–1902.
18. Jones 1961, Freud's biographer, says that though Freud dropped the anatomical terms, the principles that guided his psychological theories were underneath it all based on his early training in anatomy and physiology.
19. Sherrington 1897.
20. For a summary of Sherrington's early work on reflexes, see Sherrington 1906.
21. Shepherd 1988, p. 65.
22. Shepherd 1988, p. 42.
23. Rozental et al. 2000.
24. Kuffler and Nicholls 1976.
25. Zigmond et al. 1999; Kandel et al. 2000.
26. Chen et al. 2000.
27. Muscles don't have dendrites, but have their own special kind of receptive area that is contacted by the axon terminal.

28. Based on Winson 1985.
29. Boring 1950.
30. Boring 1950.
31. Gregory 1981.
32. Shepherd 1988.
33. Shepherd 1988.
34. Shepherd 1988.
35. From Jacobson 1993.
36. The space between neurons is filled with fluids that are in essence part of a vast continuous sea of liquid in which all the neurons of the nervous system are bathed. This sea is made up of so-called cerebrospinal fluid, and it occupies the so-called extracellular space.
37. Based on Kuffler and Nicholls 1976.
38. In fact, the postsynaptic cell has to receive convergent inputs within a matter of milliseconds, otherwise the inputs will not sum together and will not produce an action potential. Since the inputs are added up in the cell body, they can arrive from many different dendrites, as long as they produce electrical responses that reach the cell body at about the same time.
39. Electrical transmission is made possible by the existence of special contacts between cells called gap junctions (Rozental et al. 2000). These are actually physical contacts and are exceptions to the notion promoted by the neuron theory that cells are physically separate. These turn out to be important in synchronizing hippocampal GABA cells (Fukuda and Kosaka 2000).
40. Based on Bloom and Laserson 1985.
41. Cooper et al. 1978.
42. GABA cells sometimes have long axons and communicate between brain regions, but mostly they have short axons that end on nearby cells.
43. But even the time-course distinction between fast transmitters and modulators can be blurred. Most transmitters work with a variety of receptors. GABA, for example, has A and B receptors. While the A receptor mediates the fast effects we've been talking about, when GABA binds to B receptors its action is slower and more prolonged. Glutamate, too, has some late, longer-lasting effects when it binds to some of its receptors. Another fast transmitter is acetylcholine. When it binds to its nicotinic receptor, it does its fast transmitter thing, but when it binds to its muscarinic receptor, it works slowly. So it is often best to think of transmitters and receptors together when drawing conclusions about the kind of transmission involved.
44. Shepherd 1998.
45. See note 43 above.
46. The main exception involves the cholinergic neurons of the basal forebrain, which complement the brain stem cholinergic systems.
47. This will be discussed in chapter 10.
48. Shepherd 1998; Cooper et al. 1978.
49. Selkoe and Kosik 1983.
50. Babic 1999; Yamada et al. 1999.
51. This will be discussed in detail in later chapters, especially chapters 8 and 10.

52. Stutzmann et al. 1998; Stutzmann and LeDoux 1999.
53. Gibbs 2000; Dell and Stewart 2000.
54. See note 39.
55. Quirk et al. 1995; Rolls 1999; Ono and Nishijo 1992; Collins and Pare 2000; Maren 2000.
56. Breiter et al. 1996; Morris et al. 1996; Morris et al. 1998; Whalen et al. 1998; LaBar et al. 1998.
57. Li et al. 1996; Lang and Pare 1997; Collins and Pare 1999.
58. Chapman et al. 1990; Weisskopf and LeDoux 1999.
59. Quirk et al. 1995; Collins and Pare 2000; Maren 2000.
60. Woodson et al. 2000; Szinyei et al. 2000; Smith et al. 2000.
61. Li et al. 1996; Collins and Pare 1999.
62. Stutzmann et al. 1998; Stutzmann and LeDoux 1999.
63. McEwen and Sapolsky 1995.
64. Stutzmann et al. 1998.
65. Bogerts et al. 1993; Convit et al. 1995; de Leon et al. 1988; Fukuzako et al. 1996; Sheline et al. 1996; Starkman et al. 1992; Yehuda et al. 2000; Coplan et al. 1998; Young et al. 1994.
66. Corodimas et al. 1994; Conrad et al. 1999; Makino et al. 1994; Shors et al. 1992.
67. Llinas 1988.

CHAPTER FOUR BUILDING THE BRAIN

1. This section on early development is based on Purves et al. 1996.
2. Nottebohm 1989; Gould et al. 1997; Gould et al. 1999; Fuchs and Gould 2000.
3. Rodier 2000.
4. Chan and Jan 1999; Reichert and Simeone 1999.
5. Schlaggar and O'Leary 1991.
6. Rakic 1995.
7. Schlaggar and O'Leary 1991; Shatz 1992; Rakic 1992.
8. Miyashita-Lin et al. 1999.
9. Based on Raper and Tessier-Lavigne 1998.
10. Terman and Kolodkin 1999.
11. Edelman 1987; Changeux and Danchin 1976.
12. Jerne 1967; see also Gazzaniga 1992.
13. Changeux and Dehaene 1989.
14. Edelman 1987.
15. Edelman 1987.
16. Based on text from the home page of the Neuroscience Institute, of which Gerald Edelman is the director (*www.nsi.edu*), and from a summary of Edelman's views by Flanagan 1994.
17. Changeux and Danchin 1976; Innocenti 1991.
18. For a summary, see Oppenheim 1998.
19. For review of regressive events, see O'Leary 1992.
20. Rakic et al. 1986.

21. Bourgeois et al. 1994.
22. Quartz and Sejnowski 1997.
23. Huttenlocher 1979.
24. See Quartz and Sejnowski 1997 and Katz and Shatz 1996. For one thing, it is very difficult to measure accurately the density of synapses in a brain region given that the region itself is changing size over time. Also, unless the synapse changes are related to specific cell types, it is hard to know what the implications would be. Finally, the relation of structural measures (like the number of synapses) to functional ones (is the synapse working?) is hard to assess. In early development, synapses are functional before they have the "look" of synapses, and these would go uncounted.
25. O'Leary 1992.
26. For further discussion, see Quartz and Sejnowski 1997.
27. Hubel and Wiesel 1962; Hubel and Wiesel 1963; Hubel and Wiesel 1965; Hubel and Wiesel 1972.
28. Apologies to visual scientists for this simplistic description of visual pathways.
29. For a summary see: Katz and Shatz 1996; Shatz 1996; Stryker 1991.
30. Antonini and Stryker 1993.
31. The experiment actually involved the injection of the tracer into cells in the visual thalamus area called the lateral geniculate nucleus. In this region, cells are organized in layers devoted to one eye or the other. By recording the action potentials elicited by stimulation of one eye, it is possible to find the layers and then to inject a cell in that layer with the chemical.
32. Actually, the tracer is actively transported to the terminal by natural processes that go on in the cell all the time. These are taking things made by the cell body and shipping them throughout the cell.
33. Quartz and Sejnowski 1997.
34. Neville 1990.
35. Neville and Lawson 1987.
36. Based on Katz and Shatz 1996.
37. Rakic 1977; Horton and Hocking 1996.
38. Galli and Maffei 1988; Wong et al. 1993.
39. Even when endogenous activity is blocked, the clusters develop if the nerves headed for the brain from the two eyes are electrically stimulated separately. This kind of stimulation simultaneously activates many fibers from a given eye to the brain, tricking the brain into thinking that it received lots of activity at the same time from one eye (see Stryker and Harris 1986; Crair 1998).
40. Chiaia et al. 1992.
41. Crair 1999.
42. Hebb 1949.
43. This phrase comes from Carla Shatz.
44. Katz and Shatz 1996; Shatz 1992; Shatz 1996; Stryker 1991; Purves 1994.
45. However, recall that cortical cells initially receive inputs from both eyes. So the cortical cell will actually receive correlated input from each eye, but at different times. How then can one eye come to dominate? Although each cell gets inputs from both eyes, the two eyes never quite have equal inputs, leading one eye to dominate slightly.

Hebbian plasticity builds upon this preexisting bias, wiring the connection between the cortical cell and its more efficient inputs. While Hebbian plasticity may be enough to wire up a particular cell, more is needed to establish the cell-specific clusters, the so-called ocular dominance columns, in the cortex. Ken Miller of UCSF has some interesting proposals on this; see Miller 1994 and Wimbauer et al. 1997. As a result, one eye or the other will come to be more efficient in driving a cortical cell. Miller's work on this was pointed out to me by Tony Movshon of NYU. Hebbian plasticity thus takes care of the problem of how inputs from one eye come to control an individual cell, but leaves open the question of how cells that are responsive to one eye come to cluster together. For this, it is generally assumed that there are factors that allow presynaptic inputs that are nearby and that are active at the same time to link up.

46. Glanzman et al. 1990; Martin and Kandel 1996.
47. Tsien 2000; Bliss and Collingridge 1993; Purves et al. 1996; Brown et al. 1988.
48. Katz and Shatz 1996.
49. Katz and Shatz 1996; Johnson 1998; Schuman 1999.
50. Oppenheim 1998.
51. Lorenz and Tinbergen 1938; Lorenz 1950; Tinbergen 1951.
52. Lehrman 1953.
53. Terrace 1984.
54. Terrace 1984.
55. Watson 1925; Skinner 1938; Hull 1943.
56. Chomsky 1957.
57. Gardner 1987.
58. Keil 1999.
59. Garcia and Koelling 1966.
60. For discussion, see chapters by H. S. Terrace, P. P. G. Bateson, and J. L. Gould and P. Marler in the book edited by Marler and Terrace 1984.
61. Pinker 1994.
62. Pinker 1997.
63. Pinker 1997.
64. Bickerton 1980.
65. Elman et al. 1997; Quartz and Sejnowski 1997.
66. See Gopnik 1997; Korenberg et al. 2000; Ridley 1999.
67. Pinker 1994; Gopnik 1997; Korenberg et al. 2000; Bickerton 1980; Ridley 1999.
68. Ekman 1999.
69. Cosmides and Tooby 1999; Barkow et al. 1992.
70. Cosmides and Tooby 1999; Barkow et al. 1992; Spelke 1994; Carey and Spelke 1994; Povinelli and Preuss 1995.
71. Cosmides and Tooby 1999; Barkow et al. 1992.
72. Gould 1997.
73. Gould 1991.
74. Gould quoted in Gazzaniga 1992.
75. Premack 1985.
76. Pinker 1997; Pinker and Bloom 1990; Cosmides and Tooby 1999.
77. Rose and Rose 2000.

78. For example, see Edwards and Pap 1959.
79. Spelke 1994; Carey and Spelke 1994; Marcus 1999; Pinker 1994, 1997; Piattelli-Palmarini 1989.
80. Wexler 1999.
81. Fodor 1983; Gazzaniga 1992; Tooby and Cosmides 2000; Mody et al. 1997; Denenberg 1999.
82. Keil 1999.
83. There is also a domain-independent learning system in the brain (the explicit or declarative memory system). However, this system is involved in recording facts and experiences independent of rewards and punishment, and though it might be thought of as a universal learning system, it does not appear to play an essential role in the kinds of learned behaviors that the behaviorists studied.
84. Elman et al. 1997.
85. Quartz and Sejnowski 1997.
86. Barton 1997.
87. Brothers 1997.
88. Barton 1997.
89. Neisser 1998.
90. Alcock 1998.
91. Arnold 1980.
92. Alcock 1998; Wimer and Wimer 1985.
93. Described in Alcock 1998; based on Holden 1980.
94. Tellegen et al. 1988.
95. Described in Harris 1998. Harris also argues that genetic influences are underestimated by heritability scores, noting that the correlation between parents and children on personality traits is sufficiently weak that genes they share might fully account for any similarities that exist.
96. Gardner 1998.
97. For summary, see Schuster and Ashburn 1992; Jacobson 1993.
98. For summary, see Jacobson 1993.
99. Harris 1998; Gardner 1998.
100. See Hall et al. 1998.
101. See Bruer 1999.
102. Mooney 1999; Gould and Marler 1984; Doupe and Kuhl 1999; Bottejer and Johnson 1997; Singh et al. 2000; Jarvis et al. 1998.
103. Elman et al. 1997.
104. Bruer 1999.
105. See Tallal 2000; Tallal et al. 1998.
106. Bruer 1999.
107. Gopnik et al. 1999.

CHAPTER FIVE ADVENTURES IN TIME

1. Bartlett 1932; Schacter 1999.
2. Each time the brain learns something, it is changed.

3. Semon 1904; Schacter 1982.
4. Lashley 1929.
5. Lashley 1950.
6. Scoville and Milner 1957.
7. Scoville and Milner 1957; Milner 1962; Milner 1965; Milner 1967; Milner 1972.
8. Squire 1987; Cohen and Eichenbaum 1993.
9. Scoville and Milner 1957.
10. MacLean 1949; MacLean 1952; MacLean 1970.
11. Milner 1962; Corkin 1968.
12. Cohen 1980; Cohen and Squire 1980; Cohen and Corkin 1981.
13. Warrington and Weiskrantz 1973; Graf et al. 1984.
14. Weiskrantz and Warrington 1979.
15. Cohen and Squire 1980.
16. Schacter and Graf 1986.
17. The parahippocampal region consists of the entorhinal cortex, perirhinal cortex, and the parahippocampal cortex, as defined by Witter et al. 1989.
18. See Amaral et al. 1987; Suzuki and Amaral 1994; Witter et al. 1989; Burwell et al. 1995; Van Hoesen and Pandya 1975.
19. Entorhinal cortex, perirhinal cortex, and parahippocampal cortex are included in the parahippocampal region, as defined by Witter et al. 1989.
20. Jones and Powell 1970; Damasio 1989.
21. Mesulam et al. 1977.
22. This section is based on Squire and Kandel 1999.
23. The reason ECT produces memory disturbance is related to the fact that the conversion of short- to long-term memory is disturbed. For discussion, see Squire 1987.
24. McClelland et al. 1995.
25. Winson 1985; Buzsaki 1989; McNaughton 1998; Wilson and McNaughton 1994.
26. Wilson and McNaughton 1994; Nadasdy et al. 1999; Poe et al. 2000; Louie and Wilson 2001.
27. Nadel and Moscovitch 1997.
28. For discussion, see Nadel and Moscovitch 1997; Knowlton and Fanselow 1998.
29. Bontempi et al. 1999.
30. See Tulving 1983.
31. Vargha-Khadem et al. 1997.
32. Squire and Zola 1998.
33. Milner 1970.
34. For summary, see Mishkin and Murray 1994; Murray and Richmond 2001; Squire and Zola 1996, 1998.
35. Eichenbaum et al. 1994.
36. Section title adapted from Nadel and Willner 1980.
37. O'Keefe and Nadel 1978.
38. Olton et al. 1979.
39. O'Keefe and Nadel 1978.
40. Ranck 1973.
41. Muller et al. 1999.

42. See McNaughton 1998.
43. Morris 1984.
44. Cohen and Eichenbaum 1993.
45. Eichenbaum 2000.
46. Wicklegren 1979; Rolls 1990; Schmajuk and DiCarlo 1992; Gluck and Myers 1993; McClelland et al. 1995; Rudy and Sutherland 1992; Rudy and O'Reilly 1999.
47. O'Reilly and Rudy 2001.
48. McClelland et al. 1995.
49. Rudy and Sutherland 1992.
50. Skinner 1938; Skinner 1972; Hull 1943; Hull 1954.
51. Kandel and Spencer 1968.
52. Kandel and Spencer also noted that any approach to cellular physiology that could illuminate synaptic plasticity, even if irrelevant to behavior, might be reasonable at this stage since so little was known.
53. Cowan 1998.
54. Cohen 1974.
55. Pavlov 1927.
56. McAllister and McAllister 1971; Brown et al. 1951; Bolles and Fanselow 1980; Blanchard and Blanchard 1969.
57. Kluver and Bucy 1937; Weiskrantz 1956; Blanchard and Blanchard 1972.
58. LeDoux 1996; LeDoux 2000; Davis 1992; Davis et al. 1997; Kapp et al. 1992; Maren and Fanselow 1996; Maren 2001; Fendt and Fanselow 1999; Weinberger 1995.
59. Pitkänen et al. 1997.
60. LeDoux et al. 1990; Amaral et al. 1992; Herzog and Van Hoesen 1976.
61. Pitkänen et al. 1997.
62. Amorapanth et al. 2000.
63. Romanski and LeDoux 1992a; Doron and LeDoux 2000.
64. Bordi and LeDoux 1992.
65. Quirk et al. 1995; Quirk et al. 1997.
66. Romanski et al. 1993.
67. Quirk et al. 1995; Quirk et al. 1997; Repa et al. 2001.
68. Collins and Pare 2000; Maren 2000.
69. LeDoux 1990; Campeau and Davis 1995; LeDoux et al. 1990; Amorapanth et al. 2000.
70. Muller et al. 1997; Wilensky et al. 1999; Wilensky et al. 2000; Schafe and LeDoux 2000; Bailey et al. 1999; Helmstetter and Bellgowan 1994; Fanselow et al. 1994; Lee and Kim 1998; Maren et al. 1996; Miserendino et al. 1990; Gewirtz and Davis 1997.
71. Fanselow and LeDoux 1999; LeDoux 2000.
72. Thompson and Spencer 1966.
73. Gormezano 1972.
74. Weinberger 1995; Weinberger 1998.
75. For summary, see Thompson et al. 1983; Steinmetz and Thompson 1991; Hesslow and Yeo 1998; Medina et al. 2000.
76. Desmond and Moore 1982.
77. Thompson et al. 1983; Steinmetz and Thompson 1991; Thompson 1986; Thompson and Kim 1996; Hesslow and Yeo 1998.

78. Ito 1984.
79. Marr 1969; Eccles 1977; Ito 1984; Ito 1989.
80. Llinas and Welsh 1993.
81. Lisberger 1996; Lisberger 1998.
82. Steiner 1973.
83. Steiner 1973.
84. Grill and Norgren 1978.
85. As with any so-called innate function, a role for environmental factors cannot be ruled out (see discussion in chapter 4).
86. Garcia and Koelling 1966.
87. Garcia 1990.
88. See Chambers 1990; Yamamoto et al. 1994; Lamprecht and Dudai 2000.
89. Although some CS-US integration may occur in the hindbrain, any integration here is not believed to be sufficient to mediate CTA.
90. Berrige 1999.
91. Lamprecht and Dudai 1996; Lamprecht et al. 1997; Dunn and Everitt 1988; Lamprecht and Dudai 2000.
92. Schafe et al. 1998.
93. Manns et al. 2000.
94. Moyer et al. 1990; Moyer et al. 1996; LaBar and Disterhoft 1998; Huerta et al. 2000.
95. Chun and Phelps 1999.
96. Blanchard and Blanchard 1969; Bolles and Fanselow 1980.
97. Kim and Fanselow 1992; Phillips and LeDoux 1992; Maren and Fanselow 1996; Frankland et al. 1998; Selden et al. 1991.
98. Schacter 2001.
99. De Leon et al. 1995.
100. Damage to implicit systems does not destroy personality solely by affecting implicit memory functions. All systems operate on the basis of synaptic connections that are epigenetically specified during early development (that is, that are constructed from genetic and environmental influences) and then altered each time the neural system involved is active and engages in some form of learning. For more information on development, see chapter 4. For more information on learning-induced synaptic changes, see chapter 6.

CHAPTER SIX SMALL CHANGE

1. Ramón y Cajal 1909–1911.
2. Hartley 1749.
3. James 1890.
4. Freud 1887–1902.
5. Hebb 1949.
6. Comment made by Bruce McNaughton at a Neurobiology of Learning and Memory Conference in Park City, Utah, sometime in the late 1990s. McNaughton had been a student at McGill.
7. The Hebb learning rule was formalized by Stent 1973.
8. Jacobson 1993.

9. Ramón y Cajal 1911 proposed a mechanism called neurobiotaxis, but this notion has long been viewed as unrealistic (see Kandel and Spencer 1968).
10. Konorski 1948.
11. For summary, see Kandel 1976. Particularly important were early studies by Larrabee and Bronk 1947; Lloyd 1949; Brock et al. 1952.
12. Brock et al. 1952.
13. Eccles 1953.
14. This reflex is studied by measuring muscle twitches in response to electrical stimulation of the skin or nerves carrying sensory information from the skin to the spinal cord.
15. Thompson and Spencer 1966.
16. Hawkins et al. 1987.
17. Kandel and Spencer 1968.
18. For a review of some of the major invertebrate systems that have been studied over the years, see Beggs et al. 1999.
19. Lømo 1966.
20. Interview with T. Lømo in the *Journal of NIH Research* 1995.
21. Bliss and Lømo 1973.
22. Interview with Bliss in the *Journal of NIH Research* 1995.
23. It's not exactly right to say they failed. Spinal cord researchers found that they could produce changes in synaptic transmission lasting up to several hours. However, it took very strong and prolonged stimuli to produce these effects, and the potentiation then drifted away. In contrast, in the hippocampus, much milder stimuli did the trick and the responses remained for very long periods, as long as weeks in some studies where LTP was recorded in awake behaving animals.
24. Iriki et al. 1987; Castro-Alamancos et al. 1995; Bear and Malenka 1994; Huang and Kandel 1998; Weisskopf et al. 1999; Randic et al. 1993; Pennartz et al. 1993; Kombian and Malenka 1994.
25. Sanes and Lichtman 1999.
26. Bliss and Collingridge 1993; Lynch 1986; McNaughton and Barnes 1990.
27. McNaughton et al. 1978.
28. Levy and Steward 1979.
29. Kelso et al. 1986; Malinow and Miller 1986; Wigström et al. 1986. It's important to note that the postsynaptic cell does not have to fire action potentials to be active in the Hebbian sense. Its membrane potential just needs to become less negative.
30. For summaries of early research on LTP mechanisms, see Lynch 1986; Bliss and Collingridge 1993; Nicoll and Malenka 1995; Malenka and Nicoll 1999.
31. Nowak et al. 1984; Mayer and Westbrook 1987; Bliss and Collingridge 1993; Malenka and Nicoll 1999; Nicoll and Malenka 1995.
32. Brown et al. 1988.
33. Husi and Grant 2001.
34. The actual duration is not known since brain slices can't be kept alive much longer than several hours in a dish.
35. Huang et al. 1996.
36. For a discussion of different forms of LTP, see: Bliss and Collingridge 1993; Nicoll and Malenka 1995; Johnston et al. 1999; Morgan and Teyler 1999.

37. For a discussion of the relation of short-term memory in psychology and biology, see Dudai 1989, 1996, 1997; Squire and Kandel 1999.
38. Davis and Squire 1984.
39. Huang et al. 1996.
40. Squire and Kandel 1999.
41. Some of the relevant kinases include calcium/calmodulin protein kinases (CaMKII), protein kinase C, and tyrosine kinase. The actions of CaMKII are particularly well worked out. When calcium enters through the NMDA receptor, it forms a chemical complex with calmodulin (calcium/calmodulin) that then activates the alpha form of calcium/calmodulin kinase. This kinase, which is inactive until it interacts with the calcium/calmodulin complex, remains active even when calcium levels return to normal, since it can engage in autophosphorylation (it activates itself). See Lisman 1994; Elgersma and Silva 1999; Mayford et al. 1996; Mayford and Kandel 1999.
42. Soderling 1996; Shi et al. 1999.
43. Huang et al. 1996; Malgaroli and Tsien 1992; Bekkers and Stevens 1990.
44. Malgaroli and Tsien 1992; Bekkers and Stevens 1990.
45. Hawkins et al. 1994; O'Dell et al. 1994.
46. Malenka and Nicoll 1999.
47. Included are protein kinase A (PKA) and mitogen-activated protein kinase (MAP kinase). The steps involved in PKA activation are believed to include the following. First, calcium entry through NMDA receptors leads to the formation of the calcium/calmodulin complex. In early LTP, this leads to activation of CaKII, but for late LTP, calcium/calmodulin has to activate the cAMP cascade by initiating the production of cAMP by adenyl cyclase (this is part of the classic cell cycle, an important metabolic process that occurs in all cells in the body—that is, energy is released when adenyl cyclase makes cAMP from ATP). cAMP then activates PKA. Once activated, the regulatory subunit is removed from the catalytic subunit, and PKA moves to the cell nucleus. See Huang et al. 1996; Elgersma and Silva 1999; Mayford et al. 1996; Mayford and Kandel 1999.
48. For summary of genetic tools, see: Silva et al. 1997; Mayford et al. 1995; Mansuy 1998; Mayford and Kandel 1999; Tsien 2000.
49. For example, see McHugh et al. 1996.
50. Silva et al. 1997; Mayford et al. 1995; Gerlai 2000; Mayford and Kandel 1999; Tsien 2000; Tsien et al. 1996; Tang et al. 1999; Huerta et al. 2000.
51. Elgersma and Silva 1999; Silva et al. 1998; Kandel and Pittenger 1999; Mayford and Kandel 1999; Tsien 2000.
52. See Sanes and Lichtman 1999; Sweatt 1999; Kennedy 1999.
53. Sanes and Lichtman 1999.
54. Sweatt 1999; Kennedy 1999.
55. Frey and Morris 1997; Martin et al. 1997.
56. Lee et al. 1980; Chang et al. 1991; Desmond and Levy 1986; Engert and Bonhoeffer 1999; Toni et al. 1999.
57. Steward and Schuman 2001.
58. Morris et al. 1986.
59. For review, see Martin et al. 2000.

60. Staubli et al. 1989; Shapiro and Caramanos 1990; Bannerman et al. 1995; Cain et al. 1996; Shors and Matzel 1997; Keith and Rudy 1990.
61. Shors and Matzel 1997; Keith and Rudy 1990; Gallistel 1995.
62. Martin et al. 2000.
63. Tsien et al. 1996.
64. Tang et al. 1999.
65. Martin et al. 2000; Silva et al. 1997; Mayford et al. 1995; Gerlai 2000; Mayford and Kandel 1999.
66. Silva et al. 1998.
67. Abel et al. 1997; Mayford et al. 1996.
68. For discussion, see Shors and Matzel 1997.
69. Martin et al. 2000; Nicoll and Malenka 1995; Malenka and Nicoll 1999; Bliss and Collingridge 1993; Grover and Tyler 1990; Bortolotto et al. 1999; Bortolotto et al. 1999; Bekkers and Stevens 1990; Staubli et al. 1990; Weisskopf and Nicoll 1995.
70. Linden 1994; Bear and Malenka 1994; Ito 1996.
71. Shors and Matzel 1997; Gallistel 1995; Keith and Rudy 1990.
72. For review, see Teyler and DiScenna 1987; Teyler 1992; McNaughton and Barnes 1990; Moser 1995; Staubli 1995; Martinez and Derrick 1996; Morris et al. 1989; Morris 1992; Morris 1994; Martin 2000.
73. See Shors and Matzel 1997; Martinez and Derrick 1996; Martin et al. 2000; Barnes 1995; Eichenbaum 1995; Eichenbaum 1996.
74. Barnes 1995; Eichenbaum 1995.
75. Romanski et al. 1993.
76. An award given by the Society for Neuroscience.
77. Stevens 1998.
78. Barnes 1995; Eichenbaum 1995; Eichenbaum 1996.
79. Brown et al. 1988.
80. Rodrigues 2001; Miserendino et al. 1990; Gewirtz and Davis 1997; Walker and Davis 2000; Tang et al. 1999; Fanselow and Kim 1994; Maren et al. 1996; Lee and Kim 1998.
81. Schafe and LeDoux 2000; Schafe et al. 2000.
82. Schafe et al. 1999; Bourtchouladze 1998.
83. Josselyn et al. 2001.
84. Silva et al. 1992; Abel et al. 1997; Brambilla et al. 1997; Mayford and Kandel 1999; Tsien et al. 1996; Tang et al. 1999; Huerta et al. 2000; Silva et al. 1998.
85. Huang and Kandel 1998; Huang et al. 2000.
86. For a summary of other studies of amygdala LTP, see Maren 1999; Chapman et al. 1990; Chapman 2001.
87. Lamprecht et al. 1997; Rosenblum et al. 1997; Berman et al. 1998; Berman et al. 2000.
88. Weisskopf et al. 1999.
89. Schafe et al. 2001; Blair et al. 2001.
90. Tsien 2000; Paulsen and Sejnowski 2000; Magee and Johnston 1997; Johnston et al. 1999.
91. Huang and Kandel 1998; Huang et al. 2000; McKernan and Shinnick-Gallagher 1997.

92. Fanselow and LeDoux 1999; Cahill et al. 1999; Schafe et al. 2001; Blair et al. 2001.
93. Nader et al. 2000. For the history of reconsolidation research, see Sara 2000.
94. For review, see: Beggs et al. 1999; Sahley 1995; Crow 1988; Alkon 1989; Jing and Gillette 1995.
95. Some of Kandel's main colleagues and collaborators in the *Aplysia* work have included James Schwartz, Irving Kupferman, Vince Castellucci, Tom Carew, Robert Hawkins, Tom Abrams, Jack Byrne, Sam Schacter, Steve Sigelbaum, David Glanzman, Craig Bailey, Mary Chen, and Kelsey Martin. For summaries of the Kandel lab research on *Aplysia,* see Hawkins and Kandel 1984; Hawkins et al. 1987; Kandel 1989; Bailey et al. 1996; Kandel 1997.
96. Byrne et al. 1993; Cleary et al. 1998; Lechner and Byrne 1998.
97. Summarized in Dudai 1989.
98. For a discussion of why the learning in the *Aplysia* may not really be so simple, see Glanzman 1995.
99. In spite of the enormous progress, though, there's a gap between the relation of the research on the reduced experimental setups and the whole organism and its ability to learn. For a discussion, see Dudai 1989. Recent attempts have been made to close this gap by making the test paradigms for the simplified experimental setup more like real-life learning (see Hawkins et al. 1998).
100. The actual steps are that when serotonin binds to its receptors, adenyl cyclase is activated. cAMP is then made from ATP. The cAMP in turn activates PKA.
101. One of PKA's jobs is to phosphorylate an ion channel (a place on the cell membrane where ions, like calcium, sodium, and potassium, flow in and out). The ion channel in question is a special potassium channel. When phosphorylated by PKA, this channel closes, keeping potassium trapped inside the cell. The effect of this is that an action potential traveling down the sensory axon to the terminal lasts a bit longer. The reason for this is related to the fact that the exit of potassium out of the cell plays a key role in resetting the cell's electrical properties after an action potential, and this process is prolonged by phosphorylation of the potassium channel. Since calcium flows into the cell during action potentials, more calcium flows into the terminal after shock than before (since the action potential lasts longer). A longer action potential thus means more calcium coming in, which means more transmitter going out. And the more transmitter that goes out of the sensory terminal, the bigger will be the postsynaptic response of the motor neuron.
102. Specifically, what happens is that repeated shocks lead to the removal of regulatory subunits of PKA, leaving the catalytic subunit to translocate to the cell nucleus.
103. Involved are cell adhesion molecules.
104. Glanzman 1995; Murphy and Glanzman 1999; Lechner and Byrne 1998; Bao et al. 1998.
105. McKernan and Shinnick-Gallagher 1997.
106. Huang and Kandel 1998.
107. Huang et al. 1996; Malgaroli and Tsien 1992; Bekkers and Stevens 1990; Staubli et al. 1990; Weisskopf and Nicoll 1995.
108. This section is based on Dudai 1989.
109. Quinn et al. 1974; Tully and Quinn 1985.

110. For summaries of fly studies and memory molecules, see Dubnau and Tully 2001 and Yin and Tully 1996.
111. Davis 1996.
112. Staubli et al. 1994.
113. Rogan et al. 1997.
114. Tang et al. 1999.

CHAPTER SEVEN THE MENTAL TRILOGY

1. The phrase "The Mental Trilogy" comes from Hilgard 1980, who traced the three-way partition of mind to eighteenth-century Germany. However, a forerunner of this view is clearly found in Plato's tripartite soul.
2. See Hilgard 1980.
3. It's not completely fair to say that the trilogy was forgotten about during behaviorism. Instead, the processes involved were subsumed under behavioral interpretations. For example, behaviorists were interested in motivation as a driving force in behavior, but not as a mental state.
4. For cognitivists, emotions and motivations became kinds of cognitions, namely, thoughts about one's self in certain kinds of challenging situations. For more on the cognitive theory of emotion, see chapters 2 and 3 in *The Emotional Brain.*
5. Boring 1950.
6. Watson 1913; Watson 1925.
7. Minsky 1985.
8. It is well known that you can basically attend to one thing at a time, except under special circumstances. See Hirst et al. 1980.
9. Baddeley and Hitch 1974; Baddeley 1982; Baddeley 1992.
10. Bartlett 1932.
11. Gardner 1987.
12. Hilgard 1980.
13. Miller 1956.
14. Norman and Shallice 1980.
15. Johnson-Laird 1988.
16. Smith and Jonides 1999.
17. Hirst et al. 1980.
18. Luria 1973.
19. Stuss and Benson 1986; Nauta 1971; Fuster 1997; Lhermitte et al. 1972; Teuber 1964; Goldman-Rakic 1987; D'Esposito et al. 1995; Smith and Jonides 1999; Albright et al. 2000.
20. D'Esposito et al. 1995; Smith and Jonides 1999; Albright et al. 2000.
21. Smith and Jonides 1999.
22. Preuss 1995.
23. Jacobsen and Nissen 1937.
24. Fuster 1973; Fuster 1997; Fuster 2000; Fuster 1993; Levy and Goldman-Rakic 2000; Goldman-Rakic 1999; Goldman-Rakic 1987.
25. For a summary of prefrontal connections, see: Fuster 1989; Goldman-Rakic 1987;

Passingham 1995; Groenewegen et al. 1990; Petrides and Pandya 1999; Fuster 1997; Maioli et al. 1998.
26. Ungerleider and Mishkin 1982; Zeki 1993; Van Essen et al. 1992; Van Essen 1985.
27. Van Essen 1995.
28. Ungerleider and Mishkin 1982; Ungerleider and Haxby 1994; Goodale 1998.
29. There is not universal agreement that the parietal area is involved in processing spatial information. Some prefer the view that the parietal region is more involved in planning and making decisions about movements than in the perception of where objects are located. Some of those who dispute this are Paul Glimcher, Carol Colby, Michael Goldberg, and Richard Andersen. See: Platt and Glimcher 1999; Colby and Goldberg 1999; Xing and Andersen 2000.
30. Mesulam et al. 1977; Pandya and Seltzer 1982.
31. Fuster 1973; Fuster 1997; Funahashi et al. 1989.
32. Gnadt and Andersen 1988; Koch and Fuster 1989.
33. Mesulam et al. 1977; Pandya and Seltzer 1982.
34. Miller et al. 1993; Miller and Desimone 1994.
35. Romanski et al. 1999.
36. This simple story involving specialized short-term buffers in the sensory systems and a general purpose working memory mechanism in the prefrontal cortex is somewhat more complicated than the way I have presented it. As discussed later, the prefrontal cortex itself seems to have regions that are specialized, at least to some degree, for specific kinds of working memory functions.
37. Cohen and Servan-Schreiber 1992; Cohen et al. 1999.
38. Vendrell et al. 1995.
39. Cohen and Servan-Schreiber 1992; Cohen et al. 1999.
40. Leung et al. 2000; Peterson et al. 1999; Epstein et al. 1999.
41. Cohen and Servan-Schreiber 1992; Cohen et al. 1999.
42. Goldman-Rakic 1999.
43. D'Esposito et al. 1995.
44. Smith and Jonides 1999.
45. Jacobsen 1935.
46. Reynolds and Desimone 1999.
47. Jiang et al. 2000; Haxby et al. 2000.
48. Asaad et al. 1998; Miller 1999.
49. Chelazzi et al. 1993.
50. Miller 1999; Desimone and Duncan 1995.
51. Knight 1997; Barcelo et al. 2000.
52. Tomita et al. 1999.
53. Kim and Shadlen 1999.
54. Shadlen et al. 1996.
55. Platt and Glimcher 1999; Colby and Goldberg 1999; Batista and Andersen 2001.
56. Batista and Andersen 2001.
57. Fuster et al. 1982.
58. Levy and Goldman-Rakic 2000.
59. Wilson et al. 1993.

60. Romanski et al. 1999.
61. Smith and Jonides 1999.
62. Smith and Jonides 1999.
63. This region is often called the dorso-lateral prefrontal cortex, but we'll refer to it as lateral prefrontal to keep things simple.
64. Smith and Jonides 1999; Bush et al. 2000.
65. Pandya and Yeterian 1996; Fuster 1997; Passingham 1995; Petrides and Pandya 1999; Maioli et al. 1998.
66. Posner 1992.
67. Berger and Posner 2000; Badgaiyan and Posner 1998; Bush et al. 2000; Botvinick et al. 1999; Carter et al. 2000.
68. Another important area is the orbital cortex, as we will discuss in chapter 8.
69. Smith and Jonides 1999; Bechara et al. 1998; Robbins 1996; Owen et al. 1999; Botvinick et al. 1999; Carter et al. 2000.
70. Circuit description below is based on: Douglas and Martin 1998; Durstewitz et al. 1999; Markram et al. 1997; Kritzer and Goldman-Rakic 1995; Cauller et al. 1998; Jones 1984.
71. Arnsten 1998.
72. Arnsten 1998; Robbins 2000; Arnsten et al. 1994; Cai and Arnsten 1997; Muller et al. 1998.
73. Nieuwenhuys 1985.
74. Sawaguchi et al. 1988; Sawaguchi et al. 1990. These effects may well be related to the fact that infusion of dopamine enhances the responses of cells to cue stimuli that signal reward, and also enhances the activity of cells during a delay period between the cue and the reward.
75. Lindvall et al. 1978; Berger et al. 1976; Lewis et al. 1987.
76. Arnsten 1998; Robbins 2000.
77. Yang and Seamans 1996.
78. Thierry et al. 1993.
79. Arnsten 1998.
80. Woolf 1925.
81. Baars 1997; Johnson-Laird 1993; Kihlstrom 1987; Marcel and Bisiach 1988; Norman and Shallice 1980; Shallice 1988; Kosslyn and Koenig 1992.
82. Lashley 1950.
83. For example, see: Stuss 1991; Luria 1969; Ackerly and Benton 1947.
84. Luria 1969.
85. Ackerly and Benton 1947.
86. Stuss 1991.
87. Damasio 1999; Panksepp 1998.
88. Milner 1982; Shimamura 1995.
89. Buckner and Koutstaal 1998; Wagner 1999; Cabeza and Nyberg 2000; Lepage et al. 2000; Schacter et al. 1998.
90. Wagner 1999.
91. Crick and Koch 1990, 1995.
92. He et al. 1996; Tootell et al. 1995; Damasio 1995.

93. Tootell et al. 1995.
94. For more discussion of the implications of the Tootell study for the Crick/Koch hypothesis, see Damasio 1995.
95. Milner 1974; von der Malsburg 1995.
96. von der Malsburg 1995; Roskies 1999; Treisman 1996.
97. Engel and Singer 2001; Crick and Koch 1990; Tononi and Edelman 1998; Damasio 1990; Llinas and Ribary 1994; Grossberg 1999.
98. Engel and Singer 2001.
99. Shadlen and Movshon 1999.
100. Smith and Jonides 1999; Owen et al. 1999.
101. Some have argued that there is a rudimentary lateral prefrontal cortex in the rat, but this is debatable. See: Kolb and Tees 1990; Preuss 1995; Preuss 1995.
102. Gray 1987.
103. Aston-Jones et al. 1999.
104. Dillard 1974.
105. Gallup 1991; Kennan et al. 2000.
106. Delfour and Marten 2001; Reiss and Marino 2001. Also see the journal *Consciousness and Cognition,* volume 4, issue 2, 1995, for a discussion of this work and its implications.
107. Hauser et al. 1995.
108. Weiskrantz 1996; Kihlstrom 1987; Erdelyi 1985; LeDoux 1996; Wilson et al. 2000; Wilson (in press); Bargh 1990; Bargh and Chartrand 1999; Greenwald and Banaji 1995; Zajonc 1984; Loftus and Klinger 1992; Bowers 1984; Bowers and Meichenbaum 1984; Öhman 2000; Debner and Jacoby 1994; de Gelder et al. 1999.
109. Gazzaniga 1985; Gazzaniga 1992; Gazzaniga 1998.
110. Gazzaniga and LeDoux 1978.
111. Gazzaniga 1985; Gazzaniga 1992; Gazzaniga 1998.

CHAPTER EIGHT THE EMOTIONAL BRAIN REVISITED

1. James 1890.
2. Bard 1928; Cannon 1929; Herrick 1933; Papez 1937; Kluver and Bucy 1937; Hess and Brugger 1943; MacLean 1949; MacLean 1952; MacLean 1970; MacLean 1990.
3. The time I'm referring to is between the early 1960s and the early 1980s. Researchers like Mort Mishkin, Edmund Rolls, Jeffrey Gray, Jaak Panksepp, John Flynn, and Alan Siegel each did work on emotions and the brain for some of this period. For summaries, see: Flynn 1967; Mishkin and Aggleton 1981; Gray 1982; Rolls 1986; Panksepp 1982; Siegel and Edinger 1981.
4. See Neisser 1967; Gardner 1987.
5. Bard 1928; Cannon 1929; Herrick 1933; Papez 1937; MacLean 1949; MacLean 1952.
6. For example, during the behaviorist time, emotional and other mental states were reinterpreted in terms of verbal behavior. What had previously been called "feelings" by the introspectionists became verbal descriptions of one's emotional response tendencies. During the cognitive period, emotions were also recast, but this time in terms of consciously accessible and verbally describable thought processes called appraisals. Unconscious emotions have long been emphasized in the Freudian tradi-

tion, but verbal descriptions of mental states also play a key role in the psychoanalytic process, which seeks to bring repressed emotions to the forefront of consciousness, where they can be talked about.

7. Larsen and Fredrickson 1999; Stone et al. 1999; Schwarz and Strack 1999.
8. Plutchik 1980.
9. Plutchik 1980.
10. Hebb 1946.
11. Kahneman 1999.
12. Loftus 1986; Loftus and Hoffman 1989.
13. Bartlett 1932.
14. Kahneman 1999; Stone et al. 1999.
15. Larsen and Fredrickson 1999; Stone et al. 1999; Schwarz and Strack 1999.
16. Schacter and Singer 1962; Schacter 1975.
17. Panksepp 1998; Masson and McCarthy 1995.
18. Lamb et al. 1991.
19. Tinbergen 1951.
20. Neisser 1967; Gardner 1987.
21. Panksepp 1998.
22. LeDoux 1984; LeDoux 1987; LeDoux 1990.
23. Murphy and Zajonc 1993; Soares and Öhman 1993; Morris et al. 1998; Morris et al. 1999.
24. This does not, strictly speaking, imply that only humans are conscious. Instead, it means that humans are an example of a species where consciousness clearly exists. I discussed some issues about consciousness in other animals in chapter 7.
25. Behaviorists often used the term *black box* to refer to the fact that psychological processes in the mind were invisible to the experimenter. The cognitive and brain sciences have changed that.
26. Schacter and Singer 1962; Frijda 1986; Lazarus 1991; Smith and Lazarus 1990; Frijda 1993; Scherer 1988; Scherer 1993; Smith and Ellsworth 1985; Ellsworth 1991; Averill 1994; Oatley and Johnson-Laird 1987; Ortony et al. 1988.
27. MacLean 1949; MacLean 1952; MacLean 1970; MacLean 1990; Isaacson 1982.
28. Scoville and Milner 1957.
29. MacLean 1949; MacLean 1952.
30. Swanson 1983.
31. Nauta 1979.
32. Kaada 1960.
33. Isaacson 1982; Swanson 1983; Livingston and Escobar 1971.
34. Brodal 1982; Kotter and Meyer 1992; LeDoux 1987; LeDoux 1991.
35. Some researchers had been using fear conditioning to study behavior, and these were important developments in the ultimate application of this work to the brain. Included were: Estes and Skinner 1941; Mowrer and Lamoreaux 1946; Mowrer 1947; Miller 1948; Miller 1951; Brady and Hunt 1951; Solomon and Wynne 1954; Kamin 1963; Rescorla and Solomon 1967; Annau and Kamin 1961; Stebbins and Smith 1964; Brown et al. 1951; LoLordo 1967; Blanchard and Blanchard 1969; McAllister and McAllister 1971; Bolles et al. 1966; Bolles and Fanselow 1980; Bouton and Bolles 1980.

36. Several researchers used conditioned fear to study brain mechanisms of fear but did not pursue the circuits in detail, including John Harvey, Orville Smith, and Neal Schneiderman. See: Harvey et al. 1965; Marshall and Smith 1975; Schneiderman et al. 1974.
37. Blanchard and Blanchard 1972; Kapp et al. 1979; Kapp et al. 1984; Kapp et al. 1992; Hitchcock and Davis 1986; Davis et al. 1987; Davis 1992; Davis et al. 1997; Kim et al. 1993; Fanselow 1994; Maren and Fanselow 1996; Maren et al. 1996; Maren 2001; Fendt and Fanselow 1999; LeDoux 1984; LeDoux et al. 1985; Iwata et al. 1986; LeDoux et al. 1990; LeDoux 1986; LeDoux et al. 1989; LeDoux 1990; LeDoux 1992; Romanski and LeDoux 1992; LeDoux 1994; LeDoux 1995; LeDoux 1996; LeDoux 2000.
38. Cohen 1974; Cohen 1980.
39. Other approaches are available for studying fear, including various forms of avoidance conditioning, studies of behavior in an open field, reactions to a shock probe, "elevated plus maze" performance, and so on. Of these, avoidance conditioning has been used most extensively to study the neural basis of fear. This will be discussed further in chapter 9.
40. Some studies have found differences in the function of the left and right amygdala in the human brain, but the extent and significance of the differences are not understood at this point. For a recent interesting example, see Cahill et al. 2001.
41. Anagnostaras et al. 1999; Kim and Fanselow 1992; Maren and Fanselow 1996; Phillips and LeDoux 1992; Everitt and Robbins 1992.
42. Darwin 1872.
43. For example, some have suggested that hippocampal lesions make animals more active and so they freeze less not because they have not processed the context but simply because they are hyperactive. But if this were true, rats with hippocampal lesions should freeze less to a tone as well, but they do not. See: McNish et al. 1997; Maren et al. 1998.
44. Michael Davis and colleagues failed to find this effect, but they studied fear conditioning using a different procedure, which may account for the difference. See Gewirtz et al. 1997.
45. Quirk et al. 2000.
46. Garcia et al. 1999.
47. Damasio 1994; Bechara et al. 1999.
48. LaBar et al. 1995.
49. Bechara et al. 1995.
50. O'Connor et al. 1999.
51. LaBar et al. 1998; Morris et al. 1998.
52. Rolls 1999.
53. Actually, these studies don't measure neural activity but instead infer it from such measures as blood oxygenation in fMRI studies or blood flow in PET studies.
54. Morris et al. 1999.
55. Morris et al. 1996.
56. Breiter et al. 1996.
57. Whalen et al. 1998.
58. Adolphs et al. 1994; Calder et al. 1996; Young et al. 1996; Hamann et al. 1996; Scott et al. 1997.

59. Adolphs et al. 1998.
60. Kluver and Bucy 1937.
61. Rolls 1999; Ono and Nishijo 1992.
62. Phelps et al. 2000; Hart et al. 2000.
63. Bargh 1992; Jacoby and Toth 1992.
64. Brown and Kulik 1977; Christianson 1989; Neisser and Harsch 1992.
65. See McGaugh 2000; Cahill and McGaugh 1998; McGaugh 1990; McGaugh and Gold 1989; Gold and Zornetzer 1983; Gold 1995.
66. Bower and Cohen 1982; Bower 1992.
67. Sapolsky 1996; McEwen and Sapolsky 1995; Sapolsky 1998.
68. Diamond and Rose 1994; Shors and Dryver 1992; Conrad et al. 1999; Conrad et al. 1999; McEwen 1999; Kim and Yoon 1998.
69. Diamond and Rose 1994; Shors et al. 1989; Pavlides et al. 1993; Pavlides et al. 1996; McEwen 1999; Kim and Yoon 1998.
70. McEwen 1999; Sapolsky 1996; Sapolsky 1998.
71. Diorio et al. 1993.
72. Makino et al. 1994; Corodimas et al. 1994; Conrad et al. 1999; Shors et al. 1992.
73. Corodimas et al. 1994.
74. Scherer 2000; Maturana and Varela 1987; LeDoux 1996.
75. Amaral et al. 1992.
76. Weinberger 1995; Weinberger 1998.
77. Armony et al. 1998.
78. Groenewegen et al. 1990; McDonald 1998.
79. Rolls 1999; Gaffan 1992; Everitt and Robbins 1992; Rogers et al. 1999.
80. Pandya and Yeterian 1996; Petrides and Pandya 1999; Maioli et al. 1998; Passingham 1995; Fuster 1997.
81. Damasio 1994.
82. Anderson and Phelps (in press).
83. James 1884; James 1890; Schacter 1975; Berntson et al. 1993; Levenson 1992; Damasio 1994; Damasio 1999.
84. Insel 1997; Carter 1998.
85. Bowlby 1969; Bartholomew and Perlman 1994; Sternberg and Barnes 1988; Kraemer 1992; Carter 1998.
86. For summaries of their work, see Insel 1997; Carter 1998.
87. Insel 1997.
88. This description of the role of oxytocin and vasopressin is based on Insel 1997.
89. Schulkin 1999.
90. See Pfaff 1999; Meisel and Sachs 1994.
91. Canteras et al. 1995.
92. Veinante and Freund-Mercier 1997; Veinante and Freund-Mercier 1995.
93. This is reminiscent of William James's theory and its modern reincarnation and extension by Antonio Damasio. See: James 1884; James 1890; Damasio 1994; Damasio 1999. Also see: Schacter 1975; Berntson et al. 1993; Levenson 1992; Porges 1998.
94. Porges 1998.
95. As this book was being completed, I learned about a study showing activation of pre-

frontal and other areas in human subjects thinking about a loved one (Bartels and Zeki 2000).

CHAPTER NINE THE LOST WORLD

1. Miller 1948.
2. Actually, not all rats learned this. Some were unable to stop freezing and learn to adapt.
3. Hull 1943; Hull 1954; also see Bolles 1967; Cofer 1972; and Weiner 1989.
4. Hobbes 1651; Bentham 1779; also see discussion of hedonism in Bolles 1967 and Cofer 1972.
5. Thorndike 1898, 1913.
6. Watson 1925; Skinner 1938; Hull 1943.
7. Bolles 1967; Cofer 1972; Weiner 1989.
8. Sheffield and Roby 1950.
9. It should be noted, though, that while drive reduction or need satisfaction is not necessary for learning to occur, drive reduction, when it occurs, may still be involved in learning and motivation. For further discussion, see Cofer 1972.
10. Young 1961; Mowrer 1960; Bindra 1969; Bolles 1967; Toates 1986; Dickinson and Balleine 1994; Trowill et al. 1969; Everitt et al. 1999; Ikemoto and Panksepp 1999.
11. See Goddard 1964; Grossman 1967; Sarter and Markowitsch 1985; Liang et al. 1982.
12. Mowrer and Lamoreaux 1946; Mowrer 1947; Mowrer 1960; Solomon and Wynne 1954; Rescorla and Solomon 1967.
13. Amorapanth et al. 2000.
14. McAllister and McAllister 1971.
15. Killcross et al. 1997 performed a similar study. Although the main conclusion was similar—that basal amygdala lesions disrupted the learning of an instrumental response in the presence of an aversive secondary incentive (reinforcer)—they failed to find an effect of lateral nucleus lesions. This discrepancy is probably due to differences in the ways the studies were performed, as discussed in Nader and LeDoux 1997.
16. Olds and Milner 1954.
17. Olds 1973.
18. Historical information based on Carlson 1994.
19. Olds 1977.
20. Shizgall 1999.
21. Olds 1956.
22. Heath 1964.
23. Crichton 1972.
24. Shizgall 1999.
25. See Trowill et al. 1969; Shizgall 1999.
26. Hess and Brugger 1943; Hess and Akert 1955; von Holst and von Saint-Paul 1962; Flynn 1967; Hilton and Zbrozyna 1963; Valenstein 1970; Glickman and Schiff 1967; Siegel and Edinger 1981; Trowill et al. 1969; Olds 1977.
27. Trowill et al. 1969; Bindra 1969; Gallistel 1966; Deutsch and Deutsch 1966.

28. Olds 1956; Olds 1958; Trowill et al. 1969; Shizgall 1999.
29. Mogenson et al. 1980; Wise 1982; Kalivas and Nakamura 1999; Ikemoto and Panksepp 1999.
30. Cooper et al. 1978.
31. Mogenson et al. 1980; Nieuwenhuys 1985.
32. Kalivas and Nakamura 1999; Ikemoto and Panksepp 1999; Berridge and Robinson 1998; Wise 1982; Everitt et al. 1999; White 1997; Spanagel and Weiss 1999; Everitt and Robbins 1999; Schultz and Dickinson 2000; Schultz 1998; Schultz et al. 1997.
33. Nader and van der Kooy 1994; Nader et al. 1997.
34. For discussion, see: Wise 1982; White 1997; Everitt and Robbins 1999.
35. Everitt et al. 1999; Ikemoto and Panksepp 1999; Berridge and Robinson 1998; Everitt and Robbins 1999.
36. Wise 1982; White 1997; Everitt and Robbins 1999.
37. Ikemoto and Panksepp 1999; Everitt and Robbins 1999.
38. Schultz and Dickinson 2000; Schultz 1998; Schultz et al. 1997.
39. Redgrave et al. 1999.
40. Mogenson et al. 1980; Nieuwenhuys 1985.
41. Kalivas and Nakamura 1999; Ikemoto and Panksepp 1999; Berridge and Robinson 1998; Everitt et al. 1999; White 1997; Spanagel and Weiss 1999.
42. Mogenson et al. 1980; Kalivas and Nakamura 1999; Ikemoto and Panksepp 1999; Everitt et al. 1999; Everitt and Robbins 1999; Spanagel and Weiss 1999.
43. Graybiel 1976.
44. Mogenson et al. 1980.
45. Kalivas and Nakamura 1999; Ikemoto and Panksepp 1999; Everitt et al. 1999; Everitt and Robbins 1999; Spanagel and Weiss 1999.
46. The summary to follow is based on: Ikemoto and Panksepp 1999; Everitt et al. 1999; Kalivas and Nakamura 1999.
47. Morgan 1943; Morgan 1957; Bindra 1969; Gallistel 1980.
48. Ikemoto and Panksepp 1999.
49. Mogenson et al. 1980; Everitt et al. 1999.
50. Much is known about cellular communication between striatal areas, such as the accumbens, and the pallidum, and the manner in which these areas, together with dopaminergic inputs to the striatum and connections with the thalamus and cortex, regulate the control of movement. Detailed consideration of these circuits is beyond our scope here. For an overview, see chapters 33 and 34 in Zigmond et al. 1999.
51. Everitt et al. 1999.
52. Kalivas and Nakamura 1999.
53. See Alexander 1995; Calabresi et al. 1992; Kombian and Malenka 1994.
54. For summary, see Everitt et al. 1999 and Everitt and Robbins 1992. They have also studied negative incentives. Killcross et al. 1997.
55. Mishkin and Aggleton 1982; Rolls 1999; Ono and Nishijo 1992; McDonald and White 1993; Gallagher and Schoenbaum 1999; Gaffan 1992.
56. Solomon and Wynne 1954; Linden 1969.
57. Ikemoto and Panksepp 1999.
58. White 1997; Wise et al. 1996.
59. Kalivas and Nakamura 1999.

60. Groenewegen et al. 1997; Groenewegen et al. 1990; Alheid and Heimer 1996; Alheid and Heimer 1988.
61. Rainville et al. 1997.
62. Rolls 1999; Gaffan et al. 1993.
63. Amaral et al. 1992; Fuster 1989; Goldman-Rakic 1987; Passingham 1995; Groenewegen et al. 1990; Fuster 1997; Petrides and Pandya 1999; Maioli et al. 1998.
64. Damasio 1994; Damasio 1999.
65. Bechara et al. 1998.
66. Anderson et al. 1999.
67. Platt and Glimcher 1999.
68. Petrides and Pandya 1999; Fuster 1997; Goldman-Rakic 1987; Passingham 1995; Maioli et al. 1998.
69. Colby and Goldberg 1999; Xing and Andersen 2000; Pare and Wurtz 1997.
70. Contemporary theories of decision-making have their roots in economic theory. Decision-making has classically been modeled by economists in terms of rationality, where individuals are assumed to make choices by computing the probability of gain in a particular situation. However, it is well known that people and animals do not simply go for maximal gain, and, in fact, do things that seem irrational when considered purely from the point of view of the probability of gain. People make career choices on the basis not just of potential income but also of lifestyle and other factors considered "intangible" within economic theory. But from other points of view, such behavior is perfectly reasonable, even rational; it's just less profitable. Behaviorists have attempted to go beyond rational choice theory in the modeling of decision-making, but have relied on one's history of reinforcement as the main determinant of choice. Cognitive scientists have also gone beyond rational choice conceptions and given the so-called economic man a psychological makeup. But the field of cognitive science, as we've seen, has traditionally ignored the emotional and motivational sides of psychology, and has mainly viewed decision-making in terms of cognitive processes, such as plans, intentions, expectations, and beliefs. A more complete understanding of decision-making requires that the whole organism be taken into consideration—emotional and motivational factors are as important as cognitions.
71. Zajonc 1968.
72. Cantor et al. 1986; Markus and Kitayama 1991.
73. Higgins et al. 1985.
74. Adler 1931.
75. McClelland 1951.
76. Weinberger and McClelland 1990.
77. Weinberger and McClelland 1990; Strauss and Quinn 1997.
78. Strauss and Quinn 1997.
79. Squire 1992; McDonald and White 1993; Packard et al. 1994.
80. Bargh 1990; Bargh and Chartrand 1999; Greenwald and Banaji 1995; Wilson et al. 2000; Wilson (in press).
81. Botvinick et al. 1999.
82. Today, this view is less common than in the 1980s, when cognitive science came to be thought of as the science of mind. It started out as the science of thinking, but then came to be a science of mind by ignoring other aspects of mind. Now, emotion and

motivation are returning in spite of the fact that the field is still called cognitive science. Maybe "mind science" would be a better designation for this field.

CHAPTER TEN SYNAPTIC SICKNESS

1. Barondes 1993.
2. Critics of biological psychiatry: Valenstein 1999; Breggin and Breggin 1995; Breggin 1995; Glenmullen 2000; Ross and Pam 1995.
3. Grossman 1960; Grossman 1967; Miller 1965; Myers 1974.
4. Abood 1960.
5. The smart missile analogy was used by Valenstein 1999.
6. Hyman 2000.
7. For an overview of the biological basis of mental disorders, see Charney et al. 1999.
8. Barondes 1993; Valenstein 1999.
9. Jacobson 1993.
10. See chapter 3 for more about chemical and electrical transmission.
11. The history of LSD research is based on Valenstein 1999.
12. The history of drug treatment for schizophrenia is based on Barondes 1993 and Valenstein 1999.
13. Kline 1954.
14. Delay et al. 1952.
15. Carlsson et al. 1957; Cotzias et al. 1967.
16. Deniker 1983.
17. Carlsson 1983.
18. Van Roussum 1967.
19. Seeman et al. 1975; Creese et al. 1976.
20. See Seeman 1992.
21. Seeman et al. 1975; Creese et al. 1976.
22. Seeman 1992.
23. Seeman and Kapur 2000.
24. Andreasen et al. 1990.
25. Davis et al. 1991; Friedman et al. 1999; Lindstrom 2000; Keltner et al. 1998.
26. Friedman et al. 1999; Lindstrom 2000; Keltner et al. 1998.
27. Andreasen et al. 1986; Weinberger et al. 1980; Morihisa and McAnulty 1985; Selemon and Goldman-Rakic 1999; Tamminga 1991; Shelton et al. 1988; Akil and Lewis 1994; Arnold et al. 1995; Bogerts 1993; Bogerts et al. 1990; Bogerts et al. 1993; Breier et al. 1992; Bruton et al. 1990; Casanova et al. 1993; Eastwood et al. 1995; Flaum et al. 1995; Howard et al. 1995; Jakob and Beckmann 1994; Nopoulos et al. 1995; Pakkenberg 1987; Petty et al. 1995; Roberts et al. 1993; Selemon et al. 1995.
28. Seeman 1992; Abi-Dargham et al. 2000; Sedvall and Farde 1996; Okubo et al. 1997; Joyce and Meador-Woodruff 1997.
29. Andreasen et al. 1986; Pettegrew et al. 1991; Buchsbaum 1990; Ingvar and Franzen 1974; Franzen and Ingvar 1975; Farkas et al. 1984; Berman and Weinberger 1991; Berman et al. 1988; Carter et al. 1998; Stevens et al. 1998; Isenberg et al. 1999; Epstein et al. 1999; Silbersweig et al. 1995; Liddle et al. 1992.

30. Reith et al. 1994; Hietala et al. 1995; Dao-Castellana 1997; Hietala et al. 1999; Lindstrom et al. 1999.
31. Abi-Dargham et al. 2000; Seeman and Kapur 2000.
32. Friedman et al. 1999.
33. Cohen and Servan-Schreiber 1992; Weinberger and Gallhofer 1997; Braver et al. 1999; Arnsten 1998; Goldman-Rakic et al. 1992.
34. Goldman-Rakic et al. 1992.
35. Cortes et al. 1989; Arnsten et al. 1994; Goldman-Rakic et al. 1992; Arnsten 1998.
36. Okubo et al. 1997.
37. Carter et al. 1998; Stevens et al. 1998; Weinberger et al. 1986; Berman et al. 1988.
38. Arnsten et al. 1994; Arnsten 1998; Sawaguchi and Goldman-Rakic 1994; Sawaguchi et al. 1988.
39. Davis et al. 1991; Friedman et al. 1999.
40. Goff et al. 1995.
41. Lewis et al. 1999; Benes 1999.
42. Feinberg 1982; Weinberger et al. 1992; Friston and Frith 1995; Andreasen et al. 1997; Selemon and Goldman-Rakic 1999.
43. Lewis et al. 1999; Benes 1999.
44. Silbersweig et al. 1995; Stern and Silbersweig 1998; Epstein et al. 1999.
45. Grace et al. 1998; Grace 1993; Grace et al. 1997.
46. Crow 1980.
47. As in the discussion of schizophrenia, my survey of the history of treatment for depression borrows from Barondes 1993 and Valenstein 1999.
48. Kline 1974.
49. Kramer 1993.
50. Quoted in Valenstein 1999.
51. McGrath et al. 2000.
52. Quoted in Valenstein 1999.
53. Breggin and Breggin 1995; Breggin 1995. Breggin's Website: *http://www.breggin.com/.*
54. Glenmullen 2000.
55. Jamison 1997.
56. Wurtzel 1999.
57. Quitkin et al. 2000; McGrath et al. 2000; Feighner and Overo 1999.
58. Quitkin et al. 2000; Schatzberg and Kraemer 2000; Rush 2000.
59. Hrobjartsson and Gotzsche 2001. The extent to which this critique applies to psychiatric effects is not known.
60. Nestler 1998.
61. Keltner et al. 1997; Charney et al. 1998.
62. Nelson et al. 1991.
63. Nestler 1998.
64. Berman et al. 1996; Hyman and Nestler 1996; Heninger et al. 1996.
65. Nemeroff 1998; Nestler 1998; Charney et al. 1998.
66. Nemeroff 1998.
67. Nemeroff 1998; McEwen and Sapolsky 1995.
68. Maier 1984; Porsolt 2000; Willner 1995; Sanchez and Meier 1997.

69. Sapolsky 1999.
70. McEwen and Sapolsky 1995; McEwen 1998; Sapolsky 1999.
71. Gould et al. 1998; Gould et al. 1999; Gould et al. 1999.
72. Bremner et al. 2000.
73. Lupien et al. 1998.
74. Starkman et al. 1999.
75. Sapolsky 1999.
76. Sapolsky 1999.
77. Duman et al. 1999; Duman et al. 1997.
78. Drevets 1999; Drevets 1998; Drevets et al. 1997; Davidson and Slagter 2000.
79. Diorio et al. 1993.
80. Dunkin et al. 2000; Lockwood et al. 2000.
81. Sheline et al. 1998; Drevets 1999.
82. Keltner et al. 1998.
83. Barondes 1993; Taylor 1998.
84. The following history of drug treatment of anxiety is based on Valenstein 1999.
85. Gray 1982.
86. Stein et al. 1973; Thiebot et al. 1980; Graeff and Schoenfeld 1970; Redmond 1979; Gallager 1978.
87. See File et al. 2000; Gonzalez et al. 1998; Andrews et al. 1994; Plaznik et al. 1994.
88. Gray and McNaughton 2000.
89. LeDoux 1996.
90. See File 2000; Treit et al. 1993; Treit and Menard 1997; Pesold and Treit 1995; Gonzalez et al. 1996; Harris and Westbrook 1995; Hodges et al. 1987; Shibata et al. 1989; Scheel-Kruger and Petersen 1982; Graeff et al. 1993; Gonzalez et al. 1998; Sanders and Shekhar 1995.
91. Niehoff and Kuhar 1983; Onoe et al. 1996.
92. File 2000.
93. Davis and Lee 1998; Davis et al. 1997.
94. De Olmos and Heimer 1999.
95. The following summary of treatments for anxiety disorders is based on Keltner et al. 1998; Foa et al. 1999; Taylor 1998.
96. Bouton 2000; Öhman et al. 2000; Seligman 1971; Marks 1987; Mineka and Cook 1993; Mineka 1979; Shalev et al. 1992; Pitman et al. 2000; Charney et al. 1995; Klein 1993; Jacobs and Nadel 1985; Barlow et al. 1996.
97. Reiman et al. 1984; Stewart et al. 1988; Woods et al. 1988; Reiman et al. 1989; Feistel 1993; Javanmard et al. 1999; Kuikka et al. 1995; Nordahl et al. 1998; Bremner et al. 1997; Bremner et al. 1999; Fischer et al. 1996; Gurvits et al. 1996; Liberzon et al. 1999; Rauch et al. 1997; Rauch et al. 1996; Shin et al. 1997; Shin et al. 1999; Rauch et al. 1995; Fredrikson et al. 1995; Wik et al. 1996; Bell et al. 1999; Schneider et al. 1999; Birbaumer et al. 1998.
98. This paragraph was taken from our center's grant proposal.
99. LaBar et al. 1998; Buchel et al. 1998; Morris et al. 1998; Morris et al. 1999.
100. O'Connor et al. 1999.
101. Ross and Pam 1995.

102. Gottesman 1991.
103. This paragraph is based on Barondes 1993.
104. This paragraph is based on Barondes 1993.
105. Butcher 2000.
106. DePaulo 2000.
107. Kandel 1998; Kandel 1999.
108. Gorman 1996.

CHAPTER ELEVEN WHO ARE YOU?

1. The brain's processors can be described at a variety of levels, from neurons to synapses to circuits to systems. I'll roughly equate the various processors of a connection machine to the various functional systems of the brain.
2. This is multiparallelism, since several tasks are performed in parallel and each is performed using multiple processors.
3. Jerison 1973.
4. LeDoux 1982.
5. Weinberger 1998; Weinberger 1995; Gilbert 1998; Merzenich et al. 1996.
6. Sanes and Donoghue 2000.
7. The artificial intelligence pioneer Marvin Minsky wrote an interesting book called *The Society of Mind* that also discusses the multiplicity of systems that contribute to mental function. He seems to have been mainly interested in pointing out the diversity and complexity of mental life, rather than in explaining how the diversity is coped with in the process of keeping one's self together.
8. This is an effort of last resort when all other methods fail. For more information on split-brain patients, see Gazzaniga 1970; Gazzaniga and LeDoux 1978.
9. Geschwind 1965.
10. Damasio 1994.
11. Harlow 1868.
12. Damasio 1994.
13. Damasio 1989; Llinas et al. 1994; von der Malsburg 1995; Roskies 1999; Reynolds and Desimone 1999; Singer 2001; for a critique of binding and synchrony as explanations of perceptual phenomena, see Shadlen and Movshon 1999.
14. Engel and Singer 2001; Gray et al. 1989; Singer 2001; Engel et al. 1992; Phillips and Singer 1997.
15. Nargeot 2001; O'Reilly and Munakata 2000; Grossberg 2000.
16. Moruzzi and Magoun 1949.
17. Nieuwenhuys 1985.
18. Izumi and Zorumski 1999; Kobayashi et al. 1997; Katsuki et al. 1997; Brocher et al. 1992; Harley 1991.
19. Stutzmann et al. 1998; Stutzmann and LeDoux 1999.
20. Many animals have subcortical convergence zones that are involved in the regulation of simple behaviors in response to sensory events, but it is in the mammalian cortex that these come to have their full significance.
21. This summary is based on Albright et al. 2000. Some of the key findings over the

years have come from David Hubel and Torsten Weisel, Horace Barlow, Semir Zeki, Charles Gross, Mortimer Mishkin, Leslie Ungerleider, David Van Essen, and many others.

22. For example, see: Rolls 1992; Gochin et al. 1994; Nicolelis and Chapin 1994; Deadwyler and Hampson 1995; Buzsaki and Chrobak 1995; Wilson and McNaughton 1993; Young and Yamane 1992; Shadlen et al. 1996.
23. Martin 1994.
24. Shadlen et al. 1996.
25. Shadlen et al. 1996; Parker and Newsome 1998; Shadlen and Newsome 1994; Zohary et al. 1994.
26. Jones and Powell 1970.
27. As defined in chapter 5, this includes the perirhinal, entorhinal, and parahippocampal cortices.
28. Platt and Glimcher 1999; Colby and Goldberg 1999; Xing and Andersen 2000; Pare and Wurtz 1997.
29. Geschwind 1965; Bhatnagar et al. 2000; Vicari et al. 2000.
30. Mesulam et al. 1977.
31. Kim and Baxter 2001.
32. Reynolds and Desimone 1999; Desimone and Duncan 1995; Kastner et al. 1999; Kastner et al. 1998; Tomita et al. 1999; D'Esposito et al. 1995; Smith and Jonides 1999.
33. Szentagothai 1984.
34. Scherer 2000; Maturana and Varela 1987.
35. Gray 1987.

WORKS CITED

ALL SOURCES CITED IN THE CHAPTER NOTES ARE INCLUDED HERE, BUT IN ABBREVIATED FORM. FOR FULL BIBLIOGRAPHIC CITATIONS, PLEASE SEE WWW.CNS.NYU.EDU/HOME/LEDOUX/SYNSELF/WORKSCITED

Abel, T., et al. 1997. *Cell* 88:615–26.

Abi-Dargham, A., et al. 2000. *Proc. Natl. Acad. Sci. USA* 97:8104–9.

Abood, L. 1960. In *The Etiology of Schizophrenia,* edited by D. Jackson, 99–119. New York: Basic Books.

Ackerly, S. S., and A. L. Benton. 1947. *Res. Publ. Assoc. Res. Nerv. Ment. Dis.* 27:479–504.

Adler, A. 1931. *What Life Should Mean to You.* Boston: Little, Brown.

Adolphs, R., et al. 1994. *Nature* 372:669–72.

Adolphs, R., et al. 1998. *Nature* 393:470–74.

Aggleton, J. 2000. *The Amygdala.* Oxford: Oxford University Press.

Akil, M., and D. A. Lewis. 1994. *Neurosci.* 60:857–74.

Albright, T. D., et al. 2000. *Neuron* 25 Suppl.:S1–55.

Alcock, J. 1998. *Animal Behavior.* Sunderland, MA: Sinauer.

Alexander, G. E. 1995. In *Handbook of Brain Theory,* edited by M. Arbib. Cambridge: MIT Press.

Alheid, G. F., and L. Heimer. 1988. *Neurosci.* 27:1–39.

Alheid, G. F., and L. Heimer. 1996. *Prog. Brain Res.* 107:461–84.

Alkon, D. L. 1989. *Sci. Am.* 261:42–50.

Amaral, D. G., et al. 1987. *J. Comp. Neurol.* 264:326–55.

Amaral, D. G., et al. 1992. In *The Amygdala,* edited by J. P. Aggleton, 1–66. New York: Wiley-Liss.

Amorapanth, P., et al. 2000. *Nat. Neurosci.* 3:74–79.

Anagnostaras, S. G., et al. 1999. *J. Neurosci.* 19:1106–14.

Anderson, A., and E. A. Phelps. *Nature* (pending revision).

Anderson, S. W., et al. 1999. *Nat. Neurosci.* 2:1032–37.

Andreasen, N. C., et al. 1986. *Arch. Gen. Psychiat.* 43:136–44.

Andreasen, N. C., et al. 1986. *Arch. Gen. Psychiat.* 43:421–29.

Andreasen, N. C., et al. 1990. *Arch. Gen. Psychiat.* 47:615–21.

Andreasen, N. C., et al. 1997. *Lancet* 349:1730–34.

Andrews, N., et al. 1994. *Eur. J. Pharmacol.* 264:259–64.

Annau, Z., and L. J. Kamin. 1961. *J. Comp. Physiol. Psychol.* 54:428–32.

Antonini, A., and M. P. Stryker. 1993. *Science* 260:1819–21.

Arbib, M. 1999. In *Neuroscience and the Person,* edited by R. J. Russell et al. Berkeley: Vatican Observatory Publications, Vatican City State, Center for Theology and the Natural Sciences.
Ariëns Kappers, C. U. 1909. *Arch. Neurol. Psychiat.* 4:161–73.
Armony, J. L., et al. 1998. *J. Neurosci.* 18:2592–2601.
Arnold, S. E., et al. 1995. *Am. J. Psychiat.* 152:738–48.
Arnold, S. J. 1980. In *Foraging Behavior,* edited by A. Kamil and T. Sargent. New York: Garland STPM Press.
Arnsten, A. F. 1998. *Trends Cogn. Sci.* 2:419–63.
Arnsten, A. F., et al. 1994. *Psychopharmacol.* 116:143–51.
Asaad, W. F., et al. 1998. *Neuron* 21:1399–1407.
Aston-Jones, G., et al. 1999. *Biol. Psychiat.* 46:1309–20.
Averill, J. R. 1994. *Cogn. Emo.* 8:73–92.
Baars, B. J. 1997. *J. Conscious. Stud.* 4:292–309.
Babic, T. 1999. *J. Neurol. Neurosurg. Psychiat.* 67:558.
Baddeley, A. 1982. *Your Memory.* New York: Macmillan.
Baddeley, A. 1992. *Science* 255:556–59.
Baddeley, A., and G. J. Hitch. 1974. In *The Psychology of Learning and Motivation,* edited by G. Bower. New York: Academic Press.
Badgaiyan, R. D., and M. I. Posner. 1998. *Neuroimage* 7:255–60.
Bailey, C. H., et al. 1996. *Proc. Natl. Acad. Sci. USA* 93:13445–52.
Bailey, D. J., et al. 1999. *Behav. Neurosci.* 113:276–82.
Bannerman, D. M., et al. 1995. *Nature* 378:182–86.
Bao, J. X., et al. 1998. *J. Neurosci.* 18:458–66.
Barcelo, F., et al. 2000. *Nat. Neurosci.* 3:399–403.
Bard, P. 1928. *Am. J. Physiol.* 84:490–515.
Bargh, J. 1992. In *Perception Without Awareness,* edited by R. Bornstein and T. Pittman, 236–55. New York: Guilford Press.
Bargh, J. A. 1990. In *Handbook of Motivation and Cognition,* edited by T. Higgins and R. M. Sorrentino, 93–130. New York: Guilford Press.
Bargh, J. A., and K. Barndollar. 1996. In *The Psychology of Action,* edited by P. M. Gollwitzer and J. A. Bargh. New York: Guilford Press.
Bargh, J. A., and T. L. Chartrand. 1999. *Am. Psychol.* 54:462–79.
Barkow, J. H., et al., eds. 1992. *The Adapted Mind.* New York: Oxford University Press.
Barlow, D. H., et al. 1996. *Nebr. Symp. Motiv.* 43:251–328.
Barnes, C. A. 1995. *Neuron* 15:751–54.
Barondes, S. 1993. *Molecules and Mental Illness.* New York: Scientific American Library.
Bartels, A., and S. Zeki. 2000. *Neuroreport* 11:3829–34.
Bartholomew, K., and D. Perlman. 1994. *Attachment Processes in Adulthood. Advances in Personal Relationships.* London: Jessica Kingsley Publishers.
Bartlett, F. C. 1932. *Remembering.* Cambridge: Cambridge University Press.
Barton, R. A. 1997. *Behav. Brain Sci.* 20:556–57.
Batista, A. P., and R. A. Andersen. 2001. *J. Neurophysiol.* 85:539–44.
Beach, F. A. 1955. *Psychol. Rev.* 62:401–10.
Bear, M. F., and R. C. Malenka. 1994. *Curr. Opin. Neurobiol.* 4:389–99.

Bechara, A., et al. 1995. *Science* 269:1115–18.
Bechara, A., et al. 1998. *J. Neurosci.* 18:428–37.
Bechara, A., et al. 1999. *J. Neurosci.* 19:5473–81.
Beggs, J. M., et al. 1999. In *Fundamental Neuroscience,* edited by M. Zigmond. San Diego: Academic Press.
Bekkers, J. M., and C. F. Stevens. 1990. *Nature* 346:724–29.
Bell, C. J., et al. 1999. *Eur. Arch. Psychiat. Clin. Neurosci.* 249:S11–18.
Benes, F. M. 1999. *Biol. Psychiat.* 46:589–99.
Bennett, E. L., et al. 1964. *Science* 146:610–19.
Bentham, J. 1779; reprint 1948. *An Introduction to the Principle of Morals and Legislation.* New York: Hafner Publishing.
Berger, A., and M. I. Posner. 2000. *Neurosci. Biobehav. Rev.* 24:3–5.
Berger, B., et al. 1976. *Brain Res.* 106:133–45.
Berman, D. E., et al. 1998. *J. Neurosci.* 18:10037–44.
Berman, D. E., et al. 2000. *J. Neurosci.* 20:7017–23.
Berman, K. F., et al. 1988. *Arch. Gen. Psychiat.* 45:616–22.
Berman, K. F., and D. R. Weinberger. 1991. In *American Psychiatric Press Review of Psychiatry,* edited by A. Tasman and S. M. Goldfinger, 24–59. Washington, D.C.: American Psychiatric Press.
Berman, R. M., et al. 1996. In *Biology of Schizophrenia and Affective Disease,* edited by S. J. Watson, 295–368. Washington, D.C.: American Psychiatry Association Press.
Bermudez, J. 1996. *Ethics* 106:378–403.
Bernard, L. L. 1924. *Instinct.* New York: Holt, Rinehart, and Winston.
Berntson, G. G., et al. 1993. *Psychol. Bull.* 114:296–322.
Berridge, K. C. 1999. In *Well-Being,* edited by D. Kahneman et al. New York: Russell Sage Foundation.
Berridge, K. C., and T. E. Robinson. 1998. *Brain Res. Rev.* 28:309–69.
Bhatnagar, S. C., et al. 2000. *Brain Lang.* 74:238–59.
Bickerton, D. 1980. *The Roots of Language.* Ann Arbor, MI: Karoma.
Bindra, D. 1969. In *Nebraska Symposium on Motivation,* edited by W. J. Arnold and D. Levine, 1–33. Lincoln: University of Nebraska Press.
Birbaumer, N., et al. 1998. *Neuroreport* 9:1223–26.
Blair, H., et al. 2001. *Learn. Mem.* (in press).
Blackburn, J. R., et al. 1992. *Prog. Neurobiol.* 39:247–79.
Blanchard, D. C., and R. J. Blanchard. 1972. *J. Comp. Physiol. Psychol.* 81:281–90.
Blanchard, R. J., and D. C. Blanchard. 1969. *J. Comp. Physiol. Psychol.* 67:370–75.
Bliss, T. V., and G. L. Collingridge. 1993. *Nature* 361:31–39.
Bliss, T. V., and T. Lømo. 1973. *J. Physiol. (London)* 232:331–56.
Block, N. 1995. *Behav. Brain Sci.* 18:227–87.
Bloom, F. E., and A. Laserson. 1985. *Brain, Mind and Behavior.* New York: Freeman.
Bogerts, B. 1993. *Schizophr. Bull.* 19:431–45.
Bogerts, B., et al. 1990. *Schizophr. Res.* 3:295–301.
Bogerts, B., et al. 1993. *Biol. Psychiat.* 33:236–46.
Bolles, R. C. 1967. *Theory of Motivation.* New York: Harper and Row.
Bolles, R. C., and M. S. Fanselow. 1980. *Behav. Brain Sci.* 3:291–323.

Bolles, R. C., et al. 1966. *J. Comp. Physiol. Psychol.* 62:201–7.
Bontempi, B., et al. 1999. *Nature* 400:671–75.
Bordi, F., and J. LeDoux. 1992. *J. Neurosci.* 12:2493–2503.
Boring, E. G. 1950. *A History of Experimental Psychology.* New York: Appleton-Century-Crofts.
Bortolotto, Z. A., et al. 1999. *Curr. Opin. Neurobiol.* 9:299–304.
Bortolotto, Z. A., et al. 1999. *Nature* 402:297–301.
Bottjer, S. W., and F. Johnson. 1997. *J. Neurobiol.* 33:602–18.
Botvinick, M., et al. 1999. *Nature* 402:179–81.
Bourgeois, J. P., et al. 1994. *Cereb. Cortex* 4:78–96.
Bourtchouladze, R., et al. 1998. *Learn. Mem.* 5:365–74.
Bouton, M. E. 2000. *Health Psychol.* 19:57–63.
Bouton, M. E., et al. 2001. *Psychol. Rev.* 108:4–32.
Bouton, M. E., and R. C. Bolles. 1980. *Anim. Learn. Behav.* 8:429–34.
Bower, G. 1992. In *Handbook of Emotion and Memory,* edited by S. A. Christianson. Hillsdale, NJ: Lawrence Erlbaum Associates.
Bower, G. H., and P. R. Cohen. Associates 1982. In *Affect and Cognition,* edited by M. S. Clark and S. T. Fiske, 291–331. Hillsdale, NJ: Lawrence Erlbaum Associates.
Bowers, K. S. 1984. In *The Unconscious Reconsidered,* edited by K. S. Bowers and D. Meichenbaum, 227–72. New York: John Wiley & Sons.
Bowers, K. S., and D. Meichenbaum, eds. 1984. *The Unconscious Reconsidered.* New York: John Wiley & Sons.
Bowlby, J. 1969. *Attachment and Loss.* New York: Basic Books.
Brady, J. V., and H. F. Hunt. 1951. *J. Comp. Physiol. Psychol.* 44:204–9.
Brambilla, R., et al. 1997. *Nature* 390:281–86.
Braver, T. S., et al. 1999. *Biol. Psychiat.* 46:312–28.
Breggin, P. 1995. *Toxic Psychiatry.* New York: St. Martin's Press.
Breggin, P. R., and G. R. Breggin. 1995. *Talking Back to Prozac.* New York: St. Martin's Press.
Breier, A., et al. 1992. *Arch. Gen. Psychiat.* 49:921–26.
Breiter, H. C., et al. 1996. *Neuron* 17:875–87.
Bremmer, J. N. 1993. *The Early Greek Concept of the Soul.* Princeton: Princeton University Press.
Bremner, J. D., et al. 1997. *Arch. Gen. Psychiat.* 54:246–54.
Bremner, J. D., et al. 1999. *Biol. Psychiat.* 45:806–16.
Bremner, J. D., et al. 2000. *Am. J. Psychiat.* 157:115–18.
Brocher, S., et al. 1992. *Brain Res.* 573:27–36.
Brock, L. G., et al. 1952. *J. Physiol. (London)* 117:431–60.
Brodal, A. 1982. *Neurological Anatomy.* New York: Oxford University Press.
Brodmann, K. 1909. *Vergleichende Lokalisationslehre der Grosshirnrinde.* Munich: Barth.
Brothers, L. 1997. *Friday's Footprint.* New York: Oxford University Press.
Brown, J. S. 1961. *The Motivation of Behavior.* New York: McGraw-Hill.
Brown, J. S., et al. 1951. *J. Exp. Psychol.* 41:317–28.
Brown, R., and J. Kulik. 1977. *Cognition* 5:73–99.
Brown, T. H., et al. 1988. *Science* 242:724–28.
Bruer, J. *Phi Delta Kappa* May 1999:649–57.

Bruer, J. 1999. *The Myth of the First Three Years.* New York: Free Press.
Bruner, J., et al. 1956. *A Study of Thinking.* New York: John Wiley & Sons.
Bruton, C. J., et al. 1990. *Psychol. Med.* 20:285–304.
Buchel, C., et al. 1998. *Neuron* 20:947–57.
Buchsbaum, M. S. 1990. *Schizophr. Bull.* 16:379–89.
Buckner, R. L., and W. Koutstaal. 1998. *Proc. Natl. Acad. Sci. USA* 95:891–98.
Bunney, W. 1977. *Ann. Intern. Med.* 87:319–35.
Buñuel, L. 1983. *My Last Sigh.* New York: Knopf.
Burwell, R. D., et al. 1995. *Hippocampus* 5:390–408.
Bush, G., et al. 2000. *Trends Cogn. Sci.* 4:215–22.
Butcher, J. 2000. *Lancet* 356:47.
Butler, J. 1990. *Gender Trouble.* New York: Routledge.
Buzsaki, G. 1989. *Neurosci.* 31:551–70.
Buzsaki, G. 1998. *J. Sleep Res.* 7:17–23.
Buzsaki, G., and J. J. Chrobak. 1995. *Curr. Opin. Neurobiol.* 5:504–10.
Byrne, J. H., et al. 1993. *Adv. Second Messenger Phosphoprotein Res.* 27:47–108.
Cabeza, R., and L. Nyberg. 2000. *Curr. Opin. Neurol.* 13:415–21.
Cahill, L., and J. L. McGaugh. 1998. *Trends Neurosci.* 21:294–99.
Cahill, L., et al. 1999. *Neuron* 23:227–28.
Cahill, L., et al. 2001. *Neurobiol. Learn. Mem.* 75:1–9.
Cai, J. X., and A. F. Arnsten. 1997. *J. Pharmacol. Exp. Ther.* 283:183–89.
Cain, D. P., et al. 1996. *Behav. Neurosci.* 110:86–102.
Calabresi, P., et al. 1992. *Eur. J. Neurosci.* 4:929–35.
Calder, A. J., et al. 1996. *Cogn. Neuropsychol.* 13:699–745.
Camhi, J. M. 1984. *Neuroethology.* Sunderland, MA: Sinauer.
Campbell, A. W. 1905. *Histological Studies on the Localization of Cerebral Functions.* Cambridge: Cambridge University Press.
Campeau, S., and M. Davis. 1990. *J. Neurosci. Methods* 32:25–35.
Campeau, S., and M. Davis. 1995. *J. Neurosci.* 15:2301–11.
Cannon, W. B. 1927. *Am. J. Psychol.* 39:106–24.
Cannon, W. B. 1929. *Bodily Changes in Pain, Hunger, Fear, and Rage.* New York: Appleton.
Canteras, N. S., et al. 1995. *J. Comp. Neurol.* 360:213–45.
Cantor, N., et al. 1986. In *Handbook of Motivation and Cognition,* edited by R. M. Sorrentino and E. T. Higgins. New York: Guilford.
Carey, S., and E. Spelke. 1994. In *Mapping the Mind,* edited by L. A. Hirschfield and S. A. Gelman. Cambridge: Cambridge University Press.
Carlson, N. R. 1993. *Psychology.* Boston: Allyn and Bacon.
Carlson, N. R. 1994. *Physiology of Behavior.* Boston: Allyn and Bacon.
Carlsson, A. 1983. In *Discoveries in Pharmacology,* edited by M. J. Parnham and J. Bruinvels. New York: Elsevier.
Carlsson, A., et al. 1957. *Nature* 180:1200.
Carter, C. S. 1998. *Psychoneuroendocrinology* 23:779–818.
Carter, C. S., et al. 1998. *Am. J. Psychiat.* 155:1285–87.
Carter, C. S., et al. 2000. *Proc. Natl. Acad. Sci. USA* 97:1944–48.
Casanova, M. F., et al. 1993. *Psychiat. Res.* 49:41–62.

Castro-Alamancos, M. A., et al. 1995. *J. Neurosci.* 15:5324–33.
Cauller, L. J., et al. 1998. *J. Comp. Neurol.* 390:297–310.
Chalmers, D. 1996. *The Conscious Mind.* New York: Oxford University Press.
Chambers, K. C. 1990. *Annu. Rev. Neurosci.* 13:373–85.
Chan, Y. M., and Y. N. Jan. 1999. *Curr. Opin. Neurobiol.* 9:582–88.
Chang, P. L., et al. 1991. *Neurobiol. Aging* 12:517–22.
Changeux, J. P., and A. Danchin. 1976. *Nature* 264:705–12.
Changeux, J. P., and S. Dehaene. 1989. *Cognition* 33:63–109.
Chapman, P. F. 2001. *Nat. Neurosci.* 4:556–58.
Chapman, P. F., et al. 1990. *Synapse* 6:271–78.
Charney, D., et al. 1999. *Neurobiology of Mental Illness.* New York: Oxford University Press.
Charney, D. S., et al. 1995. In *Neurobiological and Clinical Consequences of Stress,* edited by M. J. Friedman, 271–87. Philadelphia: Lippincott-Raven.
Charney, D. S., et al. 1998. In *Textbook of Psychopharmacology,* 2nd ed., edited by A. F. Schatzberg and C. B. Nemeroff. Washington, D.C.: American Psychiatric Press.
Chelazzi, L., et al. 1993. *Nature* 363:345–47.
Chen, W. R., et al. 2000. *Neuron* 25:625–33.
Chiaia, N. L., et al. 1992. *Devel. Brain Res.* 66:244–50.
Chomsky, N. 1957. *Syntactic Structures.* The Hague, Netherlands: Mouton.
Christian, J. L. 1977. *Philosophy.* New York: Holt, Rinehart, and Winston.
Christianson, S.-A. 1989. *Mem. Cogn.* 17:435–43.
Chun, M. M., and E. A. Phelps. 1999. *Nat. Neurosci.* 2:844–47.
Churchland, P. 1984. *Matter and Consciousness.* Cambridge: MIT Press.
Churchland, P. S. 1986. *Neurophilosophy.* Cambridge: MIT Press.
Churchland, P. S., and T. J. Sejnowski. 1992. *The Computational Brain.* Cambridge: MIT Press.
Clark, A. 1998. *Being There.* Cambridge: MIT Press.
Cleary, L. J., et al. 1998. *J. Neurosci.* 18:5988–98.
Clugnet, M. C., and J. E. LeDoux. 1990. *J. Neurosci.* 10:2818–24.
Cofer, C. N. 1972. *Motivation and Emotion.* Glenview, IL: Scott, Foresman.
Cohen, D. H. 1974. In *Limbic and Autonomic Nervous System Research,* edited by L. V. Di Cara. New York: Plenum Press.
Cohen, D. H. 1980. In *Neural Mechanisms of Goal-Directed Behavior and Learning,* edited by R. F. Thompson et al., 283–302. New York: Academic Press.
Cohen, J. D., and D. Servan-Schreiber. 1992. *Psychol. Rev.* 99:45–77.
Cohen, J. D., et al. 1999. *J. Abnorm. Psychol.* 108:120–33.
Cohen, N. J. 1980. Unpublished doctoral dissertation. University of California at San Diego.
Cohen, N. J., and S. Corkin. 1981. *Soc. Neurosci. Abstr.* 7:517–18.
Cohen, N. J., and H. Eichenbaum. 1993. *Memory, Amnesia, and the Hippocampal System.* Cambridge: MIT Press.
Cohen, N. J., and L. Squire. 1980. *Science* 210:207–9.
Colby, C. L., and M. E. Goldberg. 1999. *Annu. Rev. Neurosci.* 22:319–49.
Collins, D. R., and D. Pare. 1999. *Eur. J. Neurosci.* 11:3441–48.
Collins, D. R., and D. Pare. 1999. *J. Neurosci.* 15:836–44.

Collins, D. R., and D. Pare. 2000. *Learn. Mem.* 7:97–103.
Conrad, C. D., et al. 1999. *Neurobiol. Learn. Mem.* 72:39–46.
Conrad, C. D., et al. 1999. *Behav. Neurosci.* 113:902–13.
Convit, A., et al. 1995. *Lancet* 345:266.
Cooper, J. R., et al. 1978. *The Biochemical Basis of Neuropharmacology.* New York: Oxford University Press.
Coplan, J. D., et al. 1998. *Arch. Gen. Psychiat.* 55:130–36.
Corkin, S. 1968. *Neuropsychol.* 6:255–65.
Corodimas, K. P., et al. 1994. *Ann. NY Acad. Sci.* 746:392–93.
Cortes, R., et al. 1989. *Neurosci.* 28:263–73.
Cosmides, L., and J. Tooby. 1999. In *Encyclopedia of Cognitive Science,* 295–97. Cambridge: MIT Press.
Cotzias, G. C., et al. 1967. *N. Engl. J. Med.* 276:374–79.
Cowan, W. M. 1998. *Neuron* 20:413–26.
Crair, M. C. 1999. *Curr. Opin. Neurobiol.* 9:88–93.
Crair, M. C., et al. 1998. *Science* 279:566–70.
Creese, I., et al. 1976. *Science* 192:481–83.
Crespi, L. P. 1942. *Am. J. Psychol.* 467–517.
Crichton, M. 1972. *The Terminal Man.* New York: Knopf.
Crick, F. 1995. *The Astonishing Hypothesis.* New York: Touchstone Books.
Crick, F., and C. Koch. 1990. *Semin. Neurosci.* 2:263–75.
Crick, F., and C. Koch. 1995. *Nature* 375:121–23.
Crow, T. 1988. *Trends Neurosci.* 11:136–47.
Crow, T. J. 1980. *Br. Med. J.* 280:66–68.
Damasio, A. R. 1994. *Descartes' Error.* New York: Grosset/Putnam.
Damasio, A. R. 1999. *The Feeling of What Happens.* New York: Harcourt, Brace.
Damasio, A. R. 1989. *Neural Comput.* 1:123–32.
Damasio, A. R. 1990. *Semin. Neurosci.* 2:287–96.
Damasio, A. R. 1995. *Nature* 375:106–7.
Dao-Castellana, M. H., et al. 1997. *Schizophr. Res.* 23:167–74.
Darwin, C. 1872; reprint 1965. *The Expression of the Emotions in Man and Animals.* Chicago: University of Chicago Press.
Davidson, R. J. 1992. *Psychol. Rev.* 3:39–43.
Davidson, R. J., and H. A. Slagter. 2000. *Ment. Retard. Dev. Disabil. Res. Rev.* 6:166–70.
Davis, H. P., and L. R. Squire. 1984. *Psychol. Bull.* 96:518–59.
Davis, K. L., et al. 1991. *Am. J. Psychiat.* 148:1474–86.
Davis, M. 1992. *Trends Pharmacol. Sci.* 13:35–41.
Davis, M., and Y. Lee. 1998. *Cogn. Emo.* 12:277–305.
Davis, M., et al. 1987. In *The Psychology of Learning and Motivation,* edited by G. H. Bower, 263–305. San Diego: Academic Press.
Davis, M., et al. 1997. *Philos. Trans. R. Soc. Lond. B Biol. Sci.* 352:1675–87.
Davis, R. L. 1996. *Physiol. Rev.* 76:299–317.
Dawkins, R. 1996. *The Blind Watchmaker.* New York: Norton.
Deadwyler, S. A., and R. E. Hampson. 1995. *Science* 270:1316–18.
Debner, J. A., and L. L. Jacoby. 1994. *J. Exp. Psychol. Learn. Mem. Cogn.* 20:304–17.

de Gelder, B., et al. 1999. *Neuroreport* 10:3759–63.
Delay, J., et al. 1952. *Ann. Med. Psychol. (Paris)* 110:112–17.
de Leon, M. J., et al. 1988. *Lancet* 2:391–92.
de Leon, M. J., et al. 1995. *Neuroimaging Clin. N. Am.* 5(1):1–17.
Delfour, F., and K. Marten. 2001. *Behav. Processes* 53:181–90.
Dell, D. L., and D. E. Stewart. 2000. *Postgrad. Med.* 108:34–36, 39–43.
Denenberg, V. H. 1999. *J. Learn. Disabil.* 32:379–83.
Deniker, P. 1983. In *Discoveries in Pharmacology,* edited by M. J. Parnham and J. Bruinvels, 163–80. Amsterdam: Elsevier.
Dennett, D. C. 1976. In *The Identities of Persons,* edited by A. O. Rorty. Berkeley: University of California Press.
Dennett, D. C. 1988. *Times Literary Supplement,* September 16–22; 1016, 1028–29.
Dennett, D. C. 1991. *Consciousness Explained.* Boston: Little, Brown.
Dennett, D. C. 1996. *Darwin's Dangerous Idea.* New York: Touchstone.
de Olmos, J. S., and L. Heimer. 1999. *Ann. NY Acad. Sci.* 877:1–32.
DePaulo, J. R. 2000. *Cerebrum* 2:43–70.
Desimone, R., and J. Duncan. 1995. *Annu. Rev. Neurosci.* 18:193–222.
Desmond, J. E., and J. W. Moore. 1982. *Physiol. Behav.* 28:1029–33.
Desmond, N. L., and W. B. Levy. 1986. *J. Comp. Neurol.* 253:476–82.
D'Esposito, M., et al. 1995. *Nature* 378:279–81.
Deutsch, J. A., and D. Deutsch. 1966. *Physiological Psychology.* Homewood, IL: Dorsey Press.
Diamond, D. M., and G. Rose. 1994. *Ann. NY Acad. Sci.* 746:411–14.
Dickinson, A. 1980. *Contemporary Animal Learning Theory.* Cambridge University Press.
Dickinson, A., and B. W. Balleine. 1994. *Anim. Learn. Behav.* 22:1–18.
Dillard, A. 1974. *Pilgrim at Tinker Creek.* New York: Harper's Magazine Press.
Diorio, D., et al. 1993. *J. Neurosci.* 13:3839–47.
Dollard, J. C., and N. E. Miller. 1950. *Personality and Psychotherapy.* New York: McGraw-Hill.
Doron, N. N., and J. E. LeDoux. 2000. *J. Comp. Neurol.* 425:257–74.
Douglas, R., and K. Martin. 1998. In *The Synaptic Organization of the Brain,* edited by G. Shepherd. New York: Oxford University Press.
Doupe, A. J., and P. K. Kuhl. 1999. *Annu. Rev. Neurosci.* 22:567–631.
Drevets, W. C. 1998. *Annu. Rev. Med.* 49:341–61.
Drevets, W. C. 1999. *Ann. NY Acad. Sci.* 877:614–37.
Drevets, W. C., et al. 1997. *Nature* 386:824–27.
Dubnau, J., and T. Tully. 2001. *Curr. Biol.* 11:R240–43.
Dudai, Y. 1989. *Neurobiology of Memory.* New York: Oxford University Press.
Dudai, Y. 1996. *Neuron* 17:367–70.
Dudai, Y. 1997. *Neuron* 18:179–82.
Duman, R. S., et al. 1997. *Arch. Gen. Psychiat.* 54:597–606.
Duman, R. S., et al. 1999. *Biol. Psychiat.* 46:1181–91.
Dunkin, J. J., et al. 2000. *J. Affect. Disord.* 60:13–23.
Dunn, L. T., and B. J. Everitt. 1988. *Behav. Neurosci.* 102:3–23.
Durstewitz, D., et al. 1999. *J. Neurosci.* 19:2807–22.
Eastwood, S. L., et al. 1995. *Neurosci.* 66:309–19.

Eccles, J. C. 1953. *The Neurophysiological Basis of Mind.* Oxford: Clarendon Press.

Eccles, J. C. 1977. *Brain Res.* 127:327–52.

Economo, C. V., and G. N. Koskinas. 1925. *Die Cytoarchitektonik der Hirnrinde des erwaschsenen Menschen.* Berlin: Julius Springer.

Edelman, G. 1987. *Neural Darwinism.* New York: Basic Books.

Edelman, G. 1993. *Bright Air, Brilliant Fire.* New York: Basic Books.

Edelman, G., and G. Tononi. 2000. *A Universe of Consciousness. How Matter Becomes Imagination.* New York: Basic Books.

Edwards, P., and A. Pap. 1959. *A Modern Introduction to Philosophy.* Glencoe, IL: Free Press.

Eichenbaum, H. 1995. *Nature* 378:131–32.

Eichenbaum, H. 1996. *Learn. Mem.* 3:61–73.

Eichenbaum, H. 2000. *Nat. Rev. Neurosci.* 1:41–50.

Eichenbaum, H., et al. 1994. *Behav. Brain Sci.* 17:449–518.

Ekman, P. 1980. In *Explaining Emotions,* edited by A. O. Rorty. Berkeley: University of California Press.

Ekman, P. 1992. *Cogn. Emo.* 6:169–200.

Ekman, P. 1999. *Annotated Update of Charles Darwin's "The Expression of the Emotions in Man and Animals."* New York: HarperCollins.

Ekman, P., and R. J. Davidson. 1994. *The Nature of Emotion.* New York: Oxford University Press.

Elgersma, Y., and A. J. Silva. 1999. *Curr. Opin. Neurobiol.* 9:209–13.

Ellsworth, P. 1991. In *International Review of Studies on Emotion,* edited by K. T. Strongman, 143–61. Chichester and New York: Wiley.

Elman, J., et al. 1997. *Rethinking Innateness.* Cambridge: MIT Press.

Elster, J. 1985. *The Multiple Self.* New York: Cambridge University Press.

Engel, A. K., and W. Singer. 2001. *Trends Cogn. Sci.* 5:16–25.

Engel, A. K., et al. 1992. *Trends Neurosci.* 15:218–26.

Engert, F., and T. Bonhoeffer. 1999. *Nature* 399:66–70.

Epstein, J., et al. 1999. *Ann. NY Acad. Sci.* 877:562–74.

Epstein, M. 1995. *Thoughts Without a Thinker.* New York: Basic Books.

Erdelyi, M. H. 1985. *Psychoanalysis.* New York: Freeman.

Estes, W. K., and B. F. Skinner. 1941. *J. Exp. Psychol.* 29:390–400.

Everitt, B. J., and T. W. Robbins. 1992. In *The Amygdala,* edited by J. P. Aggleton, 401–29. New York: Wiley-Liss.

Everitt, B. J., and T. Robbins. 1999. In *Fundamental Neuroscience,* edited by M. J. Zigmond et al. San Diego: Academic Press.

Everitt, B. J., et al. 1999. In *Advancing from the Ventral Striatum to the Extended Amygdala,* edited by J. McGintry, 412–38. New York: New York Academy of Sciences.

Eysenck, H. J., and M. W. Eysenck. 1985. *Personality and Individual Differences.* New York: Plenum.

Fanselow, M. S. 1994. *Psychon. Bull. Rev.* 1:429–38.

Fanselow, M. S., and J. J. Kim. 1994. *Behav. Neurosci.* 108:210–12.

Fanselow, M. S., and J. E. LeDoux. 1999. *Neuron* 23:229–32.

Fanselow, M. S., et al. 1994. *Behav. Neurosci.* 108:235–40.

Farkas, T., et al. 1984. *Arch. Gen. Psychiat.* 41:293–300.
Feighner, J. P., and K. Overo. 1999. *J. Clin. Psychiat.* 60:824–30.
Feinberg, I. 1982. *J. Psychiatr. Res.* 17:319–34.
Feinberg, T. 2000. *Altered Egos.* New York: Oxford University Press.
Feinberg, T. E., and M. J. Farah. 1998. *Behavioral Neurology and Neuropsychology.* New York: McGraw-Hill.
Feistel, H. 1993. *J. Nucl. Med.* 34:47
Fendt, M., and M. S. Fanselow. 1999. *Neurosci. Biobehav. Rev.* 23:743–60.
File, S. E. 2000. In *The Amygdala,* edited by J. Aggleton. Oxford: Oxford University Press.
File, S. E., et al. 2000. *Pharmacol. Biochem. Behav.* 66:65–72.
Finlay, B. L., and R. B. Darlington. 1995. *Science* 268:1578–84.
Fischer, H., et al. 1996. *Neuroreport* 7:2081–86.
Flanagan, O. 1994. *Consciousness Reconsidered.* Cambridge: Bradford Books/MIT Press.
Flaum, M., et al. 1995. *Am. J. Psychiat.* 152:704–14.
Flew, A. 1964. *Body, Mind and Death.* New York: Macmillan.
Flew, A. 1972. In *The Encyclopedia of Philosophy,* 4th ed., edited by P. Edwards, 139–50. New York: Macmillan.
Flynn, J. P. 1967. In *Biology and Behavior,* edited by D. G. Glass, 40–60. New York: Rockefeller University Press and Russell Sage Foundation.
Foa, E. G. 1999. *J. Clin. Psychiat.* 60:69–76.
Fodor, J. 1983. *Modularity of Mind.* Cambridge: MIT Press.
Foucault, M. 1978. *The History of Sexuality.* New York: Random House.
Frankland, P. W., et al. 1998. *Behav. Neurosci.* 112:863–74.
Franzen, G., and D. H. Ingvar. 1975. *J. Psychiatr. Res.* 12:199–214.
Fredrikson, M., et al. 1995. *Psychophysiology* 32:43–48.
Freud, S. 1887–1902. In *The Origins of Psychoanalysis, Letters to Wilhelm Fliess, Drafts and Notes: 1887–1902,* edited by M. Bonaparte et al. New York: Basic Books.
Freud, S. 1915. *The Standard Edition of the Complete Psychological Works of Sigmund Freud.* London: Hogarth.
Freud, S. 1938. In *The Basic Writings of Sigmund Freud,* edited by A. A Brill. New York: Modern Library.
Frey, U., and R. G. Morris. 1997. *Nature* 385:533–36.
Friedman, J. I., et al. 1999. *Biol. Psychiat.* 45:1–16.
Frijda, N. 1986. *The Emotions.* Cambridge: Cambridge University Press.
Frijda, N. H. 1993. *Cogn. Emo.* 7:357–87.
Friston, K. J., and C. D. Frith. 1995. *Clin. Neurosci.* 3:89–97.
Fuchs, E., and E. Gould. 2000. *Eur. J. Neurosci.* 12:2211–14.
Fukuda, T., and T. Kosaka. 2000. *J. Neurosci.* 20:1519–28.
Fukuzako, H., et al. 1996. *Biol. Psychiat.* 39:938–45.
Funahashi, S., et al. 1989. *J. Neurophysiol.* 61:331–49.
Fuster, J. 1997. *The Prefrontal Cortex,* 3rd ed. Philadelphia: Lippincott-Raven.
Fuster, J. 2000. *Neuron* 26:51–53.
Fuster, J. M. 1973. *J. Neurophysiol.* 36:61–78.
Fuster, J. M. 1989. *The Prefrontal Cortex.* New York: Raven.
Fuster, J. M. 1993. *Curr. Opin. Neurobiol.* 3:160–65.

Fuster, J. M. 2000. *Brain Res. Bull.* 52:331–36.
Fuster, J. M., et al. 1982. *Exp. Neurol.* 77:679–94.
Gaffan, D. 1992. In *The Amygdala,* edited by J. P. Aggleton, 471–83. New York: Wiley-Liss.
Gaffan, D., et al. 1993. *Eur. J. Neurosci.* 5:968–75.
Gallager, D. W. 1978. *Eur. J. Pharmacol.* 49:133–43.
Gallagher, I. 2000. *Trends Cogn. Sci.* 4:14–21.
Gallagher, M., and G. Schoenbaum. 1999. *Ann. NY Acad. Sci.* 877:397–411.
Gallagher, I. 1996. *Ethics* 107:129–40.
Galli, L., and L. Maffei. 1988. *Science* 242:90–91.
Gallistel, C. R. 1966. *J. Comp. Physiol. Psychol.* 62:95–101.
Gallistel, C. R. 1995. In *Brain and Memory,* edited by J. L. McGaugh et al., 328–37. New York: Oxford University Press.
Gallistel, R. 1980. *The Organization of Action.* Hillsdale, NJ: Lawrence Erlbaum Associates.
Gallup, G. 1991. In *The Self,* edited by J. Strauss and G. R. Goethals. New York: Springer.
Garcia, J. 1990. *J. Cognit. Neurosci.* 2(4):287–305.
Garcia, J., and R. A. Koelling. 1966. *Psychon. Sci.* 4:123–24.
Garcia, R., et al. 1999. *Nature* 402:294–96.
Gardner, H. 1987. *The Mind's New Science.* New York: Basic Books.
Gardner, H. 1998. *New York Review of Books,* November 5.
Gazzaniga, M. S. 1970. *The Bisected Brain.* New York: Appleton-Century-Crofts.
Gazzaniga, M. S. 1985. *The Social Brain.* New York: Basic Books.
Gazzaniga, M. S. 1988. In *Consciousness in Contemporary Science,* edited by A. Marcel and E. Bisiach. Oxford: Clarendon Press.
Gazzaniga, M. S. 1992. *Nature's Mind.* New York: Basic Books.
Gazzaniga, M. S. 1995. *The Cognitive Neurosciences.* Cambridge: MIT Press.
Gazzaniga, M. S. 1998. *The Mind's Past.* Berkeley: University of California Press.
Gazzaniga, M. S., and J. E. LeDoux. 1978. *The Integrated Mind.* New York: Plenum.
Gazzaniga, M. S., et al. 1996. *Cognitive Neuroscience.* New York: Norton.
Gergen, K. J. 1990. In *Cultural Psychology,* edited by J. W. Stigler et al. New York: Cambridge University Press.
Gergen, K. J. 1997. *Theory Psychol.* 7:723–46.
Gerlai, R. 2000. *Rev. Neurosci.* 11:15–26.
Geschwind, N. 1965. *Brain* 88:237–94.
Gewirtz, J. C., and M. Davis. 1997. *Nature* 388:471–74.
Gewirtz, J. C., et al. 1997. *Behav. Neurosci.* 111:1–15.
Gibbs, R. B. 2000. *Novartis Found. Symp.* 230:94–107.
Gilbert, C. D. 1998. *Physiol. Rev.* 78:467–85.
Glanzman, D. L. 1995. *Trends Neurosci.* 18:30–36.
Glanzman, D. L., et al. 1990. *Science* 249:799–802.
Glenmullen, J. 2000. *Prozac Backlash.* New York: Simon and Schuster.
Glickman, S. E., and B. B. Schiff. 1967. *Psychol. Rev.* 74:81–109.
Gluck, M. A., and C. E. Myers. 1993. *Hippocampus* 3:491–516.
Gnadt, J. W., and R. A. Andersen. 1988. *Exp. Brain Res.* 70:216–20.
Gochin, P. M., et al. 1994. *J. Neurophysiol.* 71:2325–37.

Goddard, G. 1964. *Psychol. Rev.* 62:89–109.
Goff, D. C., et al. 1995. *Am. J. Psychiat.* 152:1213–15.
Gold, P. E. 1995. In *Brain and Memory,* edited by J. L. McGaugh et al., 41–74. New York: Oxford University Press.
Gold, P. E., and S. F. Zornetzer. 1983. *Behav. Neural Biol.* 38:151–89.
Goldman-Rakic, P. S. 1987. In *Handbook of Physiology,* edited by F. Plum, 373–418. Bethesda: American Physiological Society.
Goldman-Rakic, P. S. 1994. *J. Neuropsychiat. Clin. Neurosci.* 6:348–57.
Goldman-Rakic, P. S. 1999. *Biol. Psychiat.* 46:650–61.
Goldman-Rakic, P. S. 1999. In *MIT Encyclopedia of Cognitive Sciences,* edited by R. A. Wilson and F. C. Keil. Cambridge: MIT Press.
Goldman-Rakic, P. S., et al. 1992. *J. Neural Transm. Suppl.* 36:163–77.
Gonzalez, L. E., et al. 1996. *Brain Res.* 732:145–53.
Gonzalez, L. E., et al. 1998. *Eur. J. Neurosci.* 10:3673–80.
Goodale, M. A. 1998. *Curr. Biol.* 8:R489–91.
Goodman, C. S., and C. J. Shatz. 1993. *Cell* 72 Suppl:77–98.
Gopnik, A., et al. 1999. *The Scientist in the Crib.* New York: Morrow.
Gopnik, M., ed. 1997. *The Inheritance and Innateness of Grammar.* Oxford: Oxford University Press.
Gorman, J. 1996. *The New Psychiatry.* New York: St. Martin's Press.
Gormezano, I. 1972. In *Classical Conditioning II,* edited by A. H. Black and W. F. Prokasy, 151–81. New York: Appleton-Century-Crofts.
Gottesman, I. I. 1991. *Schizophrenia Genesis.* New York: W. H. Freeman.
Gould, E., et al. 1997. *J. Neurosci.* 17:2492–98.
Gould, E., et al. 1998. *Proc. Natl. Acad. Sci. USA* 95:3168–71.
Gould, E., et al. 1999. *Nat. Neurosci.* 2:260–65.
Gould, E., et al. 1999. *Science* 286:548–52.
Gould, J. L. 1982. *Ethology.* New York: Norton.
Gould, J. L., and P. Marler. 1984. In *The Biology of Learning,* edited by P. Marler and H. S. Terrace, 47–74. Berlin: Springer-Verlag.
Gould, S. J. 1991. *J. Social Issues* 47:43–65.
Gould, S. J. 1997. *New York Review of Books,* June 26.
Grace, A. A. 1993. *J. Neural Transm. Gen. Sect.* 91:111–34.
Grace, A. A., et al. 1997. *Trends Neurosci.* 20:31–37.
Grace, A. A., et al. 1998. *Adv. Pharmacol.* 42:721–24.
Graeff, F. G., and R. I. Schoenfeld. 1970. *J. Pharmacol. Exp. Ther.* 173:277–83.
Graeff, F. G., et al. 1993. *Behav. Brain Res.* 58:123–31.
Graf, P., et al. 1984. *J. Exp. Psychol. Learn. Mem. Cogn.* 10:164–78.
Gray, C. M. 1999. *Neuron* 24:31–47, 111–25.
Gray, C. M., et al. 1989. *Nature* 338:334–37.
Gray, J. A. 1982. *The Neuropsychology of Anxiety.* New York: Oxford University Press.
Gray, J. A. 1987. *The Psychology of Fear and Stress.* New York: Cambridge University Press.
Gray, J. A. 1991. In *Explorations in Temperament,* edited by J. Strelau and A. P. Angleitner. New York: Plenum.
Gray, J. A., and N. McNaughton. 2000. *The Neuropsychology of Anxiety,* 2nd ed. Oxford: Oxford University Press.

Graybiel, A. 1976. Lecture at the Society for Neuroscience in Toronto, Canada.
Greenough, W. T., et al. 1985. *Proc. Natl. Acad. Sci. USA* 82:4549–52.
Greenwald, A. G. 1992. *Am. Psychol.* 47:766–79.
Greenwald, A. G., and M. R. Banaji. 1995. *Psychol. Rev.* 102:4–27.
Gregory, R. 1981. *Mind in Science.* Cambridge: Cambridge University Press.
Grill, H. J., and R. Norgren. 1978. *Science* 201:267–69.
Groenewegen, H. J., et al. 1990. In *Progress in Brain Research,* edited by H. B. M. Uylings et al., 95–118. Amsterdam: Elsevier Science Publishers B.V. (Biomedical Division).
Groenewegen, H. J., et al. 1997. *J. Psychopharmacol.* 11:99–106.
Gross, P. R., et al. 1996. *The Flight from Science and Reason. Proceedings of a Conference.* New York, New York, May 31–June 2, 1995. New York: New York Academy of Science.
Grossberg, S. 1999. *Conscious. Cogn.* 8:1–44.
Grossberg, S. 2000. *Trends Cogn. Sci.* 4:233–46.
Grossman, S. P. 1960. *Science* 132:301–2.
Grossman, S. P. 1967. *A Textbook of Physiological Psychology.* New York: Wiley.
Grover, L. M., and T. J. Tyler. 1990. *Nature* 347:477–79.
Gurvits, T. V., et al. 1996. *Biol. Psychiat.* 40:1091–99.
Guyton, A. C. 1972. *Structure and Function of the Nervous System.* Philadelphia: W. B. Saunders.
Hall, C. S., et al. 1998. *Theories of Personality.* New York: John Wiley & Sons.
Hamann, S. B., et al. 1996. *Nature* 379:497.
Happel, S. 2000. In *Neuroscience and the Person,* edited by R. J. Russell et al. Berkeley: Vatican Observatory Publications, Vatican City State, Center for Theology and the Natural Sciences.
Harley, C. 1991. *Prog. Brain Res.* 88:307–21.
Harlow, J. M. 1868. *Bull. Mass. Med. Soc.* 2:3–20.
Harre, R. 1986. *The Social Construction of Emotions.* New York: Blackwell.
Harris, J. R. 1998. *The Nurture Assumption.* New York: The Free Press.
Harris, J. A., and R. F. Westbrook., 1995. *Behav. Neurosci.* 109:295–304.
Hart, A. J., et al. 2000. *Neuroreport* 11:2351–55.
Hartley, D. 1749. *Observations on Man.* London: Leake and Frederick.
Harvey, J. A., et al. 1965. *J. Comp. Physiol. Psychol.* 59:37–48.
Hauser, M. D., et al. 1995. *Proc. Natl. Acad. Sci. USA* 92:10811–14.
Hawkins, R. D., and E. R. Kandel. 1984. *Psychol. Rev.* 91:375–91.
Hawkins, R. D., et al. 1987. In *Handbook of Physiology,* edited by F. Plum, 25–83. Bethesda: American Physiological Society.
Hawkins, R. D., et al. 1994. *J. Neurobiol.* 25:652–65.
Hawkins, R. D., et al. 1998. *Behav. Neurosci.* 112:636–45.
Haxby, J. V., et al. 2000. *Neuroimage* 11:145–56.
He, S., et al. 1996. *Nature* 383:334–37.
Heath, R. G. 1964. In *The Role of Pleasure in Behavior,* edited by R. G. Heath. New York: Harper and Row.
Hebb, D. O. 1946. *Psychol. Rev.* 53:88–106.
Hebb, D. O. 1949. *The Organization of Behavior.* New York: John Wiley & Sons.
Helmstetter, F. J., and P. S. Bellgowan. 1994. *Behav. Neurosci.* 108:1005–9.
Heninger, G. R., et al. 1996. *Pharmacopsychiatry* 29:2–11.

Heraclitus (c.540-c.480 B.C.). 1925. From D. Laertius, *Lives of Eminent Philosophers.* London: G. P. Putman.
Herrick, C. J. 1933. *Proc. Natl. Acad. Sci. USA* 19:7–14.
Herzog, A. G., and G. W. Van Hoesen. 1976. *Brain Res.* 115:57–69.
Hess, W. R. 1954. *Functional Organization of the Diencephalon,* New York: Grune and Stratton.
Hess, W. R., and K. Akert. 1955. *Arch. Neurol. Psychiat.* 73:127–29.
Hess, W. R., and M. Brugger. 1943. *Helv. Physiol. Pharmacol. Acta* 1:35–52.
Hesslow, G., and C. Yeo. 1998. *Science* 280:1817–19.
Hietala, J., et al. 1995. *Lancet* 346:1130–31.
Hietala, J., et al. 1999. *Schizophr. Res.* 35:41–50.
Higgins, E. T., et al. 1985. *Soc. Cogn.* 3:51–76.
Hilgard, E. R. 1980. *J. Hist. Behav. Sci.* 16:107–17.
Hill, W. F. 1977. *Learning.* New York: Crowell/Harper and Row.
Hilton, S. M., and A. W. Zbrozyna. 1963. *J. Physiol.* 165:160–73.
Hinde, R. A. 1966. *Animal Behavior.* New York: McGraw-Hill.
Hirst, W., et al. 1980. *J. Exp. Psychol.* 109:98–117.
Hitchcock, J., and M. Davis. 1986. *Behav. Neurosci.* 100:11–22.
Hobbes, T. 1651. *Leviathan.* London, printed for Andrew Crooke at The Green Dragon in St. Paul's Churchyard.
Hodges, H., et al. 1987. *Psychopharmacology* 92:491–504.
Holden, C. 1980. *Science* 207:1323–25, 1327–28.
Holland, P. C. 1993. *Curr. Opin. Neurobiol.* 3:230–36.
Horgan, J. 1996. *The End of Science.* New York: Broadway Books.
Horton, J. C., and D. R. Hocking. 1996. *J. Neurosci.* 16:1791–1807.
Howard, R., et al. 1995. *Psychol. Med.* 25:495–503.
Hrobjartsson, A., and P. C. Gotzsche. 2001. *N. Engl. J. Med.* 344:1594–1602.
Huang, Y. Y., and E. R. Kandel. 1998. *Neuron* 21:169–78.
Huang, Y. Y., et al. 1996. *Learn. Mem.* 3:74–85.
Huang, Y. Y., et al. 2000. *J. Neurosci.* 20:6317–25.
Hubel, D., and T. Wiesel. 1962. *J. Physiol.* 160:106–54.
Hubel, D., and T. Wiesel. 1963. *J. Neurophysiol.* 26:994–1002.
Hubel, D., and T. Wiesel. 1965. *J. Neurophysiol.* 28:1041–59.
Hubel, D. H., and T. N. Wiesel. 1972. *J. Comp. Neurol.* 146:421–50.
Huerta, P. T., et al. 2000. *Neuron* 25:473–80.
Hull, C. 1943. *Principles of Behavior.* New York: Appleton-Century-Crofts.
Hull, C. L. 1954. *A Behavior System.* New Haven: Yale University Press.
Humphrey, N. 1992. *A History of the Mind.* New York: Simon and Schuster.
Humphrey, N. 2000. *How to Solve the Mind-Body Problem.* Thorverton, UK: Imprint Academic.
Husi, H., and S. G. Grant. 2001. *Trends Neurosci.* 24:259–66.
Huttenlocher, P. R. 1979. *Brain Res.* 163:195–205.
Hyman, S. E. 2000. *Arch. Gen. Psychiat.* 57:88–89.
Hyman, S. E., and E. J. Nestler. 1996. *Am. J. Psychiat.* 153:151–62.
Ikemoto, S., and J. Panksepp. 1999. *Brain Res. Rev.* 31:6–41.

Ingvar, D. H., and G. Franzen. 1974. *Acta Psychiatr. Scand.* 50:425–62.
Innocenti, G. M. 1991. *Prog. Sens. Physiol.* 12.
Insel, T. R. 1997. *Am. J. Psychiat.* 154:726–35.
Iriki, A., et al. 1987. *Science* 245:1385–87.
Isaacson, R. L. 1982. *The Limbic System.* New York: Plenum Press.
Isenberg, N., et al. 1999. *Proc. Natl. Acad. Sci. USA* 96:10456–59.
Ito, M. 1984. *The Cerebellum and Neural Control.* New York: Raven.
Ito, M. 1989. *Annu. Rev. Neurosci.* 12:85–102.
Ito, M. 1996. *Trends Neurosci.* 19:11–12.
Iwata, J., et al. 1986. *Brain Res.* 383:195–214.
Izard, C. E. 1971. *The Face of Emotion.* New York: Appleton-Century-Crofts.
Izard, C. E. 1977. *Human Emotions.* New York: Plenum.
Izard, C. E. 1992. *Psychol. Rev.* 99:561–65.
Izumi, Y., and C. F. Zorumski. 1999. *Synapse* 31:196–202.
Jacobs, W. J., and L. Nadel. 1985. *Psychol. Rev.* 92:512–31.
Jacobsen, C. F. 1935. *Arch. Neurol. Psychiat.* 33:558–69.
Jacobsen, C. F., and H. W. Nissen. 1937. *J. Comp. Physiol. Psychol.* 23:101–12.
Jacobson, M. 1993. *Foundations of Neuroscience.* New York: Plenum.
Jacoby, L., and J. Toth. 1992. In *Perception Without Awareness,* edited by R. Bornstein and T. Pittman, 81–120. New York: Guilford Press.
Jacoby, L. L., and V. Woloshyn. 1989. *J. Exp. Psychol.: Gen.* 118:115–25.
Jakob, H., and H. Beckmann. 1994. *J. Neural Transm. Gen. Sect.* 98:83–106.
James, W. 1884. *Mind* 9:188–205.
James, W. 1890. *Principles of Psychology.* New York: Holt.
Jamison, K. R. 1997. *An Unquiet Mind.* New York: Random House.
Jarvis, E. D., et al. 1998. *Neuron* 21:775–88.
Javanmard, M., et al. 1999. *Biol. Psychiat.* 45:872–82.
Jerison, H. 1973. *Evolution of Brain and Intelligence.* New York: Academic Press.
Jerne, N. 1967. In *The Neurosciences,* edited by F. O. Schmitt. New York: Rockefeller University Press.
Jessell, T. M., and J. R. Sanes. 2000. *Curr. Opin. Neurobiol.* 10:599–611.
Jiang, Y., et al. 2000. *Science* 287:643–46.
Jing, J., and R. Gillette. 1995. *J. Neurophysiol.* 74:1900–1910.
Johnson, J. 1998. In *Fundamental Neuroscience,* edited by M. Zigmond. San Diego: Academic Press.
Johnson-Laird, P. N. 1988. *The Computer and the Mind.* Cambridge: Harvard University Press.
Johnson-Laird, P. N. 1993. In *Consciousness in Contemporary Science,* edited by A. J. Marcel and E. Bisiach, 357–68. Oxford: Oxford University Press.
Johnston, D., et al. 1999. *Curr. Opin. Neurobiol.* 9:288–92.
Jones, E. 1961. *The Life and Work of Sigmund Freud.* New York: Basic Books.
Jones, E. G. 1984. In *Cerebral Cortex,* edited by A. Peters and E. G. Jones. New York: Plenum.
Jones, E. G., and T. P. S. Powell. 1970. *Brain* 93:793–820.
Josselyn, S. A., et al. 2001. *J. Neurosci.* 21:2404–12.
Joyce, J. N., and J. H. Meador-Woodruff. 1997. *Neuropsychopharmacology* 16:375–84.

Kaada, B. R. 1960. In *Handbook of Physiology*, edited by J. Field et al., 1345–72. Washington, D.C.: American Physiological Society.
Kagan, J. 1992. *Pediatrics* 90:510–13.
Kagan, J. 1994. *Galen's Prophecy.* New York: Basic Books.
Kagan, J. 1998. In *Handbook of Child Psychology*, edited by N. Eisenberg, 177–236. New York: Wiley.
Kagan, J. 1999. *Pediatrics* 104:164–67.
Kahneman, D. 1999. In *Well-Being*, edited by D. Kahneman et al. New York: Russell Sage Foundation.
Kalivas, P. W., and M. Nakamura. 1999. *Curr. Opin. Neurobiol.* 9:223–27.
Kamin, C. J., et al. 1963. *J. Comp. Physiol. Psychol.* 56:497–501.
Kandel, E. R. 1976. *Cellular Basis of Behavior.* San Francisco: W. H. Freeman.
Kandel, E. R. 1989. *J. Neuropsychiat. Clin. Neurosci.* 1:103–25.
Kandel, E. R. 1997. *J. Cell. Physiol.* 173:124–25.
Kandel, E. R. 1998. *Am. J. Psychiat.* 155:457–69.
Kandel, E. R. 1999. *Am. J. Psychiat.* 156:505–24.
Kandel, E. R., and C. Pittenger. 1999. *Philos. Trans. R. Soc. Lond. B Biol. Sci.* 354:2027–52.
Kandel, E. R., and W. A. Spencer. 1968. *Physiol. Rev.* 48:65–134.
Kandel, E. R., et al. 2000. *Principles of Neuroscience.* New York: McGraw-Hill.
Kapp, B. S., et al. 1979. *Physiol. Behav.* 23:1109–17.
Kapp, B. S., et al. 1984. In *Neuropsychology of Memory*, edited by N. Buttlers and L. R. Squire, 473–88. New York: Guilford.
Kapp, B. S., et al. 1992. In *The Amygdala*, edited by J. P. Aggleton, 229–54. New York: Wiley-Liss.
Karten, H. J., and T. Shimizu. 1991. *J. Cogn. Neurosci.* 1:291–301.
Kastner, S., et al. 1998. *Science* 282:108–11.
Kastner, S., et al. 1999. *Neuron* 22:751–61.
Kastner, S., and L. G. Ungerleider. 2000. *Annu. Rev. Neurosci.* 23:315–41.
Katsuki, H., et al. 1997. *J. Neurophysiol.* 77:3013–20.
Katz, L. C., and C. J. Shatz. 1996. *Science* 274:1133–38.
Keenan, J. P., et al. 2000. *Trends Cogn. Sci.* 4:338–44.
Keil, F. 1999. In *Encyclopedia of Cognitive Science*, 583–85. Cambridge: MIT Press.
Keith, J. R., and J. W. Rudy. 1990. *Psychobiology* 18:251–57.
Kelso, S. R., et al. 1986. *Proc. Natl. Acad. Sci. USA* 83:5326–30.
Keltner, N., and D. G. Folks. 1997. *Psychotropic Drugs*, 2nd ed. St. Louis: Mosby.
Keltner, N., et al. 1998. *Psychobiological Foundations of Psychiatric Care.* St. Louis: Mosby.
Kennedy, M. B. 1999. *Learn. Mem.* 6:417–21.
Kihlstrom, J. F. 1987. *Science* 237:1445–52.
Kihlstrom, J. F. 1990. In *Handbook of Personality*, edited by L. Pervin, 445–64. New York: Guilford.
Killackey, H. P. 1990. *J. Cogn. Neurosci.* 2:1–17.
Killcross, S., et al. 1997. *Nature* 388:377–80.
Kim, J. J., and M. G. Baxter. 2001. *Trends Neurosci.* 24:324–30.
Kim, J. J., and M. S. Fanselow. 1992. *Science* 256:675–77.
Kim, J. J., and K. S. Yoon. 1998. *Trends Neurosci.* 21:505–9.

Kim, J. J., et al. 1993. *Behav. Neurosci.* 107:1–6.

Kim, J. N., and M. N. Shadlen. 1999. *Nat. Neurosci.* 2:176–85.

Klee, P. 1957. *The Diaries of Paul Klee 1898–1918.* Berkeley: University of California Press.

Klein, D. F. 1993. *Arch. Gen. Psychiat.* 50:306–17.

Kline, N. S. 1954. *Ann. NY Acad. Sci.* 59:107–32.

Kline, N. S. 1974. *From Sad to Glad.* New York: Putnam.

Kluver, H., and P. C. Bucy. 1937. *Am. J. Physiol.* 119:352–53.

Knight, R. T. 1997. *J. Cogn. Neurosci.* 9:75–91.

Knowlton, B. J., and M. S. Fanselow. 1998. *Curr. Opin. Neurobiol.* 8:293–96.

Kobayashi, M., et al. 1997. *Brain Res.* 777:242–46.

Koch, K. W., and J. M. Fuster. 1989. *Exp. Brain Res.* 76:292–306.

Kolb, B., and R. Tees. 1990. *The Cerebral Cortex of the Rat.* Cambridge: MIT Press.

Kolm, S.-C. 1985. In *The Multiple Self,* edited by J. Elster. New York: Cambridge University Press.

Kombian, S. B., and R. C. Malenka. 1994. *Nature* 368:242–46.

Konorski, J. 1948. *Conditioned Reflexes and Neuron Organization.* Cambridge: Cambridge University Press.

Konorski, J. 1967. *Integrative Activity of the Brain.* Chicago: University of Chicago Press.

Korenberg, J. R., et al. 2000. *J. Cogn. Neurosci.* 12:89–107.

Kosslyn, S. M., and O. Koenig. 1992. *Wet Mind.* New York: Macmillan.

Kotter, R., and N. Meyer. 1992. *Behav. Brain Res.* 52:105–27.

Kraemer, G. W. 1992. *Behav. Brain Sci.* 15:493–511.

Kramer, P. D. 1993. *Listening to Prozac.* New York: Viking.

Kritzer, M. F., and P. S. Goldman-Rakic. 1995. *J. Comp. Neurol.* 359:131–43.

Kuffler, S., and J. Nicholls. 1976. *From Neuron to Brain.* Sunderland, MA: Sinauer.

Kuikka, J. T., et al. 1995. *Nucl. Med. Commun.* 16:273–80.

LaBar, K. S., and J. F. Disterhoft. 1998. *Hippocampus* 8:620–26.

LaBar, K. S., et al. 1995. *J. Neurosci.* 15:6846–55.

LaBar, K. S., et al. 1998. *Neuron* 20:937–45.

Lamb, R. J., et al. 1991. *J. Pharmacol. Exp. Ther.* 259:1165–73.

Lamprecht, R., and Y. Dudai. 1996. *Learn. Mem.* 3:31–41.

Lamprecht, R., and Y. Dudai. 2000. In *The Amygdala,* edited by J. Aggleton. Oxford: Oxford University Press.

Lamprecht, R., et al. 1997. *J. Neurosci.* 17:8443–50.

Lang, E. J., and D. Pare. 1997. *J. Neurophysiol.* 77:353–63.

Larrabee, M. G., and D. W. Bronk. 1947. *J. Neurophysiol.* 10:139–54.

Larsen, R. J., and B. L. Fredrickson. 1999. In *Well-Being,* edited by D. Kahneman et al. New York: Russell Sage Foundation.

Lashley, K. 1950. In *Cerebral Mechanisms in Behavior,* edited by L. A. Jeffers. New York: Wiley.

Lashley, K. S. 1929. *Brain Mechanisms of Intelligence.* Chicago: University of Chicago Press.

Lashley, K. S. 1938. *Psychol. Rev.* 45:445–71.

Lashley, K. S. 1950. *Symp. Soc. Exp. Biol.* IV:454–82.

Lazarus, R. S. 1991. *Am. Psychol.* 46:352–67.

Lechner, H. A., and J. H. Byrne. 1998. *Neuron* 20:355–58.
LeDoux, J. E. 1982. *Brain Behav. Evol.* 20:196–212.
LeDoux, J. E. 1984. In *Handbook of Cognitive Neuroscience,* edited by M. S. Gazzaniga, 357–68. New York: Plenum.
LeDoux, J. E. 1986. *Integr. Psychiat.* 4:237–48.
LeDoux, J. E. 1987. In *Handbook of Physiology,* edited by F. Plum, 419–60. Bethesda: American Physiological Society.
LeDoux, J. E. 1990. In *Learning and Computational Neuroscience,* edited by M. Gabriel, 3–52. Cambridge: MIT Press.
LeDoux, J. E. 1991. *Concepts Neurosci.* 2:169–99.
LeDoux, J. E. 1992. In *The Amygdala,* edited by J. P. Aggleton, 339–51. New York: Wiley-Liss.
LeDoux, J. E. 1994. *Sci. Am.* 270:32–39.
LeDoux, J. E. 1995. *Annu. Rev. Psychol.* 46:209–35.
LeDoux, J. E. 1996. *The Emotional Brain,* New York: Simon & Schuster.
LeDoux, J. E. 1998. In *Chronicle of Higher Education* 45 (16) (December 11, 1998).
LeDoux, J. E. 2000. *Annu. Rev. Neurosci.* 23:155–84.
LeDoux, J. E., et al. 1984. *J. Neurosci.* 4:683–98.
LeDoux, J. E., et al. 1985. *J. Comp. Neurol.* 242:182–213.
LeDoux, J. E., et al. 1989. *J. Cogn. Neurosci.* 1:238–43.
LeDoux, J. E., et al. 1990. *J. Neurosci.* 10:1043–54.
LeDoux, J. E., et al. 1990. *J. Neurosci.* 10:1062–69.
Lee, H., and J. J. Kim. 1998. *J. Neurosci.* 18:8444–54.
Lee, K. S., et al. 1980. *J. Neurophysiol.* 44:247–58.
Lepage, M., et al. 2000. *Proc. Natl. Acad. Sci. USA* 97:506–11.
Lehrman, D. 1953. *Q. Rev. Biol.* 28:337–63.
Lettvin, J. Y., et al. 1959. *Proc. Inst. Radiol. Eng.* 41:1940–51.
Leung, H. C., et al. 2000. *Cereb. Cortex* 10:552–60.
Levenson, R. W. 1992. *Psychol. Sci.* 3:23–27.
Levy, R., and P. S. Goldman-Rakic. 2000. *Exp. Brain Res.* 133:23–32.
Levy, W. B., and O. Steward. 1979. *Brain Res.* 175:233–45.
Lewis, D. A., et al. 1987. *J. Neurosci.* 7:279–90.
Lewis, D. A., et al. 1999. *Biol. Psychiat.* 46:616–26.
Lhermitte, F., et al. 1972. *Revue Neurologique* 127:415–40.
Li, X. F., et al. 1996. *Synapse* 24:115–24.
Liang, K. C., et al. 1982. *Behav. Brain Res.* 4:237–49.
Liberzon, I., et al. 1999. *Biol. Psychiat.* 45:817–26.
Liddle, P. F., et al. 1992. *Br. J. Psychiat.* 160:179–86.
Linden, D. J. 1994. *Neuron* 12:457–72.
Linden, D. R. 1969. *J. Comp. Physiol. Psychol.* 69:573–78.
Lindstrom, L. H. 2000. *Trends Pharmacol. Sci.* 21:198–99.
Lindstrom, L. H., et al. 1999. *Biol. Psychiat.* 46:681–88.
Lindvall, O., et al. 1978. *Brain Res.* 142:1–24.
Lisberger, S. G. 1996. *Ann. NY Acad. Sci.* 781:525–31.
Lisberger, S. G. 1998. *Cell* 92:701–4.
Lisman, J. 1994. *Trends Neurosci.* 17:406–12.

Livingston, K. E., and A. Escobar. 1971. *Arch. Neurol.* 24:17–21.
Llinas, R. 1988. *Science* 242:1654–64.
Llinas, R. 2001. *I of the Vortex.* Cambridge: MIT Press.
Llinas, R., and U. Ribary. 1994. In *Large-Scale Neuronal Theories of the Brain,* edited by C. Koch and J. Davis, 111–24. Cambridge: MIT Press.
Llinas, R., and J. P. Welsh. 1993. *Curr. Opin. Neurobiol.* 3:958–65.
Llinas, R., et al. 1994. In *Temporal Coding in the Brain,* edited by G. Buzsaki et al., 251–72. Berlin: Springer-Verlag.
Lloyd, D. P. C. 1949. *J. Gen. Physiol.* 33:147–70.
Lockwood, K. A., et al. 2000. *Am. J. Geriatr. Psychiat.* 8:201–8.
Loftus, E. F. 1986. *Law Hum. Behav.* 10:241–63.
Loftus, E. F., and H. G. Hoffman. 1989. *J. Exp. Psychol.: Gen.* 118:100–104.
Loftus, E. F., and M. R. Klinger. 1992. *Am. Psychol.* 47:761–65.
LoLordo, V. M. 1967. *J. Comp. Physiol. Psychol.* 64:154–58.
Lømo, T. 1966. *Acta Physiol. Scand.* 68, Suppl. 277:128.
Lorenz, K. Z. 1950. *Symp. Soc. Exp. Biol.* 4:221–68.
Lorenz, K. Z., and N. Tinbergen. 1938. *Z. Tierpsych.* 2:1–29.
Louie, K., and M. A. Wilson. 2001. *Neuron* 29:145–56.
Lupien, S. J., et al. 1998. *Nat. Neurosci.* 1:69–73.
Luria, A. R. 1969. In *Handbook of Clinical Neurology,* edited by P. H. Vinken and G. W. Bruyn, 725–57. Amsterdam: North Holland.
Luria, A. R. 1973. *The Working Brain.* New York: Basic Books.
Lutz, C. A. 1988. *Unnatural Emotions.* Chicago: University of Chicago Press.
Lynch, G. 1986. *Synapses, Circuits, and the Beginnings of Memory.* Cambridge: MIT Press.
MacLean, P. D. 1949. *Psychosom. Med.* 11:338–53.
MacLean, P. D. 1952. *Electroencephalogr. Clin. Neurophysiol.* 4:407–18.
MacLean, P. D. 1970. In *The Neurosciences: Second Study Program,* edited by F. O. Schmitt, 336–49. New York: Rockefeller University Press.
MacLean, P. D. 1990. *The Triune Brain in Evolution.* New York: Plenum Press.
Magee, J. C., and D. Johnston. 1997. *Science* 275:209–13.
Maier, S. F. 1984. *Prog. Neuropsychopharmacol. Biol. Psychiat.* 8:435–46.
Maioli, M. G., et al. 1998. *Brain Res.* 789:118–25.
Makino, S., et al. 1994. *Brain Res.* 640:105–12.
Malenka, R. C., and R. A. Nicoll. 1999. *Science* 285:1870–74.
Malgaroli, A., and R. W. Tsien. 1992. *Nature* 357:134–39.
Malinow, R., and J. P. Miller. 1986. *Nature* 320:529–30.
Manning, A. 1967. *An Introduction to Animal Behavior.* Reading, MA: Addison-Wesley.
Manns, J. R., et al. 2000. *Hippocampus* 10:181–86.
Mansuy, I. M., et al. 1998. *Neuron* 21:257–65.
Marcel, A. J., and E. Bisiach. 1988. *Consciousness in Contemporary Science.* Oxford: Clarendon Press.
Marcus, G. F. 1999. *Cognition* 73:293–96.
Maren, S. 1999. *Trends Neurosci.* 22:561–67.
Maren, S. 2000. *Eur. J. Neurosci.* 12:4047–54.
Maren, S. 2001. *Ann. Rev. Neurosci.* 24:897–931.

Maren, S., and M. S. Fanselow. 1996. *Neuron* 16:237–40.
Maren, S., et al. 1996. *Behav. Neurosci.* 110:1365–74.
Maren, S., et al. 1998. *Trends Cogn. Sci.* 2:39–41.
Markram, H., et al. 1997. *J. Physiol. (Lond.)* 500:409–40.
Marks, I. 1987. *Fears, Phobias, and Rituals: Panic, Anxiety, and Their Disorders.* New York: Oxford University Press.
Markus, H. R., and S. Kitayama. 1991. *Psychol. Rev.* 98:224–53.
Marler, P., and H. Terrace. 1984. *The Biology of Learning.* Berlin: Springer-Verlag.
Marr, D. 1969. *J. Physiol. (Lond.)* 202:437–70.
Marshall, L. B., and O. A. Smith. 1975. *J. Comp. Physiol. Psychol.* 88:21–35.
Martin, J., and J. Sugarman. 2000. *Am. Psychol.* 55:397–406.
Martin, K. A. 1994. *Cereb. Cortex* 4:1–7.
Martin, K. C., and E. R. Kandel. 1996. *Neuron* 17:567–70.
Martin, K. C., et al. 1997. *Cell* 91:927–38.
Martin, S. J., et al. 2000. *Annu. Rev. Neurosci.* 23:649–711.
Martinez, J. L., Jr., and B. E. Derrick. 1996. *Annu. Rev. Psychol.* 47:173–203.
Masson, J. M., and S. McCarthy. 1995. *When Elephants Weep.* New York: Delacorte.
Maturana, H., and F. Varela. 1987. *The Tree of Knowledge.* Boston: New Science Library.
Mayer, M. L., and G. L. Westbrook. 1987. *Prog. Neurobiol.* 28:197–276.
Mayford, M., and E. R. Kandel. 1999. *Trends Genet.* 15:463–70.
Mayford, M., et al. 1995. *Curr. Opin. Neurobiol.* 5:141–48.
Mayford, M., et al. 1996. *Science* 274:1678–83.
McAllister, W. R., and D. E. McAllister. 1971. In *Aversive Conditioning and Learning,* edited by F. R. Brush, 105–79. New York: Academic Press.
McClelland, D. C. 1951. *Personality.* New York: Holt, Rinehart, and Winston.
McClelland, J. L., et al. 1995. *Psychol. Rev.* 102:419–57.
McDonald, A. J. 1998. *Prog. Neurobiol.* 55:257–332.
McDonald, R. J., and N. M. White. 1993. *Behav. Neurosci.* 107:3–22.
McDougall, W. 1908. *An Introduction to Social Psychology.* London: Methuen.
McEwen, B. S. 1994. *The Hostage Brain.* New York: Rockefeller University Press.
McEwen, B. S. 1998. *N. Engl. J. Med.* 338:171–79.
McEwen, B. S. 1999. *Annu. Rev. Neurosci.* 22:105–22.
McEwen, B. S., and R. M. Sapolsky. 1995. *Curr. Opin. Neurobiol.* 5:205–16.
McGaugh, J. L. 1990. *Psychol. Sci.* 1:15–25.
McGaugh, J. L. 2000. *Science* 287:248–51.
McGaugh, J. L., and P. E. Gold. 1989. In *Psychoendocrinology,* edited by R. B. Brush and S. Levine, 305–40. New York: Academic Press.
McGinn, C. 2000. *The Mysterious Flame.* New York: Basic Books.
McGrath, P. J., et al. 2000. *Am. J. Psychiat.* 157:344–50.
McHugh, T. J., et al. 1996. *Cell* 87:1339–49.
McKernan, M. G., and P. Shinnick-Gallagher. 1997. *Nature* 390:607–11.
McNaughton, B. L. 1998. *Neurobiol. Learn. Mem.* 70:252–67.
McNaughton, B. L., and C. A. Barnes. 1990. *Sem. Neurosci.* 2:403–16.
McNaughton, B. L., et al. 1978. *Brain Res.* 157:277–93.

McNish, K. A., et al. 1997. *J. Neurosci.* 17:9353–60.
Medina, J. F., et al. 2001. *Curr. Opin. Neurobiol.* 10:717–24.
Meichenbaum, D., and J. B. Gilmore. 1984. In *The Unconscious Reconsidered,* edited by K. S. Bowers and D. Meichenbaum, 273–98. New York: John Wiley & Sons.
Meisel, R. L., and B. D. Sachs. 1994. In *The Physiology of Reproduction,* edited by E. Knobil and D. Neill. New York: Raven Press.
Merikle, P. M. 1992. *Am. Psychol.* 47:792–95.
Merzenich, M., et al. 1996. *Cold Spring Harb. Symp. Quant. Biol.* 61:1–8.
Mesulam, M. M., et al. 1977. *Brain Res.* 136:393–414.
Metzinger, T. 1995. *Conscious Experience.* Thorverton, UK: Imprint Academic.
Miller, E. K. 1999. *Neuron* 22:15–17.
Miller, E. K., and R. Desimone. 1994. *Science* 263:520–22.
Miller, E. K., et al. 1993. *J. Neurosci.* 13:1460–78.
Miller, G. A. 1956. *Psychol. Rev.* 63:81–97.
Miller, K. D. 1994. *J. Neurosci.* 14:409–41.
Miller, N. E. 1948. *J. Exp. Psychol.* 38:89–101.
Miller, N. E. 1951. In *Handbook of Experimental Psychology,* edited by S. Stevens, 435–72. New York: Wiley.
Miller, N. E. 1965. *Science* 148:328–38.
Milner, B. 1962. In *Physiologie de L'Hippocampe,* edited by P. Plassouant. Paris: Centre de la Recherche Scientifique.
Milner, B. 1967. In *Brain Mechanisms Underlying Speech and Language,* edited by F. L. Darley. New York: Grune and Stratton.
Milner, B. 1972. *Clin. Neurosurg.* 19:421–46.
Milner, B. 1982. *Philos. Trans. R. Soc. Lond. B Biol. Sci.* 298:211–26.
Milner, P. 1965. In *Cognitive Processes and the Brain,* edited by P. M. Milner and S. E. Glickman. Princeton: Van Nostrand.
Milner, P. 1970. *Physiological Psychology.* New York: Holt, Rinehart, and Winston.
Milner, P. 1974. *Psychol. Rev.* 81:521–35.
Mineka, S. 1979. *Psychol. Bull.* 86:985–1010.
Mineka, S., and M. Cook. 1993. *J. Exp. Psychol. Gen.* 122:23–38.
Minsky, M. 1985. *The Society of Mind.* New York: Simon & Schuster.
Mischel, W. 1990. In *Handbook of Personality,* edited by L. A. Pervin. New York: Guilford.
Mischel, W. 1993. *Introduction to Personality.* Fort Worth: Harcourt, Brace, Jovanovich.
Miserendino, M. J. D., et al. 1990. *Nature* 345:716–18.
Mishkin, M., and J. Aggleton. 1981. In *The Amygdaloid Complex,* edited by Y. Ben-Ari, 409–20. Amsterdam: Elsevier/North-Holland Biomedical Press.
Mishkin, M., and E. A. Murray. 1994. *Curr. Opin. Neurobiol.* 4:200–206.
Miyashita-Lin, E. M., et al. 1999. *Science* 285:906–9.
Mody, M., et al. 1997. *J. Exp. Child Psychol.* 64:199–231.
Mogenson, G. J., et al. 1980. *Prog. Neurobiol.* 14:69–97.
Moltz, H. 1965. *Psychol. Rev.* 72:27–47.
Mooney, R. 1999. *Curr. Opin. Neurobiol.* 9:121–27.
Morgan, C. T. 1943. *Physiological Psychology.* New York: McGraw-Hill.
Morgan, C. T. 1957. *Nebr. Symp. Motiv.* 5:1–43.

Morgan, S. L., and T. J. Teyler. 1999. *J. Neurophysiol.* 82:736–40.
Morihisa, J. M., and G. B. McAnulty. 1985. *Biol. Psychiat.* 20:3–19.
Morris, J. S., et al. 1996. *Nature* 383:812–15.
Morris, J. S., et al. 1998. *Nature* 393:467–70.
Morris, J. S., et al. 1999. *Proc. Natl. Acad. Sci. USA* 96:1680–85.
Morris, R. 1984. *J. Neurosci. Methods* 11:47–60.
Morris, R. G. M., et al. 1986. *Nature* 319:774–76.
Morris, R. G. M., et al. 1989. *Neuropsychol.* 27:41–59.
Morris, R. G. M. 1992. In *Encyclopedia of Learning and Memory,* edited by L. R. Squire, 369–72. New York: Macmillan.
Morris, R. G. M. 1994. In *Animal Learning and Cognition,* edited by N. Mackintosh. San Diego: Academic Press.
Moruzzi, G., and H. W. Magoun. 1949. *Electroencephalogr. Clin. Neurophysiol.* 1:455–73.
Moser, E. I. 1995. *Behav. Brain Res.* 71:11–18.
Mowrer, O. H. 1939. *Psychol. Rev.* 46:553–65.
Mowrer, O. H. 1947. *Harv. Ed. Rev.* 17:102–48.
Mowrer, O. H. 1960. *Learning Theory and Behavior.* New York: Wiley.
Mowrer, O. H., and R. R. Lamoreaux. 1946. *J. Comp. Psychol.* 39:29–50.
Moyer, J. R., Jr., et al. 1990. *Behav. Neurosci.* 104:243–52.
Moyer, J. R., Jr., et al. 1996. *J. Neurosci.* 16:5536–46.
Muller, J., et al. 1997. *Behav. Neurosci.* 111:683–91.
Muller, R. U., et al. 1999. *Hippocampus* 9:413–22.
Muller, U., et al. 1998. *J. Neurosci.* 18:2720–28.
Munroe, R. L. 1955. *Schools of Psychoanalytic Thought.* New York: Holt, Rinehart, and Winston.
Murphy, G. G., and D. L. Glanzman. 1999. *J. Neurosci.* 19:10595–602.
Murphy, S., and R. Zajonc. 1993. *J. Pers. Soc. Psychol.* 64:723–39.
Murray, E. A., and B. J. Richmond. 2001. *Curr. Opin. Neurobiol.* 11:188–93.
Myers, R. D. 1974. *Handbook of Drug and Chemical Stimulation of the Brain.* New York: Van Nostrand Reinhold.
Nadasdy, Z., et al. 1999. *J. Neurosci.* 19:9497–9507.
Nadel, L., and M. Moscovitch. 1997. *Curr. Opin. Neurobiol.* 7:217–27.
Nadel, L., and J. Willner. 1980. *Physiol. Psychol.* 8:218–28.
Nader, K., and J. E. LeDoux. 1997. *Trends Cognit. Sci.* 1:241–44.
Nader, K., and D. van der Kooy. 1994. *Psychobiol.* 22:68–76.
Nader, K., et al. 1997. *Annu. Rev. Psychol.* 48:85–114.
Nader, K., et al. 2000. *Nature* 406:722–26.
Nagel, T. 1974. *Phil. Rev.* 83:435–50.
Nargeot, R. 2001. *J. Neurosci.* 21:3282–94.
Nauta, W. J. 1971. *J. Psychiatr. Res.* 8:167–87.
Nauta, W. J. H. 1979. In *Functional Neurosurgery,* edited by T. Rasmussen and R. Marino, 7–23. New York: Raven Press.
Nauta, W. J. H., and H. J. Karten. 1970. In *The Neurosciences: Second Study Program,* edited by F. O. Schmitt, 7–26. New York: Rockefeller University Press.
Neisser, U. 1967. *Cognitive Psychology.* Englewood Cliffs, NJ: Prentice Hall.

Neisser, U. 1988. *Philos. Psychol.* 1:35–39.

Neisser, U., ed. 1998. *The Rising Curve.* Washington, D.C.: American Psychological Association.

Neisser, U., and R. Fivush. 1994. *The Remembering Self.* New York: Cambridge University Press.

Neisser, U., and N. Harsch. 1992. In *Affect and Accuracy in Recall: Studies of "Flashbulb" Memories,* edited by E. Winograd and U. Neisser. New York: Cambridge University Press.

Nelson, J. C., et al. 1991. *Arch. Gen. Psychiat.* 48:303–7.

Nemeroff, C. B. 1998. *Biol. Psychiat.* 44:517–25.

Nestler, E. J. 1998. *Biol. Psychiat.* 44:526–33.

Neville, H. J. 1990. *Ann. NY Acad. Sci.* 608:71–87.

Neville, H. J., and D. Lawson. 1987. *Brain Res.* 405:268–83.

Nicolelis, M. A., and J. K. Chapin. 1994. *J. Neurosci.* 14:3511–32.

Nicoll, R. A., and R. C. Malenka. 1995. *Nature* 377:115–18.

Niehoff, D. L., and M. J. Kuhar. 1983. *J. Neurosci.* 3:2091–97.

Nieuwenhuys, R. 1985. *Chemoarchitecture of the Brain.* Berlin: Springer-Verlag.

Nisbett, R. E., and T. D. Wilson. 1977. *Psychol. Rev.* 84:23–59.

Nopoulos, P., et al. 1995. *Am. J. Psychiat.* 152:1721–23.

Nordahl, T. E., et al. 1998. *Biol. Psychiat.* 44:998–1006.

Norman, D. A., and T. Shallice. 1980. In *Consciousness and Self-Regulation,* edited by R. J. Davidson et al. New York: Plenum.

Northcutt, R. G., and J. H. Kaas. 1995. *Trends Neurosci.* 18:373–79.

Nottebohm, F. 1989. *Sci. Am.* 260:74–79.

Nowak, L., et al. 1984. *Nature* 307:462–65.

Oatley, K., and P. Johnson-Laird. 1987. *Cogn. Emo.* 1:29–50.

O'Connor, K. J., et al. 1999. Presentation at Sixth Annual Meeting of the Cognitive Neuroscience Society, Washington, D.C.

O'Connor, T. G., and M. Rutter. 2000. *J. Am. Acad. Child Adolesc. Psychiat.* 39:703–12.

O'Connor, T. G., et al. 2000. *Child Dev.* 71:376–90.

O'Dell, T. J., et al. 1994. *Science* 265:542–46.

Öhman, A. 2000. In *Handbook of Emotions,* 2nd ed., edited by M. Lewis and J. M. Haviland-Jones, 573–93. New York: Guilford Press.

Öhman, A., and J. J. Soares. 1994. *J. Abnorm. Psychol.* 103:231–40.

Öhman, A., et al. 2000. In *Cognitive Neuroscience of Emotion,* edited by R. D. Lane and L. Nadel. New York: Oxford University Press.

O'Keefe, J., and L. Nadel. 1978. *The Hippocampus as a Cognitive Map.* New York: Oxford University Press.

Okubo, Y., et al. 1997. *Nature* 385:634–36.

Olds, J. 1956. *Sci. Am.* 195:105–16.

Olds, J. 1958. *Science* 127:315–24.

Olds, J. 1973. In *Brain Stimulation and Motivation,* edited by E. Valenstein, Glenview, IL: Scott, Foresman.

Olds, J. 1977. *Drives and Reinforcement.* New York: Raven Press.

Olds, J., and P. Milner. 1954. *J. Comp. Physiol. Psychol.* 47:419–27.

O'Leary, D. D. 1992. *Curr. Opin. Neurobiol.* 2:70–77.

Olton, D., et al. 1979. *Behav. Brain Sci.* 2:313–65.
Ono, T., and H. Nishijo. 1992. In *The Amygdala,* edited by J. P. Aggleton, 167–90. New York: Wiley-Liss.
Onoe, H., et al. 1996. *Neuroreport* 8:117–22.
Oppenheim, R. W. 1998. In *Fundamental Neuroscience,* edited by M. Zigmond. San Diego: Academic Press.
O'Reilly, R. C., and Y. Munakata. 2000. *Computational Explorations in Cognitive Neuroscience,* Cambridge: MIT Press.
O'Reilly, R. C., and J. W. Rudy. 2001. *Psychol. Rev.* 108:311–45.
Ortony, A., et al. 1988. *The Cognitive Structure of Emotions.* Cambridge: Cambridge University Press.
Owen, A. M., et al. 1999. *Eur. J. Neurosci.* 11:567–74.
Packard, M. G., et al. 1994. *Proc. Natl. Acad. Sci. USA* 91:8477–81.
Pakkenberg, B. 1987. *Br. J. Psychiat.* 151:744–52.
Pandya, D. N., and B. Seltzer. 1982. *J. Comp. Neurol.* 204:196–210.
Pandya, D. N., and E. H. Yeterian. 1996. In *Neurobiology of Decision-Making,* edited by A. R. Damasio et al. Berlin: Springer-Verlag.
Panksepp, J. 1982. *Behav. Brain Sci.* 5:407–67.
Panksepp, J. 1998. *Affective Neuroscience.* New York: Oxford University Press.
Papez, J. W. 1937. *Arch. Neurol. Psychiat.* 79:217–24.
Pare, M., and R. H. Wurtz. 1997. *J. Neurophysiol.* 78:3493–97.
Parker, A. J., and W. T. Newsome. 1998. *Annu. Rev. Neurosci.* 21:227–77.
Passingham, R. 1995. *The Frontal Lobes and Voluntary Action.* Oxford: Oxford University Press.
Paulsen, O., and T. J. Sejnowski. 2000. *Curr. Opin. Neurobiol.* 10:172–79.
Pavlides, C., et al. 1993. *Hippocampus* 3:183–92.
Pavlides, C., et al. 1996. *Brain Res.* 738:229–35.
Pavlov, I. P. 1927. *Conditioned Reflexes.* New York: Dover.
Pennartz, C. M., et al. 1993. *Eur. J. Neurosci.* 5:107–17.
Penrose, R. 1989. *The Emperor's New Mind.* New York: Penguin Books.
Pesold, C., and D. Treit. 1995. *Brain Res.* 671:213–21.
Peterson, B. S., et al. 1999. *Biol. Psychiat.* 45:1237–58.
Petrides, M., and D. N. Pandya. 1999. *Eur. J. Neurosci.* 11:1011–36.
Pettegrew, J. W., et al. 1991. *Arch. Gen. Psychiat.* 48:563–68.
Petty, R. G., et al. 1995. *Am. J. Psychiat.* 152:715–21.
Pfaff, D. W. 1999. *Drive.* Cambridge: MIT Press.
Phelps, E. A., et al. 2000. *J. Cogn. Neurosci.* 12:729–38.
Phillips, R. G., and J. E. LeDoux. 1992. *Behav. Neurosci.* 106:274–85.
Phillips, W. A., and W. Singer. 1997. *Behav. Brain Sci.* 20:657–83.
Piattelli-Palmarini, M. 1989. *Cognition* 31:1–44.
Pinker, S. 1994. *The Language Instinct.* New York: William Morrow.
Pinker, S. 1997. *How the Mind Works.* New York: Norton.
Pinker, S., and P. Bloom. 1990. *Behav. Brain Sci.* 13:723–24.
Pitkänen, A., et al. 1997. *Trends Neurosci.* 20:517–23.
Pitman, R. K., et al. 2000. In *The New Cognitive Neurosciences,* edited by M. S. Gazzaniga. Cambridge: MIT Press.

Plato, P. 1964. In *Body, Mind and Death,* edited by A. Flew. New York: Macmillan.
Platt, M. L., and P. W. Glimcher. 1999. *Nature* 400:233–38.
Plaznik, A., et al. 1994. *Eur. J. Pharmacol.* 257:293–96.
Plutchik, R. 1980. *Emotion.* New York: Harper & Row.
Poe, G. R., et al. 2000. *Brain Res.* 855:176–80.
Poletti, C. E. 1986. In *The Limbic System,* edited by B. K. Doane and K. E. Livingston, 79–95. New York: Raven Press.
Popper, K. P., and J. C. Eccles. 1977. *The Self and Its Brain.* Berlin, New York: Springer.
Porges, S. W. 1998. *Psychoneuroendocrinology* 23:837–61.
Porsolt, R. D. 2000. *Rev. Neurosci.* 11:53–58.
Posner, M. 1992. *Curr. Dir. Psychol. Sci.* 1:11–14.
Povinelli, D. J., and T. M. Preuss. 1995. *Trends Neurosci.* 18:418–24.
Premack, D. 1965. In *Nebraska Symposium on Motivation,* edited by D. Levine. Lincoln: University of Nebraska Press.
Premack, D. 1985. *Cognition* 19:207–96.
Preuss, T. 1995. In *The Cognitive Neurosciences,* edited by M. S. Gazzaniga. Cambridge: MIT Press.
Preuss, T. M. 1995. *J. Cogn. Neurosci.* 7:1–24.
Prien, R. F., et al. 1973. *Arch. Gen. Psychiat.* 29:420–25.
Purves, D. 1994. *Neural Activity and the Growth of the Brain.* Cambridge: Cambridge University Press.
Purves, D., et al., eds. 1996. *Neuroscience.* Sunderland, MA: Sinauer.
Quartz, S. R., and T. J. Sejnowski. 1997. *Behav. Brain Sci.* 20:537–56.
Quinn, W. G., et al. 1974. *Proc. Natl. Acad. Sci. USA* 71:708–12.
Quirk, G. J., et al. 1995. *Neuron* 15:1029–39.
Quirk, G. J., et al. 1997. *Neuron* 19:613–24.
Quirk, G. J., et al. 2000. *J. Neurosci.* 20:6225–31.
Quitkin, F. M., et al. 2000. *Am. J. Psychiat.* 157:327–37.
Rainville, P., et al. 1997. *Science* 277:968–71.
Rakic, P. 1977. *Philos. Trans. R. Soc. Lond. B Biol. Sci.* 278:245–60.
Rakic, P. 1992. *Science* 258:1421–22.
Rakic, P. 1995. *Trends Neurosci.* 18:383–88.
Rakic, P., et al. 1986. *Science* 232:232–35.
Ramachandran, V. S., and S. Blakeslee. 1998. *Phantoms in the Brain.* New York: William Morrow.
Ramón y Cajal, S. 1909–1911. *Histologie du Systeme Nerveux de L'Homme et des Vertebres.* Paris: A. Maloine.
Ranck, J. B., Jr. 1973. *Exp. Neurol.* 41:461–531.
Randic, M., et al. 1993. *J. Neurosci.* 13:5228–41.
Raper, J. A., and M. Tessier-Lavigne. 1998. In *Fundamental Neuroscience,* edited by M. Zigmond. San Diego: Academic Press.
Rauch, S. L., et al. 1995. *Arch. Gen. Psychiat.* 52:20–28.
Rauch, S. L., et al. 1996. *Arch. Gen. Psychiat.* 53:380–87.
Rauch, S. L., et al. 1997. *Biol. Psychiat.* 42:446–52.
Recanzone, G. H., et al. 1993. *J. Neurosci.* 13:87–103.
Redgrave, P., et al. 1999. *Trends Neurosci.* 22:146–51.

Redmond, D. E. J. 1979. In *Phenomenology and Treatment of Anxiety,* edited by W. G. Fann et al., 153–202. New York: Spectrum.
Reichert, H., and A. Simeone. 1999. *Curr. Opin. Neurobiol.* 9:589–95.
Reiman, E. M., et al. 1984. *Nature* 310:683–85.
Reiman, E. M., et al. 1989. *Arch. Gen. Psychiat.* 46:493–500.
Reiss, D., and L. Marino. 2001. *Proc. Natl. Acad. Sci. USA* 98:5937–42.
Reith, J., et al. 1994. *Proc. Natl. Acad. Sci. USA* 91:11651–54.
Repa, J. C., et al. 2001. *J. Neurosci.* 4:724–31.
Rescorla, R. A., and R. L. Solomon. 1967. *Psychol. Rev.* 74:151–82.
Reynolds, J. H., and R. Desimone. 1999. *Neuron* 24:19–29, 111–25.
Richter. 1927. *Q. Rev. Biol.* 2:307–43.
Ridley, M. 1999. *Genome.* New York: HarperCollins.
Robbins, T. W. 1996. *Philos. Trans. R. Soc. Lond. B Biol. Sci.* 351:1463–70.
Robbins, T. W. 2000. *Exp. Brain Res.* 133:130–38.
Robbins, T. W., et al. 1989. *Neurosci. Biobehav. Rev.* 13:155–62.
Roberts, G. W., et al. 1993. *Neuropsychiatric Disorders.* London: Wolfe.
Rochat, P. 1995. *The Self in Infancy.* New York: Elsevier.
Rodier, P. M. 2000. *Sci. Am.* 282:56–63.
Rodrigues, S. M., et al. 2001. *J. Neurosci.* (pending).
Rogan, M. T., and J. E. LeDoux. 1995. *Neuron* 15:127–36.
Rogan, M. T., et al. 1997. *Nature* 390:604–7.
Rogan, M. T., et al. 1997. *J. Neurosci.* 17:5928–35.
Rogers, R. D., et al. 1999. *J. Neurosci.* 19:9029–38.
Rolls, E. 1990. In *An Introduction to Neural and Electronic Networks,* edited by S. F. Zornetzer et al. San Diego: Academic Press.
Rolls, E. T. 1986. In *Emotions,* edited by Y. Oomur, 325–44. Tokyo: Japan Scientific Societies Press.
Rolls, E. T. 1992. *Philos. Trans. R. Soc. Lond. B Biol. Sci.* 335:11–20.
Rolls, E. T. 1999. *The Brain and Emotion.* Oxford: Oxford University Press.
Romanski, L. M., and J. E. LeDoux. 1992. *J. Neurosci.* 12:4501–9.
Romanski, L. M., et al. 1993. *Behav. Neurosci.* 107:444–50.
Romanski, L. M., et al. 1999. *Nat. Neurosci.* 2:1131–36.
Rorty, R. 1979. *Philosophy and the Mirror of Nature.* Princeton, NJ: Princeton University Press.
Rose, H., and S. Rose. 2000. *Alas, Poor Darwin.* New York: Harmony Books.
Rosenblum, K., et al. 1997. *J. Neurosci.* 17:5129–35.
Roskies, A. L. 1999. *Neuron* 24:7–9, 111–25.
Ross, C. A., and A. Pam. 1995. *Pseudoscience in Biological Psychiatry.* New York: John Wiley.
Roth, P. 1986. *The Counterlife.* New York: Farrar, Straus, Giroux.
Routtenberg, A. 1999. *Trends Neurosci.* 22:255–56.
Rozental, R., et al. 2000. *Brain Res. Rev.* 32:11–15.
Rozin, P. 1976. In *Progress in Psychobiology and Physiological Psychology,* edited by J. M. Sprague and A. N. Epstein. New York: Academic Press.
Rudy, J. W., and R. C. O'Reilly. 1999. *Behav. Neurosci.* 113:867–80.

Rudy, J. W., and R. J. Sutherland. 1992. *J. Cognit. Neurosci.* 4:208–16.
Rush, A. J. 2000. *Biol. Psychiat.* 47:745–47.
Rushdie, S. February 4, 1990. In *Independent on Sunday.* London.
Russell, R. J., et al. 2000. *Neuroscience and the Person.* Berkeley: Vatican Observatory Publications, Vatican City State, Center for Theology and the Natural Sciences.
Ryle, G. 1949. *The Concept of Mind.* New York: Barnes and Noble.
Sahley, C. L. 1995. *J. Neurobiol.* 27:434–45.
Sanchez, C., and E. Meier. 1997. *Psychopharmacology (Berl.)* 129:197–205.
Sanders, S. K., and A. Shekhar. 1995. *Pharmacol. Biochem. Behav.* 52:701–6.
Sanes, J. N., and J. P. Donoghue. 2000. *Annu. Rev. Neurosci.* 23:393–415.
Sanes, J. R., and J. W. Lichtman. 1999. *Nat. Neurosci.* 2:597–604.
Sapolsky, R. M. 1996. *Science* 273:749–50.
Sapolsky, R. M. 1998. *Why Zebras Don't Get Ulcers.* New York: Freeman.
Sapolsky, R. M. 1999. In *Well-Being,* edited by D. Kahneman et al. New York: Russell Sage Foundation.
Sara, S. J. 2000. *Learn. Mem.* 7:73–84.
Sarter, M. F., and H. J. Markowitsch. 1985. *Behav. Neurosci.* 99:342–80.
Sass, L. A. 1992. In *Psychology and Postmodernism,* edited by S. Kvale, 166–82. Thousand Oaks, CA: Sage.
Sawaguchi, T., and P. S. Goldman-Rakic. 1991. *Science* 251:947–50.
Sawaguchi, T., and P. S. Goldman-Rakic. 1994. *J. Neurophysiol.* 71:515–28.
Sawaguchi, T., et al. 1988. *Neurosci. Res.* 5:465–73.
Sawaguchi, T., et al. 1990. *J. Neurophysiol.* 63:1385–1400.
Schacter, D. L. 1982., *Stranger Behind the Engram.* Hillsdale, NJ: Lawrence Erlbaum Associates.
Schacter, D. L. 1987. *J. Exp. Psychol.: Learn. Mem. Cogn.* 13:501–18.
Schacter, D. L. 1999. *Am. Psychol.* 54:182–203.
Schacter, D. L. 2001. *The Seven Sins of Memory.* Boston: Houghton Mifflin.
Schacter, D. L., and P. Graf. 1986. *J. Exp. Psychol.: Learn. Mem. Cogn.* 12(3):432–44.
Schacter, D. L., et al. 1998. *Philos. Trans. R. Soc. Lond. B Biol. Sci.* 353:1861–78.
Schacter, S. 1975. In *Handbook of Physiology,* edited by M. S. Gazzaniga and C. B. Blakemore, 529–63. New York: Academic Press.
Schacter, S., and J. E. Singer. 1962. *Psychol. Rev.* 69:379–99.
Schafe, G. E., and J. LeDoux. 2000. *J. Neurosci.* 20:1–5 (RC96).
Schafe, G. E., et al. 1998. *Learn. Mem.* 5:481–92.
Schafe, G. E., et al. 1999. *Learn. Mem.* 6:97–110.
Schafe, G. E., et al. 2000. *J. Neurosci.* 20:8177–87.
Schafe, G. E., et al. 2001. *Trends Neurosci.* 24:540–46.
Schatzberg, A. F., and H. C. Kraemer. 2000. *Biol. Psychiat.* 47:736–44.
Scheel-Kruger, J., and E. N. Petersen. 1982. *Eur. J. Pharmacol.* 82:115–16.
Scherer, K. 2000. In *Emotion, Development, and Self-Organization,* edited by M. Lewis and I. Granic, 70–99. New York: Cambridge University Press.
Scherer, K. R. 1988. In *Cognitive Perspectives on Emotion and Motivation,* edited by V. Hamilton et al., 89–126. Norwell, MA: Kluwer Academic Publishers.
Scherer, K. R. 1993. *Cogn. Emo.* 7:325–55.

Schlaggar, B. L., and D. D. O'Leary. 1991. *Science* 252:1556–60.
Schmajuk, N. A., and J. J. DiCarlo. 1992. *Psychol. Rev.* 99:268–305.
Schneider, F., et al. 1999. *Biol. Psychiat.* 45:863–71.
Schneiderman, N., et al. 1974. In *Limbic and Autonomic Nervous System Research,* edited by L. V. DiCara, 277–309. New York: Plenum.
Schore, A. N. 1994. *Affect Regulation and the Origin of the Self.* Hillsdale, NJ: Lawrence Erlbaum Associates.
Schulkin, J. 1999. *The Neuroendocrine Regulation of Behavior.* New York: Cambridge University Press.
Schultz, W. 1998. *J. Neurophysiol.* 80:1–27.
Schultz, W., and A. Dickinson. 2000. *Annu. Rev. Neurosci.* 23:473–500.
Schultz, W., et al. 1997. *Science* 275:1593–99.
Schuman, E. M. 1999. *Curr. Opin. Neurobiol.* 9:105–9.
Schuster, C. S., and S. S. Ashburn. 1992. *The Process of Human Development.* Philadelphia: Lippincott.
Schwartz, C. E., et al. 1999. *J. Am. Acad. Child Adolesc. Psychiat.* 38:1008–15.
Schwarz, N., and F. Strack. 1999. In *Well-Being,* edited by D. Kahneman et al., New York: Russell Sage Foundation.
Scott, S. K., et al. 1997. *Nature* 385:254–57.
Scoville, W. B., and B. Milner. 1957. *J. Neurol. Psychiat.* 20:11–21.
Searle, J. R. 1992. *The Rediscovery of the Mind.* Cambridge: MIT Press.
Searle, J. R. 2000. *Ann. Rev. Neurosci.* 23:557–78.
Sedvall, G., and L. Farde. 1996. *Lancet* 347:264.
Seeman, P. 1992. *Neuropsychopharmacology* 7:261–84.
Seeman, P., and S. Kapur. 2000. *Proc. Natl. Acad. Sci. USA* 97:7673–75.
Seeman, P., et al. 1975. *Proc. Natl. Acad. Sci. USA* 72:4376–80.
Selden, N. R., et al. 1991. *Neurosci.* 42:335–50.
Selemon, L. D., and P. S. Goldman-Rakic. 1999. *Biol. Psychiat.* 45:17–25.
Selemon, L. D., et al. 1995. *Arch. Gen. Psychiat.* 52:805–18.
Seligman, M. E. P. 1971. *Behav. Ther.* 2:307–20.
Selkoe, D., and K. Kosik. 1983. In *The Clinical Neurology of Aging,* edited by M. Albert. Oxford: Oxford University Press.
Semon, R. 1904. *Die Mneme als erhaltendes Prinzip im Wechsel des organischen Geschehens.* Leipzig: Wilhelm Engelmann.
Shadlen, M. N., and J. A. Movshon. 1999. *Neuron* 24:67–77, 111–25.
Shadlen, M. N., and W. T. Newsome. 1994. *Curr. Opin. Neurobiol.* 4:569–79.
Shadlen, M. N., et al. 1996. *J. Neurosci.* 16:1486–1510.
Shakespeare, W. 1542–1544. *The Comedy of Errors.* Act 1, scene 1, line 47.
Shalev, A. Y., et al. 1992. *Biol. Psychiat.* 31:863–65.
Shallice, T. 1988. In *Consciousness in Contemporary Science,* edited by A. Marcel and E. Bisiach, 305–33. Oxford: Oxford University Press.
Shapiro, M. L., and Z. Caramanos. 1990. *Psychobiol.* 18:231–43.
Shatz, C. J. 1992. *Science* 258:237–38.
Shatz, C. J. 1996. *Proc. Natl. Acad. Sci. USA* 93:602–8.
Sheffield, F. D., and T. B. Roby. 1950. *J. Comp. Physiol. Psychol.* 43:471–81.

Sheline, Y. I., et al. 1996. *Proc. Natl. Acad. Sci. USA* 93:3908–13.
Sheline, Y. I., et al. 1998. *Neuroreport* 9:2023–28.
Shelton, R. C., et al. 1988. *Am. J. Psychiat.* 145:154–63.
Shepherd, G. 1988. *Neurobiology.* New York: Oxford University Press.
Shepherd, G. 1998. *The Synaptic Organization of the Brain.* New York: Oxford University Press.
Sherrington, C. S. 1897. In *A Textbook of Physiology,* 7th ed., edited by M. Foster. London: Macmillan.
Sherrington, C. S. 1906. *The Integrative Action of the Nervous System.* New Haven: Yale University Press.
Shevrin, H., and S. Dickman. 1980. *Am. Psychol.* 35:421–34.
Shevrin, H., et al. 1992. *Conscious. Cogn.* 1:340–66.
Shi, S. H., et al. 1999. *Science* 284:1811–16.
Shibata, S., et al. 1989. *Psychopharmacol.* 98:38–44.
Shimamura, A. 1995. In *The Cognitive Neurosciences,* edited by M. S. Gazzaniga. Cambridge: MIT Press.
Shin, L. M., et al. 1997. *Arch. Gen. Psychiat.* 54:233–41.
Shin, L. M., et al. 1999. *Am. J. Psychiat.* 156:575–84.
Shizgall, P. 1999. In *Well-Being,* edited by D. Kahneman et al. New York: Russell Sage Foundation.
Shors, T. J., and E. Dryver. 1992. *Psychobiol.* 20:247–53.
Shors, T. J., and L. D. Matzel. 1997. *Behav. Brain Sci.* 20:597–613.
Shors, T. J., et al. 1989. *Science* 244:224–26.
Shors, T. J., et al. 1992. *Science* 257:537–39.
Siegel, A., and H. Edinger. 1981. In *Handbook of the Hypothalamus,* edited by P. J. Morganc and J. Panksepp, 203–40. New York: Marcel Dekker.
Silbersweig, D. A., et al. 1995. *Nature* 378:176–79.
Silva, A. J., et al. 1992. *Science* 257:201–6.
Silva, A. J., et al. 1997. *Annu. Rev. Genet.* 31:527–46.
Silva, A. J., et al. 1998. *Annu. Rev. Neurosci.* 21:127–48.
Singer, W. 1998. *Philos. Trans. R. Soc. Lond. B Biol. Sci.* 353:1829–40.
Singer, W. 1999. *Curr. Opin. Neurobiol.* 9:189–94.
Singer, W. 2001. *Ann. NY Acad. Sci.* 929:123–46.
Singh, T. D., et al. 2000. *J. Neurobiol.* 44:82–94.
Skinner, B. F. 1938. *The Behavior of Organisms.* New York: Appleton-Century-Crofts.
Skinner, B. F. 1953. *Science and Human Behavior.* New York: Free Press.
Skinner, B. F. 1972. *Beyond Freedom and Dignity.* New York: Knopf.
Smith, C. A., and P. C. Ellsworth. 1985. *J. Pers. Soc. Psychol.* 56:339–53.
Smith, C. A., and R. S. Lazarus. 1990. In *Handbook of Personality,* edited by L. A. Pervin, 609–37. New York: Guilford.
Smith, E. E., and J. Jonides. 1999. *Science* 283:1657–61.
Smith, Y., et al. 2000. *J. Comp. Neurol.* 24:496–508.
Snell, B. 1960. *The Discovery of the Mind.* New York: Harper & Row.
Soares, J. J. F., and A. Öhman. 1993. *Psychophysiol.* 30:460–66.
Soderling, T. R. 1996. *Neurochem. Int.* 28:359–61.

Solomon, R. L., and L. C. Wynne. 1954. *Psychol. Rev.* 61:353.
Sorabji, P. 2001. *Emotion and Peace of Mind.* Oxford: Oxford University Press.
Spanagel, R., and F. Weiss. 1999. *Trends Neurosci.* 22:521–27.
Spelke, E. 1994. *Cognition* 50:431–45.
Spencer, H. 1866. *Principles of Biology.* New York: D. Appleton and Company.
Sperry, R. 1966. In *New Views on the Nature of Man,* edited by J. R. Platt. Chicago: University of Chicago Press.
Sperry, R. 1984. *Neuropsychol.* 22:661–73.
Sperry, R., and N. Miner. 1955. *J. Comp. Physiol. Psychol.* 48:50–58.
Squire, L. 1987. *Memory and the Brain.* New York: Oxford University Press.
Squire, L. R. 1987. In *Handbook of Physiology,* edited by F. Plum, 295–371. Bethesda: American Physiological Society.
Squire, L. R. 1992. *Psychol. Rev.* 99:195–231.
Squire, L. R., and E. R. Kandel. 1999. *Memory: From Mind to Molecules.* New York: Scientific American Library.
Squire, L. R., and S. M. Zola. 1996. *Proc. Natl. Acad. Sci. USA* 93:13515–22.
Squire, L. R., and S. M. Zola. 1998. *Hippocampus* 8:205–11.
Squire, L. R., et al. 1993. *Annu. Rev. Psychol.* 44:453–95.
Starkman, M. N., et al. 1992. *Biol. Psychiat.* 32:756–65.
Starkman, M. N., et al. 1999. *Biol. Psychiat.* 46:1595–1602.
Staubli, U., et al. 1989. *Behav. Neurosci.* 103:54–60.
Staubli, U., et al. 1990. *Synapse* 5:333–35.
Staubli, U., et al. 1994. *Proc. Natl. Acad. Sci. USA* 91:777–81.
Staubli, U. V. 1995. In *Brain and Memory,* edited by J. L. McGaugh et al., 303–18. New York: Oxford University Press.
Stebbins, W. C., and O. A. Smith. 1964. *Science* 144:881–83.
Stein, L., et al. 1973. In *The Benzodiazepines,* edited by S. Garattini et al., 299–326. New York: Raven Press.
Steiner, J. E. 1973. *Symp. Oral Sens. Percept.* 4:254–78.
Steinmetz, J. E., and R. F. Thompson. 1991. In *Neurobiology of Learning, Emotion and Affect,* edited by J. I. Madden, 97–120. New York: Raven Press.
Stellar, E. 1954. *Psychol. Rev.* 61:5–22.
Stent, G. S. 1973. *Proc. Natl. Acad. Sci. USA* 70:997–1001.
Stern, E., and D. Silbersweig. 1998. In *The Pathogenesis of Schizophrenia,* edited by M. F. Lenzenweger and R. H. Dworkin, 235–46. Washington, D.C.: American Psychological Association.
Sternberg, R. J., and M. I. Barnes. 1988. *The Psychology of Love.* New Haven: Yale University Press.
Stevens, A. A., et al. 1998. *Arch. Gen. Psychiat.* 55:1097–1103.
Stevens, C. F. 1998. *Neuron* 20:1–2.
Steward, O., and E. M. Schuman. 2001. *Annu. Rev. Neurosci.* 24:299–325.
Stewart, R. S., et al. 1988. *Am. J. Psychiat.* 145:442–49.
Stoerig, P. 1996. *Trends Neurosci.* 19:401–6.
Stone, A. A., et al. 1999. In *Well-Being,* edited by D. Kahneman et al. New York: Russell Sage Foundation.

Strauss, C., and N. Quinn. 1997. *A Cognitive Theory of Cultural Meaning.* New York: Cambridge University Press.
Strawson, P. 1959. In *Individuals.* London: Methuen.
Stryker, M. 1991. In *Development of the Visual System,* edited by D. Lam and C. Shatz. Cambridge: MIT Press.
Stryker, M. P., and W. A. Harris. 1986. *J. Neurosci.* 6:2117–33.
Stuss, D. T. 1991. In *The Self,* edited by J. Strauss and G. R. Goethals. New York: Springer.
Stuss, D. T., and D. F. Benson. 1986. *The Frontal Lobes.* New York: Raven Press.
Stutzmann, G. E., and J. E. LeDoux. 1999. *J. Neurosci.* 19:RC8 (online).
Stutzmann, G. E., et al. 1998. *J. Neurosci.* 18:9529–38.
Suga, N. 1990. *Sci. Am.* 262:60–68.
Suzuki, W. A., and D. G. Amaral. 1994. *J. Comp. Neurol.* 350:497–533.
Swanson, L. W. 1983. In *Neurobiology of the Hippocampus,* edited by W. Seifert, 3–19. London: Academic Press.
Sweatt, J. D. 1999. *Learn. Mem.* 6:399–416.
Szentagothai, J. 1984. *Annu. Rev. Neurosci.* 7:1–11.
Szinyei, C., et al. 2000. *J. Neurosci.* 20:8909–15.
Talbot, M. 2000. In *New York Times Magazine,* January 9.
Tallal, P. 2000. *Proc. Natl. Acad. Sci. USA* 97:2402–4.
Tallal, P., et al. 1998. *Exp. Brain Res.* 123:210–19.
Tamminga, C. A. 1991. In *Advances in Neuropsychiatry and Psychopharmacology,* edited by C. A. Tamminga and S. C. Schulz, 99–109. New York: Raven Press.
Tang, Y. P., et al. 1999. *Nature* 401:63–69.
Taylor, C. B. 1998. In *Textbook of Psychopharmacology,* 2nd ed., edited by A. F. Schatzberg and C. B. Nemeroff. Washington, D.C.: American Psychiatric Press.
Tellegen, A., et al. 1988. *J. Pers. Soc. Psychol.* 54:1031–39.
Terman, J. R., and A. L. Kolodkin. 1999. *Neuron* 23:193–95.
Terrace, H. 1984. In *The Biology of Learning,* edited by P. Marler and H. S. Terrace, 15–45. Berlin: Springer-Verlag.
Teuber, H. L. 1964. In *The Frontal Granular Cortex and Behavior,* edited by J. M. Warren and K. Akert. New York: McGraw-Hill.
Teyler, T. 1992. In *Neuroscience Year: Supplement 2 to the Encyclopedia of Neuroscience,* edited by B. Smith and G. Adelman, 91–93. Cambridge: Brikhauser Boston.
Teyler, T., and P. DiScenna. 1987. *Annu. Rev. Neurosci.* 10:131–61.
Thiebot, M. H., et al. 1980. *Neurosci. Lett.* 16:213–17.
Thierry, A. M., et al. 1993. In *Motor and Cognitive Functions of the Prefrontal Cortex,* edited by A. M. Thierry et al. Berlin: Springer-Verlag.
Thompson, R. F. 1986. *Science* 233:941–47.
Thompson, R. F., and J. J. Kim. 1996. *Proc. Natl. Acad. Sci. USA* 93:13438–44.
Thompson, R. F., and W. A. Spencer. 1966. *Psychol. Rev.* 73:16–43.
Thompson, R. F., et al. 1983. *Prog. Psychobiol. Physiol. Psychol.* 10:167–96.
Thorndike, E. L. 1898. *Psychol. Monog.* 2:109.
Thorndike, E. L. 1913. *The Psychology of Learning.* New York: Teachers College.
Thorpe, W. H. 1963. *Learning and Instinct in Animals.* London: Methuen.
Tinbergen, N. 1951. *The Study of Instinct.* Oxford: Clarendon Press.

Toates, F. M. 1986. *Motivational Systems.* Cambridge: Cambridge University Press.
Tomita, H., et al. 1999. *Nature* 401:699–703.
Tomkins, S. S. 1962. *Affect, Imagery, Consciousness.* New York: Springer.
Toni, N., et al. 1999. *Nature* 402:421–25.
Tononi, G., and G. M. Edelman. 1998. *Science* 282:1846–51.
Tooby, J., and L. Cosmides. 2000. In *The New Cognitive Neurosciences,* edited by M. S. Gazzaniga. Cambridge: MIT Press.
Tootell, R. B., et al. 1995. *Nature* 375:139–41.
Treisman, A. 1996. *Curr. Opin. Neurobiol.* 6:171–78.
Treit, D., and J. Menard. 1997. *Behav. Neurosci.* 111:653–58.
Treit, D., et al. 1993. *Behav. Neurosci.* 107:770–85.
Trowill, J. A., et al. 1969. *Psychol. Rev.* 76:264–81.
Tsien, J. Z. 2000. *Curr. Opin. Neurobiol.* 10:266–73.
Tsien, J. Z., et al. 1996. *Cell* 87:1327–38.
Tully, T., and W. G. Quinn. 1985. *J. Comp. Physiol.* [A] 157:263–77.
Tulving, E. 1983. *Elements of Episodic Memory.* New York: Oxford University Press.
Ungerleider, L. G., and J. Haxby. 1994. *Curr. Opin. Neurobiol.* 4:157–65.
Ungerleider, L. G., and M. Mishkin. 1982. In *Analysis of Visual Behavior,* edited by D. J. Ingle et al., 549–86. Cambridge: MIT Press.
Uvnas-Moberg, K. 1998. *Psychoneuroendocrinology* 23:819–35.
Valenstein, E. 1970. In *The Neurosciences; Second Study Program,* edited by F. O. Schmitt, 207–17. New York: Rockefeller University Press.
Valenstein, E. 1999. *Blaming the Brain.* New York: Free Press.
Van Essen, D. C. 1985. In *Cerebral Cortex,* edited by A. Peters and E. G. Jones. New York: Plenum.
Van Essen, D. C. 1995. *Trans. Am. Ophthalmol. Soc.* 93:123–33.
Van Essen, D. C., et al. 1992. *Science* 255:419–23.
Van Hoesen, G., and D. N. Pandya. 1975. *Brain Res.* 95:1–24.
Van Roussum, M. 1967. In *Proceedings of the Fifth Collegium Internationale Neuropharmacologicum,* edited by H. Brill, 321–29. Amsterdam: Excerpta Medica Foundation.
Vargha-Khadem, F., et al. 1997. *Science* 277:376–80.
Veinante, P., and M. J. Freund-Mercier. 1995. *Adv. Exp. Med. Biol.* 395:347–48.
Veinante, P., and M. J. Freund-Mercier. 1997. *J. Comp. Neurol.* 383:305–25.
Vendrell, P., et al. 1995. *Neuropsychol.* 33:341–52.
Vicari, S., et al. 2000. *Cortex* 36:31–46.
von der Malsburg, C. 1995. *Curr. Opin. Neurobiol.* 5:520–26.
von Holst, E., and U. von Saint-Paul. 1962. *Sci. Am.* 206:50–59.
Wagner, A. D. 1999. *Neuron* 22:19–22.
Wagner, A. R., and S. E. Brandon. 1989. In *Contemporary Learning Theories,* edited by S. B. Klein and R. R. Mowrer, 149–89. Hillsdale, NJ: Lawrence Erlbaum Associates.
Walker, D. L., and M. Davis. 2000. *Behav. Neurosci.* 114:1019–33.
Walter, G. 1953. *The Living Brain.* New York: Norton.
Warrington, E., and L. Weiskrantz. 1973. *Neuropsychol.* 20:233–48.
Watson, J. B. 1913. *Psychol. Rev.* 20:158–77.
Watson, J. B. 1925. *Behaviorism.* New York: W. W. Norton.

Watson, J. 1938. In *The Behavior of Organisms,* edited by B. F. Skinner. New York: Appleton-Century-Crofts.
Watt, D. F. 1998. *J. Neuropsychiat. Clin. Neurosci.* 10:113–16.
Weinberger, D. R., and B. Gallhofer. 1997. *Int. Clin. Psychopharmacol.* 12 Suppl. 4:S29–36.
Weinberger, D. R., et al. 1980. *Arch. Gen. Psychiat.* 37:11–13.
Weinberger, D. R., et al. 1986. *Arch. Gen. Psychiat.* 43:114–24.
Weinberger, D. R., et al. 1992. *Am. J. Psychiat.* 149:890–97.
Weinberger, J., and D. C. McClelland. 1990. In *Handbook of Motivation and Cognition,* edited by E. T. Higgins and R. M. Sorrentino. New York: Guilford.
Weinberger, N. M. 1995. In *The Cognitive Neurosciences,* edited by M. S. Gazzaniga, 1071–90. Cambridge: MIT Press.
Weinberger, N. M. 1998. *Neurobiol. Learn. Mem.* 70:226–51.
Weiner, B. 1989. *Human Motivation.* Hillsdale, NJ: Lawrence Erlbaum Associates.
Weiskrantz, L. 1956. *J. Comp. Physiol. Psychol.* 49:381–91.
Weiskrantz, L. 1996. *Curr. Opin. Neurobiol.* 6:215–20.
Weiskrantz, L., and E. Warrington. 1979. *Neuropsychol.* 17:187–94.
Weisskopf, M. G., and J. E. LeDoux. 1999. *J. Neurophysiol.* 81:930–34.
Weisskopf, M. G., and R. A. Nicoll. 1995. *Nature* 376:256–59.
Weisskopf, M. G., et al. 1999. *J. Neurosci.* 19:10512–19.
Wexler, K. 1999. In *Encyclopedia of Cognitive Science,* 408–9. Cambridge: MIT Press.
Whalen, P. J., et al. 1998. *J. Neurosci.* 18:411–18.
White, N. M. 1997. *Curr. Opin. Neurobiol.* 7:164–69.
Wicklegren, W. A. 1979. *Psychol. Rev.* 86:44–60.
Wigström, H., et al. 1986. *Acta Physiol. Scand.* 126:317–19.
Wik, G., et al. 1996. *Int. J. Neurosci.* 87:267–76.
Wilde, O. 1891. *The Picture of Dorian Gray.* London: Ward, Lock and Co.
Wilensky, A. E., et al. 1999. *J. Neurosci.* 19:(RC48).
Wilensky, A. E., et al. 2000. *J. Neurosci.* (in press).
Willner, P. 1995. *Adv. Biochem. Psychopharmacol.* 49:19–41.
Wilson, E. O. 1999. *Consilience.* New York: Random House.
Wilson, F. A., et al. 1993. *Science* 260:1955–58.
Wilson, M. A., and B. L. McNaughton. 1993. *Science* 261:1055–58.
Wilson, M. A., and B. L. McNaughton. 1994. *Science* 265:676–79.
Wilson, T. D., et al. 2000. *Psychol. Rev.* 107:101–26.
Wilson, T. D. *Strangers to Ourselves.* Cambridge: Harvard University Press (in press).
Wimbauer, S., et al. 1997. *Biol. Cybern.* 77:453–61.
Wimer, R. E., and C. C. Wimer. 1985. *Annu. Rev. Psychol.* 36:171–218.
Winson, J. 1985. *Brain and Psyche.* Garden City, NY: Anchor/Doubleday.
Wise, R. A. 1982. *Behav. Brain Sci.* 5:39–87.
Wise, S. P., et al. 1996. *Crit. Rev. Neurobiol.* 10:317–56.
Witter, M. P., et al. 1989. *Prog. Neurobiol.* 33:161–253.
Wong, R. O., et al. 1993. *Neuron* 11:923–38.
Woods, S. W., et al. 1988. *Lancet* 2:678.
Woodson, W., et al. 2000. *Synapse* 38:124–37.
Woodworth, R. S. 1918. *Dynamic Psychology.* New York: Columbia University Press.

Woolf, V. 1928. *Orlando.* Barcelona: EDHASA.

Woolf, V. 1979. Letter, December 28, 1932. In *The Sickle Side of the Moon: Letters,* edited by N. Nicolson.

Woolf, V. S. 1925. *The Common Reader.* New York: Harcourt, Brace.

Wurtzel, E. 1999. *Prozac Nation.* Boston: Houghton Mifflin.

Xing, J., and R. A. Andersen. 2000. *J. Cogn. Neurosci.* 12:601–14.

Yamada, K., et al. 1999. *Jpn. J. Pharmacol.* 80:9–14.

Yamamoto, T., et al. 1994. *Behav. Brain Res.* 65:123–37.

Yang, C. R., and J. K. Seamans. 1996. *J. Neurosci.* 16:1922–35.

Yehuda, R., et al. 2000. *Am. J. Psychiat.* 157:1252–59.

Yin, J., and T. Tully. 1996. *Curr. Opin. Neurobiol.* 6:264–68.

Young, A. W., et al. 1996. *Neuropsychol.* 34:31–39.

Young, E. A., et al. 1994. *Arch. Gen. Psychiatry* 51:701–7.

Young, M. P., and S. Yamane. 1992. *Science* 256:1327–31.

Young, P. T. 1961. *Motivation and Emotion.* New York: Wiley.

Zajonc, R. B. 1968. In *Handbook of Social Psychology,* edited by G. Lindzey and E. Aronson. Reading, MA: Addison-Wesley.

Zajonc, R. B. 1980. *Am. Psychol.* 35:151–75.

Zajonc, R. B. 1984. *Am. Psychol.* 39:117–23.

Zeki, S. 1993. *Curr. Opin. Neurobiol.* 3:155–59.

Zeki, S., and A. Bartels. 1999. *Conscious Cogn.* 8:225–59.

Zigmond, M. J., et al. 1999. *Fundamental Neuroscience.* San Diego: Academic Press.

Zohary, E., et al. 1994. *Nature* 370:140–43.

Zola-Morgan, S., and L. R. Squire. 1993. *Annu. Rev. Neurosci.* 16:547–63.

Zuckerman, M. 1991. *Psychobiology of Personality.* Cambridge: Cambridge University Press.

INDEX

Page numbers in *italics* refer to illustration captions.